ROBERT
DECEMBER, 1972

Structural Design Guide to the ACI Building Code

Structural Design Guide to the ACI Building Code

Paul F. Rice

Edward S. Hoffman

Van Nostrand Reinhold Company

New York / Cincinnati / Toronto / London / Melbourne

Van Nostrand Reinhold Company Regional Offices:
New York Cincinnati Chicago Millbrae Dallas
Van Nostrand Reinhold Company International Offices:
London Toronto Melbourne

Library of Congress Catalog Card Number: 72-4070
ISBN: 0-442-36903-4

Manufactured in the United States of America
Published in the United States of America
450 West 33rd Street, New York, N.Y. 10001
Published simultaneously in Canada by Van Nostrand Reinhold Ltd.
15 14 13 12 11 10 9 8 7 6 5 4 3 2 1

Library of Congress Cataloging in Publication Data

Rice, Paul F 1921-
Structural design guide to the ACI building code.

Includes bibliographical references.
1. Reinforced concrete construction—Contracts and specifications—United States. I. Hoffman, Edward S., 1920 joint author. II. Title.
TA683.24.RI 624'.1834 72-4070
ISBN 0-442-36903-4

preface

This book is intended to guide practicing structural engineers familiar with earlier ACI building codes into more profitable routine designs with the ACI 1971 Building Code (ACI 318-71).

Each new ACI Building Code expresses the latest knowledge of reinforced concrete in legal language for safe design application. Beginning in 1956 with the introduction of ultimate strength design, each new code offered better utilization of high-strength reinforcement and the compressive strength of the concrete itself. Each new code thus permitted more economy as to construction material, but achieved it through more detailed and complicated design calculations. In addition to competition requiring independent structural engineers to follow the latest code for economy, the reduction of shear and bond values from those previously allowed created a professional obligation to follow the latest code for accepted levels of structural safety.

The increasing complexity of codes has encouraged the use of computers for design and has stimulated the development of computer-based handbooks. Computer time and development of such new computer design programs are costly, however, and computers are not available to all engineers. Also, it is not economical to program all design problems encountered.

This book will guide the user to the various sections of the Code pertinent to design of common reinforced concrete structural elements. A brief explanation of the significance of these sections is presented, together with limits of applicability, the range in which results may control design; and, where possible, design short cuts to insure automatic conformance to the Code without calculations.

This Guide does not duplicate nor replace the ACI Code, its Commentary, design handbooks, or use of computers. It complements the ACI Code and Commentary, shows how to take full advantage of available handbooks based on the 1963 Code, and should shorten time required to develop computer design programs. It converts some code formulas from the review form (or trial designs) to direct design. It presents some simple appropriate formulas, tabulations, and charts for conservative longhand direct design.

Specifications for materials and special Code requirements superimposed upon the ASTM Specifications for materials are explained to aid the structural engineer to avoid difficulties with use of obsolete specifications.

The overall objective of this book is simply to save the Engineer time in reinforced concrete design.

HOW TO USE THIS GUIDE

Code requirements applicable to the design of structural building elements, scattered through various Code chapters, have been assembled for the analysis and design of one-way slabs, one-way joists, beams, the various types of two-way slab systems with and without beams, prestressed flexural members, columns, walls, and footings.

Most of the numerical examples are based on normal weight concrete with $f_c' = 4{,}000$ psi for flexural members and 4,000, 5,000, or 6,000 psi for columns; and the standard Grade 60 reinforcement. For lightweight aggregate concrete, see Chapter 15. Other concrete strengths used are so indicated.

Provisions new to the Code are noted as such for especial attention. It is assumed that users of the guide are familiar with reinforced concrete design and structural analysis as well as the terms and symbols in common use. Definitions of new nomenclature and symbols follow this introduction.

This guide is intended for use *with* the Code itself. Space limitations make it impracticable to include the Code. The Guide indicates the proper sections of the Code in the order that a designer would normally require their use for design of a particular building element. In the appropriate chapter for the element being designed, the Engineer will find the applicable Code sections indicated in parenthesis thus: "(Section 00.00.00)" following the explanations of their application.

Explanations of requirements difficult to interpret are followed by numerical examples. Where several Code equations or requirements are applicable simultaneously and must be solved to determine which controls, computer solutions over the usual range have been included as convenient tables or curves. Where examination of computer solutions over a wide range show simple approximations by longhand possible, such short cuts, together with the limitations of range or accuracy are given. For especially difficult or unusual problems outside the scope of this guide, other references are cited to guide the Engineer to a quick source of information for detailed study. References to the ACI Code Commentary are indicated thus: "(Commentary 00.00.00)."

No attempt has been made to explain each individual Section of the Code in this book. A large number of Code provisions which have provoked little or no question of interpretation in past codes have been repeated essentially without change in the 1971 Code.

Other exclusions are precast concrete and composite (precast with cast-in-place concrete) design procedures which involve no separate design theory but merely consideration of different load conditions due to construction sequences. An alternate working stress design is permitted by the Code, but the working stress is uneconomical. The new Code provisions for thin shell and folded-plate design serve only to include these structures within the scope of the Code. Similarly, the new provisions for seismic design of special ductile frames are intended to provide generally accepted details of design for use with various regional seismic code requirements. The variety of plates, shells, and seismic details precludes their inclusion here. Any explanation of seismic details here would also of necessity duplicate the report on seismic details by ACI Committee 315* in which both authors participated and recommend.

Two indexes are provided, a subject index and a Code section reference index. The user wishing to locate all Code references to a particular subject, as well as the user interested in the interpretation of a particular Code section, should find this arrangement most convenient.

* "Seismic Details for Special Ductile Frames," by ACI Com. 315, *ACI Journal,* May 1970; paper title No. 67-22.

contents

1 structural materials, specifications, and testing

GENERAL

All structural materials for reinforced concrete and the standard acceptance tests for such materials are prescribed under the ACI Code as the standard specifications issued by the American Society for Testing and Materials (Section 3.8.1). The ASTM specifications, including the date of latest issue at the time each new ACI Code is itself adopted, are adopted by reference to the Code, as if set forth fully therein. Because the specifications are made an integral part of the Code, and the latter itself is intended to be incorporated into statutory general building codes as an integral part, it is legally necessary that the "outside" specifications be identified as to year of issue.

Since the ASTM, like ACI, revises its specifications periodically to reflect new materials; improvements in production methods, materials, or testing; or new requirements, some of the materials specifications made part of a current ACI Code will always be obsolete. Many users of the Code simply specify material specifications of "latest issue" for their projects designed under the ACI Code. Usually, this procedure is not only satisfactory but desirable, since most revisions to material specifications upgrade the requirements. The specifying engineer, however, is responsible that materials used under statutory codes meet the legal requirements thereof, and so should be acquainted with any variations in later issues of the specifications. Some recent variations in the latest issues of the ASTM specifications from those in the last Code, as well as variations resulting from Code requirements more restrictive than the ASTM specifications cited therein, are described in the following sections on individual materials.

REINFORCEMENT—SPECIFICATIONS

Although the applicable requirements are titled "Metal Reinforcement" (Section 3.5), only steel is permitted. Cast iron is no longer usable as part of a composite member. For reinforced concrete, deformed reinforcing bars, wire, or welded wire fabric may be used (Section 3.5). Plain (smooth) bars or smooth wire may be used only as column spirals (Sections 3.5.2 and 3.5.5). Since the ASTM specifications for reinforcing bars do not include plain bars, the Code requires that they need conform only to the strength and elongation requirements for the grade specified (Section 3.5.2). Welded wire fabric may be composed of either plain or deformed wire (Section 3.5.6 and 3.5.7). For composite steel-concrete construction, steel pipe or tubing (Section 3.5.11) or rolled shapes and built-up structural steel members may be used (Section 3.5.12). Wire, strand, or high-strength low-alloy bars may be used for prestressing (Sections 3.5.9 and 3.5.10).

REINFORCING BARS—SPECIAL REQUIREMENTS

Deformed reinforcing bars (termed "rebars" hereafter in this text) are available in three types of steel, billet (ASTM A615-68), rail (A616-68), and axle (A617-68). The 1971 ACI Code is intended specifically for the use of standard Grade 60 reinforcement, although provision is also made for the use of Grades 40, 50, and 75 which are available under the reference ASTM specifications, up to a maximum limit of 80,000 psi yield strength, except for prestressing steels (Section 9.4.1).

Axle steel rebars (Section 3.5.1-c) under A617-68 are available in sizes #3 through #11 only, and in Grades 40 and 60. There are no additional requirements imposed by the Code not required in A617-68.

Rail steel rebars (Section 3.5.1-b) are available under A616-68 in sizes #3 through #11 only, and in Grades 50 and 60. The Code requires that if rail steel rebars are to be used bent they shall meet the bend test requirements for Grade 60 billet steel rebars. A616-68 contains no provisions for bend testing.

Billet steel rebars (Section 3.5.1-a) are commonly available in sizes #3 through #18 in standard Grade 60, in sizes #3 through #11 in Grade 40, and in Sizes #11, #14, and #18 only in Grade 75. Two special requirements in the Code for billet steel rebars are *not* included in A615-68: (1) if #14 or #18 rebars are to be bent they shall be capable of being bent 90 deg at a minimum temperature of 60°F around a 10 bar diameter pin without cracking transverse to the axis of the bar, and (2) for Grade 75, the specified yield strength, f_y, shall be measured at a strain of 0.35 percent. ASTM A615-68 provides no bend test for #14 and #18, and specifies that the yield strength, f_y, for Grade 75 shall be measured at a strain of 0.6 percent (see Fig. 1-1).

One additional Code requirement more restrictive than the ASTM specifications for all tension testing on rebars larger than #8 is that tension test results shall correspond to those on full-size specimens (Section 3.5.1).

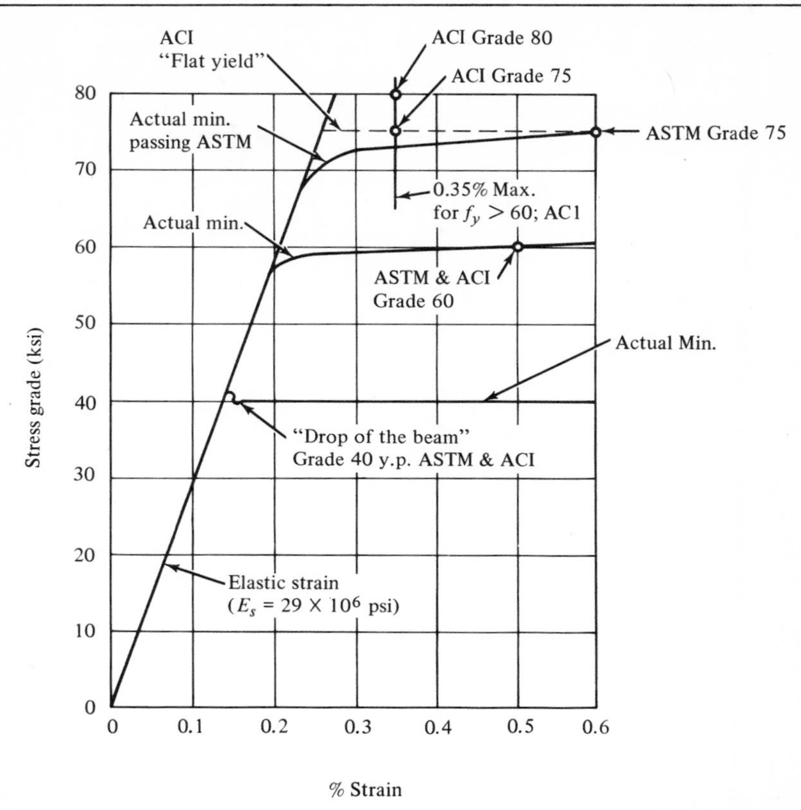

Fig. 1–1 Rebar stress-strain curves and specifications for yield strength.

The ASTM specifications permit tension tests on (1) full-size bars for an 8 in. gauge length, (2) machined-down specimens $\frac{3}{4}$ in. in diameter or larger for an 8 in. gauge length, or (3) machined-down specimens 0.500 in. in diameter for a 2 in. gauge length. In practice, machined-down specimens are commonly used on #11, #14, and #18 bars only. To the authors' knowledge, no comprehensive published data from parallel tests for the three types of specimens are available from which reliable statistical correlations could be made. The results of a pilot series of tests comparing tensile properties of various types of specimens from the same bars show somewhat higher (3 to 5 percent) average strengths for the 0.500 in. diameter machined-down specimens than for the full-size bars.* The range of variations is quite large, however, probably since the variable "stress raiser" effect of the bar deformations is eliminated in the machined speci-

* Unpublished. Presented by the American Iron and Steel Institute, Com. of Concrete Reinforcing Bar Producers, at public meetings of ACI Com. 318.

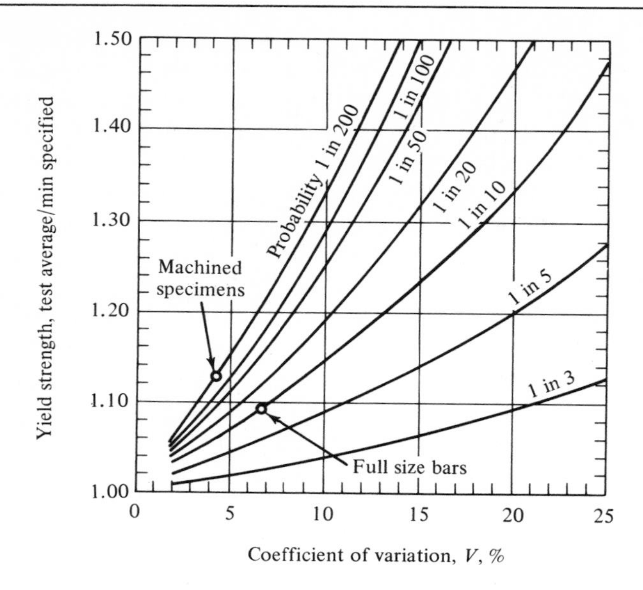

34 Heats #18, Grade 60 Bars

Machined specimens 1.128" dia.
$\overline{f_y}$ = 67,829 psi (= 1.13 f_y)
σ = 2,383 psi
V = 4.2%
Full-size specimens
$\overline{f_y}$ = 65,700 psi (= 1.095 f_y)
σ = 4,370 psi
V = 6.65%

Note: Full size specimen's taken at fabricating shop from a "population" not including any rejected on basis of prior machined specimen testing at producer's mill.

Range of differences between tests on full-size bars and the machined specimens:
-11,980 psi to +4,100 psi
R = 16,080 psi
Average difference = -2,128 psi

Near lows:

Machined	Full-size
66,350	60,750
72,730	60,750
67,530	60,500
68,000	61,000
70,220	61,500

Fig. 1–2 Probability of low-strength tests.

mens. For a single job record of parallel tests on bars from the same heat (*not* the same bars), see Fig. 1-2, in which tests on full-size field samples are compared to tests on machined specimens at the producing mill for one project.

WELDABILITY

Weldability properties are specifically excluded from ASTM specifications for standard grades of rebars. The Code requires that ASTM specifications for bars to be welded shall be supplemented by requirements to assure weldability in accordance with welding procedures specified (Section 3.5.3).* Weldability is at best a relative term. All steels are "weldable" somehow. The metallurgist defines weldability in terms of the chemical composition. His measure of weldability is the "carbon equivalent" content, for which different authorities recommend different formulas. The practical structural engineer defines weldability for rebars in terms of the strength achieved at a splice. The practical welder or contractor defines weldability in terms of cost—welding method required, amount of preheat, if any, and control of cooling, if any—required to deliver the performance normally specified by the Engineer. In this confusion all parties agree that extremely rapid cooling such as quenching in cold liquid changes the ductility at the heated area of a weld in carbon steel bars. The steel in this area becomes embrittled.

TACK WELDING (Welding of crossing rebars for assembly)

If a small portion of the cross section of a large cold rebar is rapidly heated to welding temperatures and allowed to cool equally rapidly (similar to the effect of quenching), a "metallurgical notch" effect is created (see Fig. 1-3). At this cross section the original properties of the rebar such as elongation, bendability, dynamic load resistance, fatigue strength, ultimate strength, and yield strength are reduced. This effect is most dramatically

* Welding procedures must conform to the American Welding Society specification "Recommended Practices for Welding Reinforcing Steel, Metal Inserts, and Connections in Reinforced Concrete" (AWS D12.1).

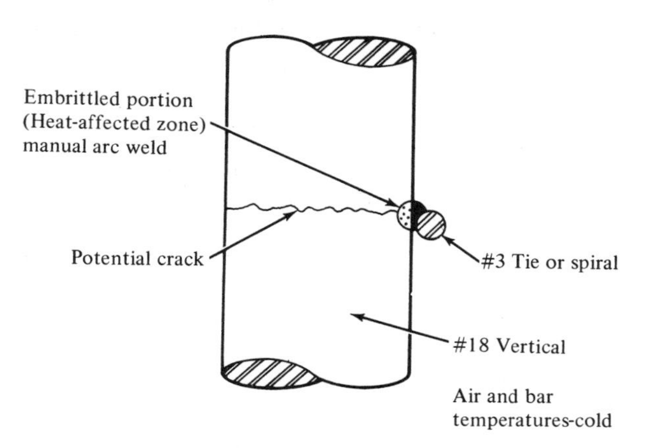

Fig. 1–3 Tack welding under most adverse conditions.

demonstrated commonly in reinforced concrete construction when a column cage is assembled by tack welding relatively small ties to large diameter vertical bars. When the assembly is performed under a combination of most adverse conditions (ambient temperature—cold; tie diameter to vertical bar diameter ratio—minimum), the cage will shatter under any accidental impact. The individual bar failures occur in the large vertical bars, which snap straight across the cross section from the welds nearest the impact.

More scientific tests on concrete beams containing beam cages with stirrups tack welded to the flexural bars (by manual arc welding under laboratory temperatures and conditions) show significant losses in the fatigue strength, etc., of the main bars. Again, the failure occurs in a brittle fashion at the welds.*

SPECIFICATIONS ON TACK WELDING

The Code permits use of only clip-connected bar mats, while the reference specification, ASTM A184-65, allows clipped or tack-welded bar mats (Section 3.5.4). (See also Section 7.3.1, which forbids field welding of crossing bars for assembly of reinforcement without permission of the Engineer.)

REINFORCEMENT—WIRE

Welded wire fabric, deformed and plain, is available under ASTM A497-69 and A185-69, respectively, in diameters up to $\frac{5}{8}$ in. The Code has two requirements in addition to those in the ASTM specifications: (1) for specified $f_y > 60{,}000$ psi, f_y shall be measured at 0.35 percent strain (Section 3.5.5), and (2) welded cross-wire spacings shall not exceed 12 in. for plain wire (Section 3.5.6) or 16 in. for deformed wire (Section 3.5.8).

EVALUATION OF TEST RESULTS ON REINFORCEMENT

It should be noted that the tension test requirements for reinforcing steel and prestressing steel are based upon the statistically impossible concept of an absolute minimum strength, except for the retest provisions of the ASTM specifications, which permit retesting when the specimen itself contains a flaw or when the testing equipment or procedure proves defective. This minimum strength concept has been abandoned for evaluat-

* *Reinforcing Bar Splices,* 2nd Edition, Concrete Reinforcing Steel Institute, Chicago, 1971.

ing the results of tests of concrete for which specific statistical methods and criteria have been provided. (See "Concrete," this chapter.)

In evaluating the test results for both the tensile and bending properties of rebars, some practical recognition of the statistical approach to variations is desirable. In practice the minimum strength concept of evaluating tests on rebars has probably developed and survived only because few projects require a sufficient number of tests for statistical evaluation; the coefficient of a variation is low, usually about 5 percent; and normally rebars are produced to develop an average yield strength equal to 1.10 or 1.15 f_y. Steel is a factory-made material, and all lots of rebars are tested at the point of production regularly. Material failing to meet specifications in this control testing is rejected, and thus removed from the "population" tested by the user from field samples. As graphically shown in Fig. 1-2, however, the indiscriminate imposition of requirements more restrictive than those of the standard ASTM specifications (followed by most producers) can greatly increase the statistical probability of understrength, or the cost necessary to avoid same.

In bend tests, results are inherently even more variable due to the mechanical effect of the deformations. The stress-raising effect can vary not only as the fresh cut roll wears, giving smoother transitions, but also with the orientation of the bar in the bending equipment and the speed of the bending operation. The final test of satisfactory bending is field inspection of the bent bars. The bendability property requirements are somewhat self-enforcing due to the producers' concern with waste involved in breakage during bending.

For small projects where too few tests are required to permit a statistical evaluation, the decision to reject material should not be made without careful retesting, and low tests close to the minimum specified should be considered within the overall framework established for safe tolerances in variables affecting the strength of the final product. In this consideration, the three philosophical criteria used for evaluating concrete tests might be of value: average (material) strength levels are intended to reduce the probability of (1) too many low tests, (2) too many consecutive low tests, and (3) individual tests being disturbingly low (for concrete, more than 500 psi below specified f_c').

CONCRETE MATERIALS

A number of new provisions in the Code regarding concrete materials should be known to the structural engineer responsible for specifications.

Chloride in concrete from any source has been found detrimental to prestressed reinforcement (stress corrosion) and to aluminum embedments (galvanic corrosion).* There is no prohibition of chloride as an admixture elsewhere, subject to the approval of the Engineer (Section 3.6.1). For prestressed concrete and concrete containing aluminum embedments of any type, the Code now limits the accidental chloride ion content of mixing water (Section 3.4.1) and of admixtures (Section 3.6.1). In applying the limit to mixing water, it is intended that free water on the aggregates be included; the maximum total chloride ion content from all sources considered allowable is about 500 ppm (Commentary 3.4.1) including any chloride contributed by the aggregates.

Nonpotable water is to be permitted only after specified comparative cube tests show that it will produce at least 90 percent of the strength achieved with potable water (Section 3.4.2).

CEMENTS

Cement specifications permitted under the Code have been broadened to include portland blast-furnace-slag cement or portland-pozzolan cement without the special curing previously required (Section 3.2.1-c.) One significant additional provision concerning cement is that the type used in the work shall correspond to that upon which the selection of proportions is based (Section 3.2.2).

With the refinements of statistical evaluation controlling concrete quality from establishment of original proportions and changes during work, the importance of variations in the cement must not be minimized. The Code provides no specific limit on the minimum amounts of cement, but limits the water-cement ratios only for concrete which will be exposed to freezing while wet or which is to be made watertight (Sections 4.2.5 and 4.2.6).

Where the original proportions for the concrete strength specified, f_c' are based upon a statistical analysis of production from an established plant (field experience) over a long period, the resulting coefficient of variation will probably include cement variations, and so the requirement that the cement used shall correspond to that upon which proportions are based becomes simply a matter of using the same *type*.

Where laboratory trial batches are the basis for the proportions (Sections 4.2.2 and 4.2.3), the standard deviation used to establish the required excess of average strength above f_c' may be based on records of strength tests within 1,000 psi on similar proportions and materials (not necessarily identical) and similar controls of proportioning (Section 4.2.2.1). To eliminate the effect of cement variations in this situation the specifier should

* *Calcium Chloride in Concrete,* 1952, Highway Research Board, Bibliography 13. Also reports of ACI Com. 222 *Corrosion of Metals in Concrete.*

require that a "corresponding" cement, usually the same type from the same source as used in the trial batches, be used for the work and provide for periodic tests on it, unless the trial batches were made with a blend of the several cements expected to be used. Where cement from several sources is anticipated, it is better practice to require trial batches separately, establishing separate mix proportions for each to allow for the considerable coefficient of variation in strength-producing properties of cement from different sources.

Although of no structural significance (to safety) where a uniform color of the exposed concrete is desired, use of cement of the same type and from the same source will improve uniformity.

AGGREGATES

The Code requires that concrete aggregates conform to ASTM C33 or C330 for normal weight or lightweight aggregate, respectively. An exception provides that nonconforming aggregate with a satisfactory service record is acceptable (Section 3.3), where authorized by the building official. Although the Code contains no specific provision parallel to that for cement requiring that the aggregate used in the structure be the same as in the trial batches, this obvious precaution is included in an overall provision (Section 4.2.2). Proportions are to be based on laboratory trial batches or field experience with the materials to be employed. The aggregate specifications are very broad and include materials within broad ranges of gradation, surface roughness, and particle shape. Conformance to C33 or C330 does not insure that all aggregates so conforming are to be used interchangeably in the same proportions under either specification.

CONCRETE QUALITY

The structural Engineer is required to specify on the plans or in the specifications the concrete strength, f_c', for which each part of the structure is designed (Section 4.1.2). Unless otherwise specified, f_c' is based on the 28-day strength (Section 4.1.4). Field samples for acceptance tests must be laboratory-cured (Section 4.3.1). If additional tests are required to measure adequacy of curing in the structure or to determine when forms may be removed, parallel specimens from the same sampling as the acceptance tests should be cured in the field in accordance with ASTM C31.

The Code provides that the field curing procedures be improved when the field-cured strengths are less than 85 percent of companion laboratory-cured specimens or less than 500 psi above f_c' when 85 percent of the laboratory-cured test results are higher than $f_c' + 500$ psi (Section 4.3.4).

The 1971 Code bases concrete proportioning on statistical considerations, except that the original proportions, where laboratory trial batches are not feasible and no experience record is available, may be based upon a table of maximum water-cement ratios (Table 4.2.4). This table is, of course, conservative and it is limited for application to maximum $f_c' =$ 5,000 psi for nonair-entrained concrete and $f_c' = 4{,}500$ psi for air-entrained concrete (Section 4.2.4).

For structural lightweight concrete, where the splitting strength is used as a design property, provision is made for tensile splitting tests (Section 4.2.9). No specific criteria are given for evaluation of the results, and the Code specifically notes that field tests are not to be used as a basis for acceptance. Unless the design engineer can be assured, by a well-documented series of splitting tests on the structural lightweight concretes available, that the average value will justify higher stresses, it should be conservative to use the values prescribed for use without tests as allowable shear (Section 11.3) and the standard coefficients for development length (Section 12.5-c).

CONCRETE TESTS AND EVALUATION OF RESULTS

The primary interest of the structural Engineer is that the concrete placed will develop the strength, f_c', used for design before the structural element concerned can receive its design load. Code provisions for initial proportioning (Section 4.2.2) and for later adjustment of the initial proportions (Section 4.2.2.2) are intended to develop concrete with an average strength which is always greater than f_c'.

The procedures for evaluating the strength of the concrete produced (Sections 4.3.1, 4.3.2, and 4.3.3) are based upon probability concepts outside the scope of this book. The basic reference for this task is *Recommended Practice for Evaluation of Compression Test Results of Field Concrete* (ACI 214-65). Another is the commentary to the Code.

When it has been determined that the strength of field samples is below acceptable limits, the Code provides for an orderly progression of further testing to evaluate the questionable concrete. If field-cured samples show deficient strengths and the companion laboratory-cured samples give acceptable results, further curing can be required. The code provides for taking core samples which can be utilized to confirm the need for further curing or the results of further curing.

CORE TESTS

If the strength results on laboratory-cured specimens are unacceptable, core tests are prescribed (Section 4.3.5). (If parallel sets of core samples are taken, one set can be tested immediately and one after intensive further curing to evaluate the potential benefit of further curing.) It will also be helpful to take numerous surface hardness (impact) readings in conjunction with any core testing. The results of the impact tests should be corrected by a calibration factor from impact readings on actual test cylinders and cores and the compression tests on same. These tests can be very helpful in locating areas of dubious strength and progress in strength gain under additional curing.

Note that core test criteria are different from criteria for the evaluation of the regular cylinder tests. There is an inherently larger scatter in core test results. Core test results are to be considered adequate if the average of three tests is equal to or greater than 0.85 f_c', and no single core is less than 0.75 f_c' (Section 4.3.5.1).

A common difficulty in securing reliable core test results is the tendency toward using small diameter cores. For faster core drilling, less damaging holes, and the standard 2:1 ratio of length to diameter, core samples in practice are frequently of smaller diameter than desirable in relation to aggregate size. Since cores are cut through, and often loosen pieces of coarse aggregate lying partly within the core, the core samples are more sensitive to maximum aggregate size because of the resulting stress concentrations than are cast cylinders. As a rule, more reliable results will be achieved if the minimum core diameter is the same as that required for test cylinders, even if a correction factor for a smaller length to diameter ratio is required.*

LOAD TESTS

The last resort before structural strengthening, reduced load rating, or removal and replacement is the standard load test (Sections 4.3.5.1 and Code Chapter 20). Even at this point the Code provides for an analytical investigation, a load test, or a combination of the two methods (Section 20.1.1). (See Chapter 17, this book.)

* ASTM C42-68, *Standard Method of Obtaining and Testing Drilled Cores and Sawed Beams of Concrete,* Sections 2.1, 5.1, and 5.7.

2 structural analysis and design—general

DESIGN METHODS

The Code is based upon design in two stages, strength and serviceability. Design for strength (safety) (Section 9.1.1) involves use of load magnification factors for ultimate conditions (Section 9.3) and capacity reduction (ϕ) factors (Section 9.2.1) to allow for various possible understrength effects. The second state of design, serviceability (Section 9.1.2), consists of controls upon the computed deflection and crack widths under service loads (Sections 9.5 and 10.6). This method of design has been called "Ultimate Strength Design" in previous codes.

An alternate design method using the actual service loads, in which all load factors and ϕ-factors = 1.0, is permitted (Sections 8.1.2 and 8.10). Under the alternate method the straight line theory of stress-strain relationship with reduced allowable stresses is employed for flexure only; stresses due to combined flexure and axial load, shear, torsion, and bearing are computed with unit load factors by the same equations as in the principal (strength) design method, but with special capacity reduction factors applied to computed capacities and allowable stresses (Sections 8.10.1 through 8.10.6). Development lengths (bond in previous codes) are the same by both methods of design (Section 8.10.4).

The Code is intended to become part of statutory general building codes by reference. Live loads with permissible reductions thereof as prescribed in the statutory code for application to all types of construction are to be used for reinforced concrete design under the 1971 Code (Section 8.2.1) (See Fig. 2-1). (Note: In the 1971 Code the term "load" means the *actual* dead weight of building elements or the *actual* minimum loads as prescribed for design by the statutory general code. "Design load" means actual load times the load magnification factor prescribed in the 1971 ACI Code.) The Code, including Appendix A, "Special Provisions for Seismic Design," provides for design to resist the lateral loads of wind or earthquake, and requires consideration of the effects of forces from vibration, impact, shrinkage, temperature, creep, and unequal settlements (Sections 8.2.2 and 8.2.3). It will be noted that, except for prestressed con-

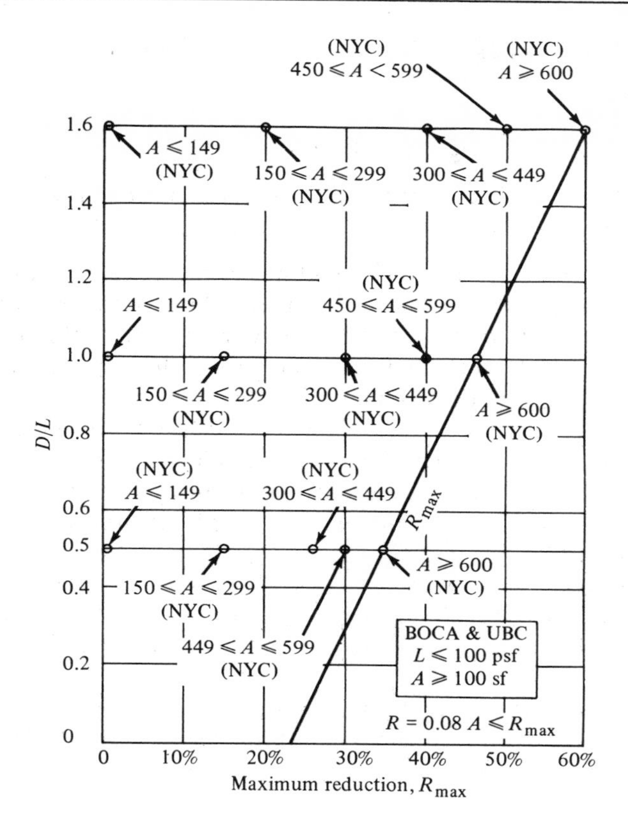

Fig. 2–1 Typical live load reductions.

crete (Section 18.10), provisions for the effect of fatigue have been omitted as not applicable to reinforced concrete in most building construction.

For both strength and alternate design methods, the modulus of elasticity in psi, E_c, for concrete is prescribed to vary with the concrete weight and compressive strength as:

$$E_c = w^{1.5}\, 33\, \sqrt{f_c'} \text{ for concretes between 90 and 155 pcf}$$

or

$$E_c = 57{,}000\, \sqrt{f_c'} \text{ for normal weight concrete (Section 8.3.1)}$$

Although "normal" is not defined, presumably concrete made with all natural aggregate would be normal weight, commonly varying between 140 to 150 pcf. Structural lightweight concrete is defined as concrete containing lightweight aggregate and not exceeding a weight of 115 pcf (Section 2.1).

DRAWINGS AND SPECIFICATIONS

For either design method, certain minimum data are required to be shown on the design drawings or specifications (Section 1.2.1):

1. The *size* and *position* of all structural elements *and* reinforcing steel. "Size and position" of steel includes splice locations, and, for lap splices, lap length; orientation of column verticals within the column; and where different lengths for staggered splices are used, the location of each within the column.
2. Provisions for dimensional changes. Strictly interpreted, such provisions include all joints. Construction joints are not only practical necessities, but the location of construction joints and the resulting sequence and size of elements cast are specific provisions to regulate shrinkage.
3. Specified strength of concrete, f_c', and grade of steel (or f_y). If different strengths of concrete or grades of steel are to be used for different elements on the same drawing, of course, the location of each should be shown or noted.
4. Magnitude, location, and, where applied in increments, sequence of prestressing forces. Magnitude must be completely described as "maximum jacking force," "initial prestress," "final prestress after losses," etc.
5. Live loads and other loads used in the design. (Although not specifically required, it is also desirable to show the design method used, since the Code permits use of two methods.)

Where design calculations are required (by the Building Official), computer design assumptions, input, and output data are acceptable. Results of model analyses are also recognized (Section 1.2.2).

FRAME ANALYSIS BY COEFFICIENTS (Section 8.4)

The approximate method of frame analysis is essentially unchanged from past codes (Section 8.4.2). Note that the load employed, *w*, is the "design load" (Section 8.0), which is the actual load times the load factors (Section 9.3), for use with the moment coefficients (Section 8.4.2). Note also that the limitations upon the use of the approximate method (Section 8.4.2) include the limit of three upon the ratio of live to dead load. Here the term "load" is the actual dead load and the actual specified live load with all reductions permitted by the general code. The limitations also restrict the use of the approximate method to nonprestressed concrete members with uniformly distributed loads.

The moment coefficients supply the maximum points of a moment envelope at the critical sections required for the design. The theoretical moment envelope consists of maximum moments within the limitations on adjacent unequal spans with live loads applied to all, alternate, or adjacent pairs of spans. In the approximate method of analysis, theoretical maximum negative moments, at centers of supports, have been reduced to allow for usual width of supports and some moment redistribution. Theoretical maximum positive moments have been increased for such redistribution. No further redistribution (limit design) is permitted with the moments calculated using the coefficients of this section (Section 8.6).

Coefficients are not prescribed for negative moments that can develop at the centers of spans, with L/D ratios equal to or greater than 1.0 in the shortest spans permitted under the limitations on the approximate method. See Fig. 2-2 for illustrations of the various coefficients of the approximate frame analysis method. Under the general "frame analysis" method (Section 8.5), it can be expected that some top reinforcement will often be required continuously across short spans. See Figs. 2-3 and 2-4 for a comparison of design moments by the two methods of analysis for an exmaple within the limitations of the approximate method.

An important advantage to using the approximate method of analysis when the loads and the frame dimensions are within the limitations prescribed, is that a "three dimensional" analysis considering torsional stiffnesses of the members at right angles to the frame is not required. Such an analysis is required with the general frame analysis whenever the flexural members are integral with their supports (Section 8.5.3.1) for beam-column frames, and (Section 13.4.1.5) two-way slab and column frames with or without beams).

EXAMPLE. To evaluate the approximate frame analysis (by coefficients) under typical, real conditions, consider a joist-slab continuous across alternate spans of 21′–0″ and 25′–0″ between centerlines of supports *not* integral with the joist-slab (worst condition for slab moments). The supports, if masonry walls, can be taken as 12 in. thick. Use a joist-slab with total depth of 13 in. (10 + 3) with 6 in. ribs @ 26″. This depth is less than the minimum (Section 9.5.2.1 and Table 9.5-a), and so computed deflections must be compared with those permitted for conditions (Table 9.5-b.). Clear span l_n = 21′–0″–12″ (support width) + 13″ depth) $\leqq$ 21′–0″ (center to center span) (Section 8.5.2.1). End span, l_n = 21′–0″; similarly, interior spans, l_n = 25′–0″

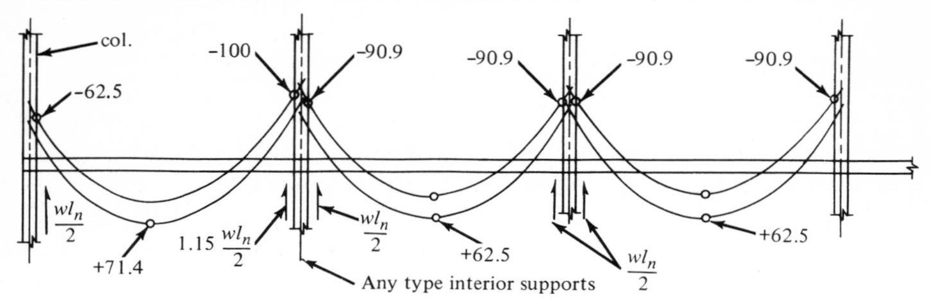

wl_n^2 = 1,000 – Discontinuous end integral with column; equal spans

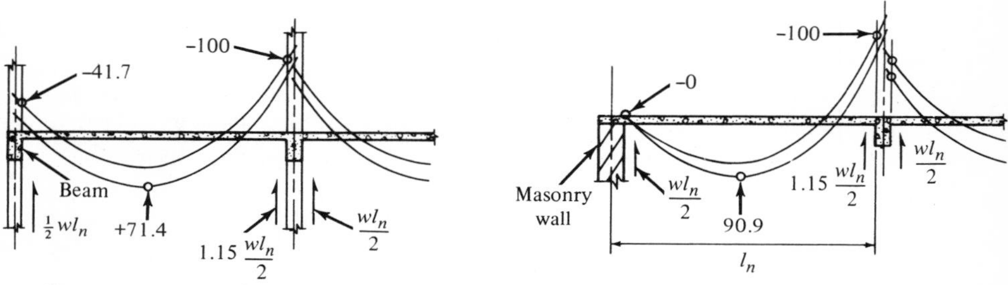

wl_n^2 = 1,000 – Discontinuous end integral with support; spans equal

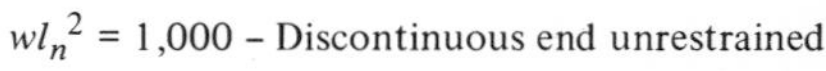

wl_n^2 = 1,000 – Discontinuous end unrestrained

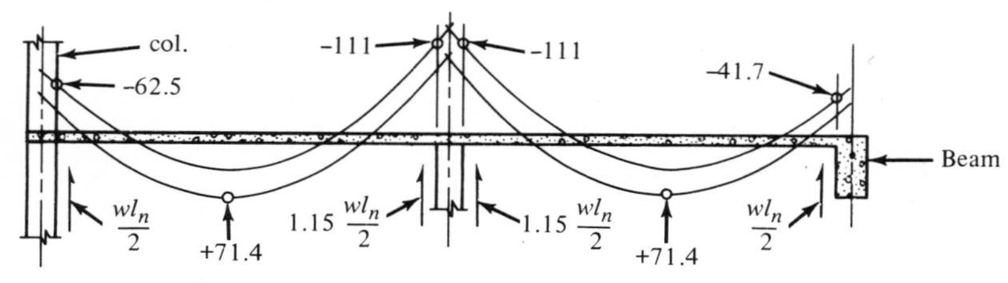

wl_n^2 = 1,000 – Two equal spans – integral supports

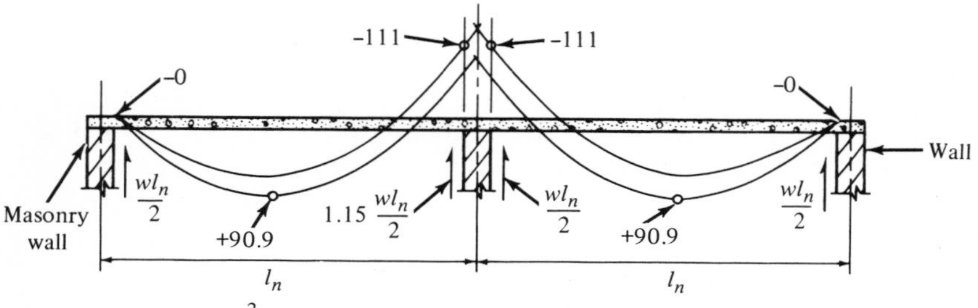

wl_n^2 = 1,000 – Two equal spans; ends unrestrained

Fig. 2–2 Approximate frame analysis coefficients.

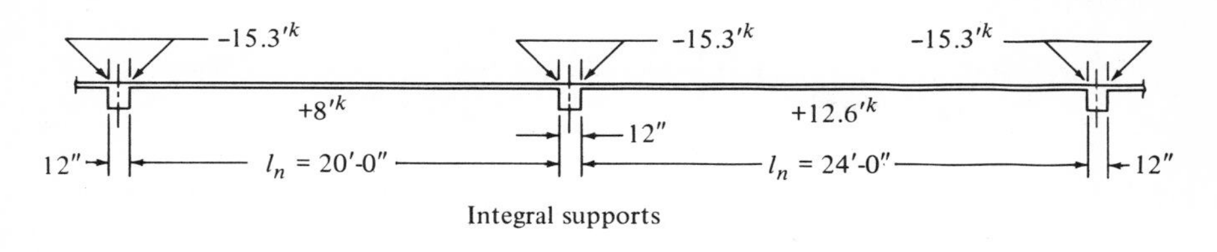

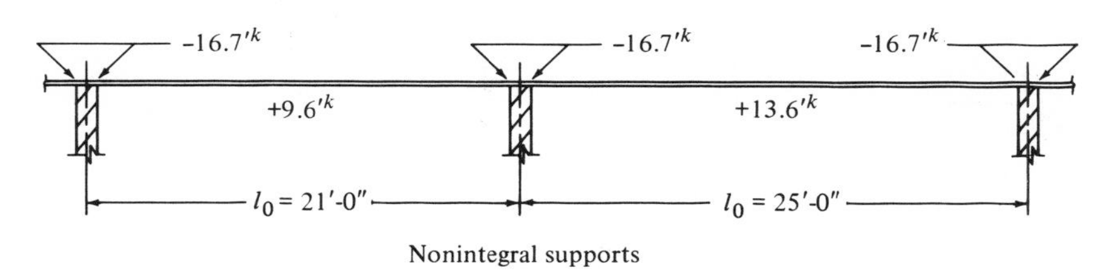

Fig. 2–3 Approximate analysis; $-M$ at faces of supports.

Let the design live load, w_l = 250 psf. (Actual specified live load after reductions permitted, L = 250/1.7 = 147 psf.) The dead weight of the joist-slab, D = 70 psf for normal weight concrete. Design dead load, w_d = 1.4 D = 98 psf. Total design load, w = 348 psf.

The approximate frame analysis provides moment coefficients to be applied to the factors, wl_n^2. Note that l_n for the negative moments is the average of the adjacent clear spans at a support (Section 8.0).

For a 1 ft wide strip, these factors are:

End, $l_n = 21'$; $wl_n^2 = (0.348)(21)^2 = 153$ ft-kips

Interior, $l_n = 25'$; $wl_n^2 = (0.348)(25)^2 = 217$ ft-kips

Average, $l_n = 23'$; $wl_n^2 = (0.348)(23)^2 = 184$ ft-kips

Maximum design moments for interior spans per strip 1 ft wide, at the interior face of the first interior support and at the faces of all other interior supports,

$$-M_u = -\frac{1}{11}(184) = -16.7 \text{ ft-kips}$$

At the center of the 21 ft interior spans,

$$+M_u = +\frac{1}{16}(153) = +9.6 \text{ ft-kips}$$

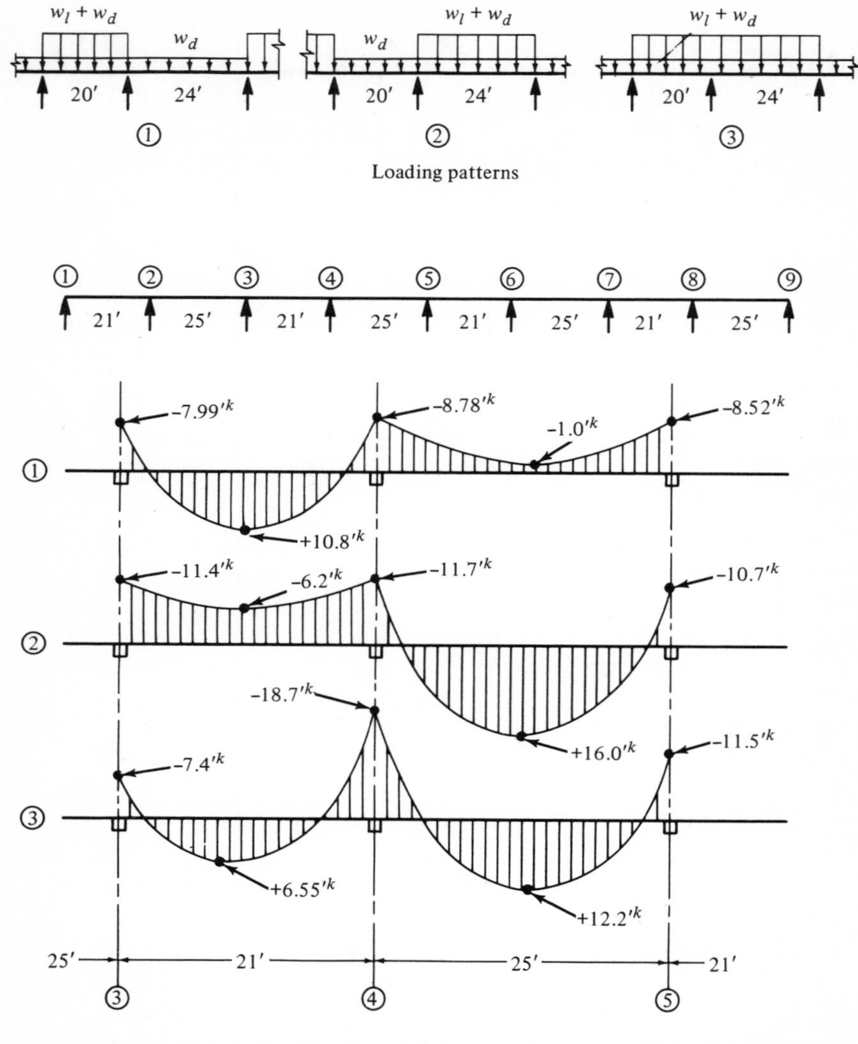

Fig. 2–4 Frame analysis moments—midspan and centers of supports.

At the center of the 25 ft interior spans,

$$+M_u = +\frac{1}{16}(217) = +13.6 \text{ ft-kips}$$

The preceding example was selected to show the application of the definition of clear span to nonintegral supports (Section 8.5.2.1). Although torsional stiffness of integral supports is not directly considered in the approximate frame analysis, clear span is measured from the face of each support. If integral supports were provided of the same width as in the preceding example, say 12 in. wide beams monolithic with the joist-slab system, the clear spans would become 20 ft and 24 ft, respectively.

For a 1 ft wide strip, wl_n^2 factors become:

$$\text{End } l_n = 20 \text{ ft}; \qquad wl_n^2 = 139 \text{ ft-kips}$$

$$\text{Interior } l_n = 24 \text{ ft}; \qquad wl_n^2 = 201 \text{ ft-kips}$$

$$\text{Average } l_n = 22 \text{ ft}; \qquad wl_n^2 = 168 \text{ ft-kips}$$

Maximum design moments for interior spans per 1 wt wide strip, at the interior face of the first interior support and at the faces of all other interior supports.

$$-M_u = \frac{1}{11}(169) = -15.3 \text{ ft-kips}$$

At the center of the 20′ interior spans,

$$+M_u = \frac{1}{16}(140) = +8.75 \text{ ft-kips}$$

At the center of the 24′ interior spans,

$$+M_u = \frac{1}{16}(202) = +12.6 \text{ ft-kips}$$

See Fig. 2-3 for a summary of these design moments.

APPROXIMATE METHOD VERSUS FRAME ANALYSIS

The maximum moment values for interior spans in Fig. 2-4 were derived from a frame analysis applied to the preceding example of approximate analysis (Section 8.5). Three loading patterns were considered as prescribed (Section 8.5.1.2):*

---

* Note that *design* loads are employed for the various loading patterns. At the limiting ratio of live to dead load, $L/D = 3.0$, for approximate analysis, the ratio of *design* loads $w_l/w_d = \frac{(1.7)}{(1.4)} 3 = 3.64$.

1. Design dead and live load, $w_d + w_l$, all spans
2. w_d all spans with w_l on adjacent spans
3. w_d all spans with w_l on alternate spans

Maximum values from these three analyses can be combined to form a moment envelope. Using the frame analysis method, if the torsional stiffness of integral supports can be neglected, the design moments will be the same for integral or nonintegral supports where depth of the member is the same or more than the width of the nonintegral supports (Sections 8.5.2.1 and 8.5.2.2.). Center-to-center spans are used for determination of the moments by frame analysis for both of the preceding examples.

For comparison to the maximum design moments by the approximate analysis, the negative moments at the center of supports in Fig. 2-4 must be reduced to the face of supports. Reduce all negative moments by (1/2) $(wl)\left(\frac{6}{12}\right)$ ft-kips using total w. Compare negative moments by the frame analysis, − 16.4 ft-kips, to the average negative moment by approximate analysis, − 16.7 ft-kips. Maximum positive moments in the long spans for frame analysis versus approximate analysis are + 16.0 ft -kips versus + 13.6 ft-kips. These values are well within the redistribution limit allowed with the general frame analysis (Section 8.6). See Fig. 2-5.

The comparison of the results from the frame analysis to the results from the approximate analysis for integral supports is also within the redistribution limit allowed for usual steel ratios, though not as close as for nonintegral supports. The greatest difference, in positive moments, + 10.8 versus +8.7 and + 16.0 versus + 12.6, would be reduced somewhat in the usual structure where the usual beam-column support stiffness would be significant (Section 8.5.3.1).

The only significant difference is the minimum midspan moment for the short span which is −6.2 ft-kips by the frame analysis and 0 by the coefficients. Rigorous application of the frame analysis procedures (Section 8.5) would require top tensile reinforcement continuous across the short spans. Decades of experience with approximate analysis have shown such reinforcement, within the limitations of 20 percent difference in spans and L/D ratios not greater than 3.0, to be usually unnecessary. Some negative moment can be resisted by concrete tension, approximately 8.5 ft-kips in this example. If tensile strength is exceeded, top cracking would occur and a significant redistribution of moments would result.

The general absence of such cracks in actual applications where $L/D \leq 3.0$ indicates that the stress is less than the concrete tensile strength, perhaps because the rigorous pattern of full ultimate live load on alternate long spans and none on alternate short spans occurs very seldom.

In order to avoid unnecessary top steel and to save time, the authors recommend use of the approximate method wherever the limitations permit and alternate pattern live loadings with full live load are not anticipated. When either the live/dead load or the span ratios approach the limitation and full live load in alternate spans can be expected, the approximate method can be employed to evaluate negative moment at the center of short spans.

For the preceding example the approximate analysis can be extended to evaluate pattern live loading as follows:

$$\text{Total load in long spans, } -M_u = \frac{(0.35)(24)^2}{11} = -17.6 \text{ ft-kips}$$

$$\text{Dead load only in short spans, } -M_u = \frac{(0.10)(20)^2}{11} = -3.6 \text{ ft-kips}$$

$$\text{Average } -M_u = \frac{17.6 + 3.6}{2} = -10.6 \text{ ft-kips}$$

$$\text{Simple span moment in short span, } +M_u = \frac{(0.10)(20)^2}{8} = +5.0 \text{ ft-kips}$$

Negative moment

$$\text{at center of short span, } -M_u = -10.6 + 5.0 = -5.6 \text{ ft-kips}$$

It will be noted that the result by this approximate analysis is within 10 percent of the frame analysis result (Fig. 2-4).

The authors recommend further that, if the frame analysis is used for conditions *within* the limitations for approximate analysis, the probability of full live load in alternate spans be evaluated and any resulting negative moments at midspan be adjusted accordingly for design.

Where full live load in alternate spans is expected and an approximate analysis indicates negative moments, the authors recommend use of the frame analysis method, particularly for elements forming the primary framing system. It will, of course, frequently be desirable to employ a more accurate analysis, even where the code permits use of the coefficients, to determine unbalanced moments to columns, to consider effects of torsional stiffness or width of wide integral supports, etc. (see Chapter 9).

REDISTRIBUTION OF NEGATIVE DESIGN MOMENTS (Limit Design)

In the previous example, frame analysis results for the negative design moments for the 25 ft. span and the reductions permitted by redistribution

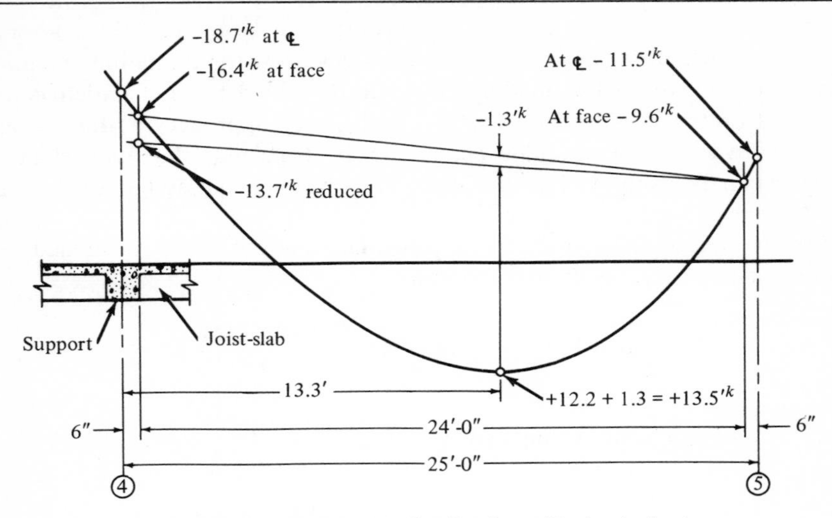

Fig. 2-5 Moment redistribution (limit design).

are shown in Fig. 2-5. Such redistribution is permitted when the net steel ratio $\rho - \rho' \leqq 0.5\ \rho_b$ (Section 8.6.1). For Grade 60 bars and $f_c' = 4{,}000$ psi,

$$\rho_b = \frac{0.85\ \beta_1 f_c'}{f_y} \times \frac{87{,}000}{87{,}000 + f_y} \qquad \text{ACI Eq. (8-1)}$$

$$\rho_b = 0.0285$$

For the joists on a span of 25 ft., reinforcement could consist of four #4 straight top bars and one #5 + one #6 straight bottom bars, as shown in Fig. 2-6.

Average width, $b = 6.5$ in.

Effective depth, $d = 12$ in.

$$\text{At the face of the supports, } \rho = \frac{4 \times 0.20}{6.5 \times 12} = 0.010$$

Net ρ:

Assuming the two bottom bars are made the same length, and the minimum required embedment of 6 in. (Section 12.2.1) is provided for both, the

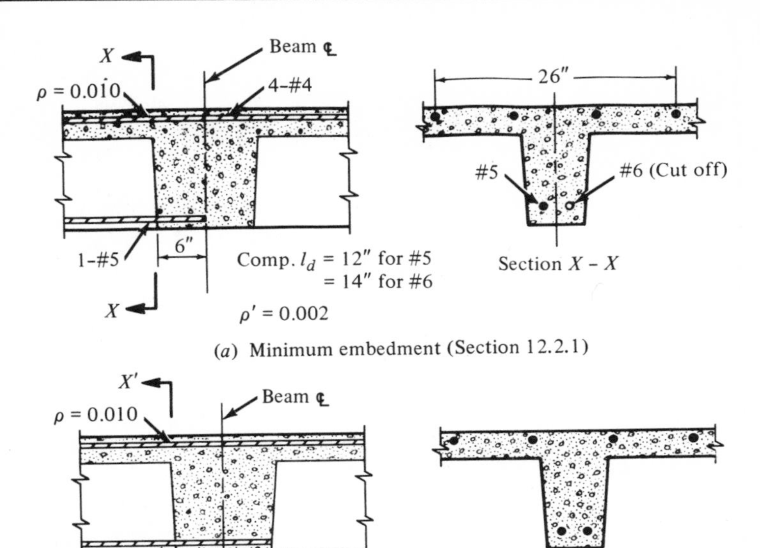

(*a*) Minimum embedment (Section 12.2.1)

(*b*) Embedment (Section 12.6.1)

Embed

1 – #5 into support 6 in. (Minimum allowed): $\rho - \rho' = 0.008$

Both bottom bars 6 in. (practical minimum): $\rho - \rho' = 0.005$

Both bottom bars 12 in. (practical maximum): $\rho - \rho' = 0.007$

Fig. 2–6 Embedment and effectiveness of bottom bars as compressive reinforcement.

bottom bars can be assumed 50 percent effective for compression at the face of the support (see Table 13-1).

$$\rho' = \frac{(0.44 + 0.31)}{6.5 \times 12}(0.50) = 0.005$$

$$\rho - \rho' = 0.010 - 0.005 = 0.005 \text{ at the face of the support}$$

Negative moment may be decreased* for design,

* $-M$ may also be increased in a like amount if desired (Section 8.6.1). The authors do not recommend increasing the negative moment for design. For practical reasons it is not desirable to increase congestion of the reinforcing steel at supports; also any error in placing steel is likely to reduce effectiveness of top steel.

$$(20)\left(1 - \frac{\rho - \rho'}{\rho_b}\right) \text{ percent (Section 8.6.1)}$$

$$(20)\left(1 - \frac{0.005}{0.0285}\right) = 16.5 \text{ percent}$$

The adjusted maximum negative moment for design becomes (16.4) (1 − 0.165) = −13.7 ft-kips. No adjustment should be made to the less-than-maximum negative moment at the other support. This adjustment, 16.4 − 13.7 = 2.7 ft-kips, is compatable with an increase in the positive design moment to + 12.2 + (1/2) (2.7) = + 13.5 ft-kips for this loading condition. No increase in the bottom steel is required since the maximum positive design moment is + 16.0 ft-kips (Fig. 2-4).

PRACTICAL USES OF MOMENT REDISTRIBUTION

Note that the maximum reduction of negative reinforcement permitted, 20 percent, is reached when $\rho - \rho' = 0$. The provision offers an opportunity for savings in the total weight of reinforcement for continuous construction, particularly for heavily reinforced long span beams. A very short extension of the bottom bars into the supports, usually 20 bar diameters for full compression embedment (Sections 12.6.1), will often permit substantial further reduction of the top bar area over a length equal to more than 0.6 span.

In this short span example, a further embedment of 7 in. for full compression would permit the full reduction of 20 percent in top steel. Added weight

$$= \frac{7}{12}(1.5 + 1.0) = 1.5 \text{ lb per joist}$$

Weight reduced =

$$(20.0 - 16.5)[1.0 + 0.3(20 + 24)](4)(0.668) = 1.4 \text{ lb per joist}$$

If all or some of the joists in this example are part of the "primary lateral load resisting system," full *tension* embedment is required (Section 12.2.2). Note that the redistribution of negative moment procedure will often enable the designer to provide this additional bottom steel with no increase in total steel weight.

STRENGTH DESIGN

The first and primary stage of design is for strength (safety). The 1971 Code requires that all structural members and structures have a minimum computed strength at each section equal to the effects of prescribed design loads and forces upon the section (Section 9.1.1). Load magnification factors and strength reduction (ϕ) factors are separately prescribed, and are intended together to provide against *overloads* above the minimum live load, wind load, etc., prescribed by the general code, *overrun* of estimated dead weight of materials, and *understrength* resulting from normal variations in the strength of reinforced concrete. These factors do not, and are not intended to, protect against gross errors or negligence in analysis, design, or construction.

The Code prescribes in general an elastic analysis of the structure (exception, "Redistribution," Section 8.6.1). Except for simple span construction, in all monolithic reinforced concrete structures or where the connections of various elements are made integral, some of the sections can yield without causing failure of the structure and thus some additional load capacity is available. With the one exception noted, this additional capacity is conservatively neglected under the Code. Critical sections are designed on the basis of an elastic analysis of the structure, under the magnified design loads, for yield stress in the reinforcement, reduced by specified ϕ-factors (Sections 9.2 and 9.3). Examples showing applications for strength design to various common structural elements and systems are offered in Chapters 3 through 14.

General Assumptions for Strength Design—Flexure with No Axial Load

Strength design is based on the assumptions diagramed in Fig. 2-7 and on the satisfaction of applicable conditions of equilibrium and compatibility of strains. The Code permits the use of any shape of stress block that will result in strength predictions that are in agreement with available tests, but only the rectangular stress block (developed by Whitney) is prescribed in detail (Section 10.2.6) and 10.2.7). The authors recommend use of the Whitney formulas exclusively for manual design, and for simplicity of programming computer design.

For a summary of the design assumptions, see Fig. 2-7. See also Fig. 10-6 for applications of these assumptions to column design. ϕ in the above expressions for flexure is 0.90 (Section 9.2.1). The effects of axial loads, compression or tension, must be included with the effects of flexure. See Chapter 10 for examples.

The most important requirement for strength design involves the definition of "balanced conditions," and the limitation of tensile reinforcement based thereon. The entire concept of strength design for flexure utilizes the capacity reduction factor, $\phi = 0.90$, and is based on a gradual yield-

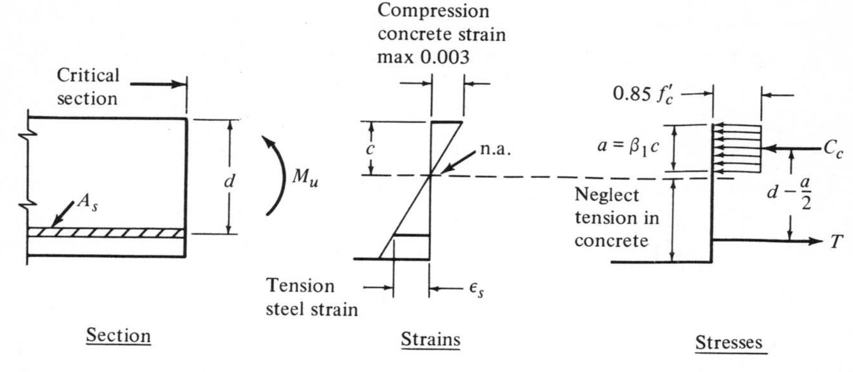

Note: Strains are directly proportional to the distance from the neutral axis, n.a.; 0.003 is the assumed maximum strain usable for the concrete in compression (Section 10.2.3).

For $f_c' \leqslant 4{,}000$ psi, $\beta_1 = 0.85$
For $f_c' \geqslant 4{,}000$ psi

$$\beta_1 = 0.85 - 0.05 \frac{(f_c' - 4{,}000)}{1{,}000}$$

(a) Singly-Reinforced Sections: Resisting moment, $M_u = \phi A_s f_y \left(d - \frac{a}{2}\right)$

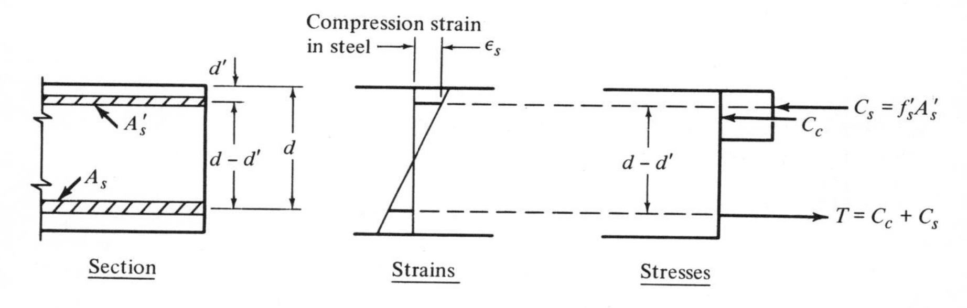

Other strains and stresses same as in (a). The stress in the compression steel $f_s' = (29 \times 10^6)(\epsilon_s) \leqslant f_y$. For all ϵ_s values greater than that corresponding to a stress, f_y, the compression stress is taken as f_y.

(b) Doubly-Reinforced Sections:

Resisting moment, $M_u = \phi C_c (d - a/2) + \phi C_s (d - d')$

$$= \phi \left(A_s - A_s' \frac{f_s'}{f_y}\right) (f_y)(d - a/2) + \phi A_s' f_s' (d - d')$$

Fig. 2–7 General assumptions for strength design (Section 10.2).

ing tensile failure. Higher ϕ-factors are prescribed where a brittle compression failure is to be expected, as in columns.

Balanced conditions are defined as simultaneously reaching a compres-

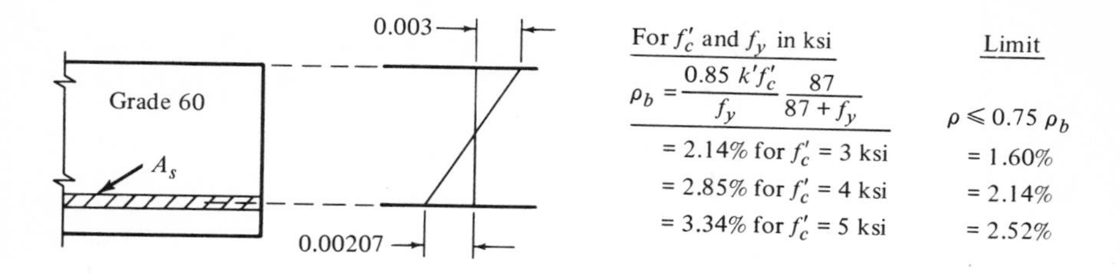

(*a*) Limits on tensile steel – singly reinforced elements

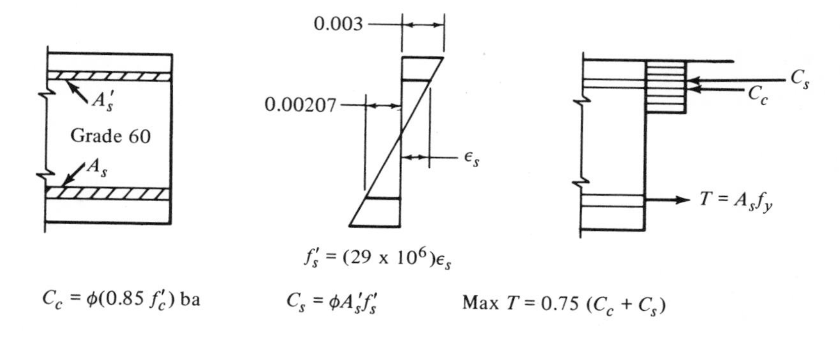

(*b*) Limits on tensile steel – doubly reinforced elements

Fig. 2–8 Limiting reinforcement in flexure (Section 10.3.2).

sion strain in the outer fiber of the concrete of 0.003 (crushing) and tensile strain in the tensile reinforcement corresponding to that for the specified yield point, f_y (Section 10.3.3) (see Fig. 2-8). The Code limits tensile reinforcement to 0.75 of that producing balanced conditions in flexure to produce ductile behavior of the structure in general and to preclude brittle failure at any point under design loads. To achieve ductile behavior, the tension steel must yield before the concrete in compression crushes. With the prescribed limit of 0.75 times balanced tensile reinforcement, a yielding failure will occur under any of the following combinations of material strengths assuming construction to the design dimensions:

Let f_{ya} = actual yield strength of steel; f_y = specified

f_{ca}' = actual concrete strength; f_c' = specified

1. $f_{ya} \leqq f_y$ and $f_{ca}' \geqq 0.75\ f_c'$

2. $f_{ya} \leqq 1.33\ f_y$ and $f_{ca}' \geqq f_c'$

3. $f_{ya} \geqq 1.33\ f_y$ and $f_{ca}' \geqq (f_{ya}/f_y - 0.33)(f_c')$

When condition (3) is exceeded, the failure will be brittle, but will not occur until the actual loading is at least one-third larger than the design loading.

The limit on the amount of tensile reinforcement applies also to columns under small axial load where the ϕ-factor is gradually reduced from $\phi = 0.90$ for zero axial load to $\phi = 0.70$ for axial load equal to 0.10 $f_c'A_g$ (Section 9.2.1.2-d). See interaction diagrams for columns from axial load zero to maximum, Chapter 10.

SERVICEABILITY

Satisfactory levels of safety (strength) do not alone ensure serviceability of a structure under its ordinary conditions of "service" loads, exposures, special requirements, etc. In this sense, serviceability is defined as acceptable conditions for the intended use of the structure, including the deflection of floors and roofs from level; crack widths for exposure conditions; stiffness against vibration, sway, or other movement; and any other structural property required.

Unnecessary additional computations for serviceability can be avoided in the design of most routine elements. For both one-way reinforced and two-way reinforced construction the Code prescribes minimum thickness limits above which deflections can be assumed to be tolerable (Section 9.5.2, one-way; Section 9.5.3, two-way).

DEFLECTION

If lesser thicknesses must be used or if deflection must be computed for any reason, standard procedures for such computations are given (Section 9.5.2) and acceptable limits on deflection so computed are given (Table 9.5.-b). The simplest illustration of possible variation between design for strength only and design for the serviceability requirement that elastic deflection under live load be less than $l_n/360$ (Table 9.5-b for floors not supporting nor attached to nonstructural elements likely to be damaged by large deflections) is a one-way slab (See Fig. 3-8).

CRACK CONTROL

Limiting crack widths were established from measured values observed to provide satisfactory resistance to various conditions of exposure. These limits have been converted into limiting parameters "z" (Section 10.6.3). An equation for the computation of z-values based on laboratory research is given (Section 10.6.2, Eq. 10-2). If reinforcement is distributed as required (Section 10.6) to result in computed z-values less than the limits prescribed for various exposures, the crack widths in one-way construction should be satisfactory.

Calculation of z-values for each element of construction could be tedious. For practical application, the authors recommend that users convert the equation for z to solve for the maximum bar spacings. The maximum bar spacings can be tabulated for each size bar, each standard cover specified, and for each z-value. Such tabulations can be prepared for any special conditions of exposure and cover desired. See Chapter 3 for bar spacings in one-way slabs, common cover, interior and exterior.

3 one-way reinforced concrete slabs

GENERAL

The simplest routine structural element for illustration of design provisions in the Code is the one-way slab. A one-way slab is defined for purposes of this book as a flexural member with depth small relative to other dimensions, supporting loads applied normal to and directly over its surface, spanning in one direction between parallel supports, and reinforced for flexure in that direction only. For purposes of analysis, one-way slabs may be restrained to any degree at the supports or may be unrestrained. A number of Code provisions refer to "flexural members," which include one- and two-way slabs, beams, girders, footings, and, where bending exists with the axial load, walls and columns. In general, when the Code provision is intended to apply to one-way slabs, the term will be used in the sense of the definition herein. For example, the Code requires stirrups in most beams, but specifically exempts slabs and footings from the requirement (Section 11.1.1). Where the Code term "structural slabs" is employed, as in the provision that the minimum reinforcement in the direction of span shall be that required for temperature and shrinkage (Section 10.5.2), the requirement applies to both one- and two-way slabs as herein defined. In this book, one-way slabs and beams are discussed separately. Beams are defined herein as one-way flexural members designed to support concentrated line loads such as slabs, girders as one-way flexural members designed to support concentrated loads such as beams. Either beams or girders or two-way slabs may be utilized as primary members of a frame. For design of beams, see Chapter 9. For design of two-way slabs, see Chapters, 5, 6, and 7.

TYPES OF ONE-WAY SLABS

One-way slabs may be solid, hollow, or ribbed. When ribbed slab dimensions conform to the standard dimensional limitations for joist-slab construction (Section 8.8) special Code provisions apply. (For design of one-way joist-slab systems, see Chapter 4.) Ribbed slabs larger than the minimum limits are designed as slabs and beams (Section 8.8.3). Various shapes of totally enclosed void volumes can be incorporated into one-way slabs. Special hollow sections (precast blocks) are commonly used in various proprietary precast planks to form one-way slabs. Proprietary longitudinal void core forms can be employed for cast-in-place construction of one-way slabs. Some of these forms are inflatable and may be reused; others, such as fiber or cardboard cylinders, are used once and left in place. Since most of the hollow slab designs are special, either proprietary or requiring unique computations, they are outside the scope of this book. One-way slabs need not be of uniform depth; they may be haunched. Where haunches are used, the effect of the haunches on stiffness, moment, flexural and shear strength, and deflection must be considered (Section 8.5.3.2.). The discussion and examples following in this chapter will be limited to the common one-way slab, solid, of uniform depth, and designed for uniformly distributed loads.

SLAB REINFORCEMENT

Reinforcement for one-way slabs commonly consists of straight bars. Welded wire fabric or combinations of straight and trussed bars are also often used where some special feature of the structure makes them advantageous. The maximum reinforcement permitted for slabs can be easily computed (see Fig. 2-8). The minimum reinforcement permitted (Section 10.5.2) is that required for temperature and shrinkage (Section 7.13). See Table 3-1 for a summary of both maximum and minimum limits on reinforcement for common slab thicknesses.

TABLE 3-1 Minimum and Maximum Areas of Steel in One-Way Slabs

LIMITING AREAS OF GRADE 60 REINFORCEMENT (IN.^{2}/12 IN. WIDTH)					
Limit	ONE-WAY SLAB THICKNESS, h				
	4 in.	5 in.	6 in.	7 in.	8 in.
Minimum $A_s = 0.0018\ bh$[a]	0.0865	0.108	0.130	0.152	0.173
Maximum $\rho = 0.75\ \rho_b$[b] where $\rho = A_s/bd$					
For $f_c' = 3{,}000$ psi[c] $0.75\ \rho_b = 0.0161$	0.578	0.761	0.955	1.15	1.34
For $f_c' = 4{,}000$ psi[c] $0.75\ \rho_b = 0.0213$	0.768	1.01	1.26	1.52	1.77

[a] As limited by Sections 10.5.2 and 7.13.1.

[b] Max $\dfrac{A_s}{bd} = 0.75\left[\dfrac{(0.85)\ (\beta_1)\ (f_c')}{f_y}\ \dfrac{(87 + f_y)}{(87)}\right]$.

[c] Applicable to bar sizes No's. 3, 4, 5 with ¾ in. clear cover for interior exposures.

SERVICEABILITY

Serviceability requirements of the Code for one-way slabs involve maximum computed crack width parameters and deflections (Sections 10.6 and 9.5, respectively). For one-way solid slabs these limitations can be expressed most easily as maximum bar spacings and minimum thicknesses to avoid need for any added calculations. See Table 3-2 for bar spacings and Table 3-3 for minimum thicknesses.

TABLE 3-2 Maximum Bar Spacings for Crack Control in One-Way Slabs

Grade 60 Bar Sizes	MAXIMUM BAR SPACINGS FOR CRACK CONTROL[a]		
	INTERIOR EXPOSURES		EXTERIOR EXPOSURES
	Clear cover[b] = 3/4 in.	Clear cover[d] = 1 in.	Clear cover[b] = 1-1/2 in. for No's 3, 4, 5; = 2 in. for No. 6
#3	(46″) $\leqq 3h \leqq$ 18″[c]	(29″) $\leqq 3h \leqq$ 18″[c]	8″
#4	(41″) $\leqq 3h \leqq$ 18″	(26″) $\leqq 3h \leqq$ 18″	7½″
#5	(36″) $\leqq 3h \leqq$ 18″	(23″) $\leqq 3h \leqq$ 18″	7″
#6	(32″) $\leqq 3h \leqq$ 18″	(21″) $\leqq 3h \leqq$ 18″	4″

[a] For Grade 60 bars, use f_s = 36 ksi (Section 10.6.2); spacings to satisfy $z = f_s \sqrt[3]{d_c A}$ (Eq. 10–2) for values of z = 156 for interior exposure; z = 129 for exterior exposure (Section 10.6.3, maximum z = 175 and 145, respectively, for beams; Commentary, z for slabs = (1.2/135) (z) for beams).
[b] To satisfy cover requirements for cast-in-place concrete (Section 7.14.1.1).
[c] Spacings are limited to $3h$ but not more than 18 in. (Section 7.4.3).
[d] Required under some general codes for maximum fire rating.

TABLE 3-3 Maximum Spans for One-Way Slabs Without Computation of Deflections

Slab Thickness (inches)	MAXIMUM CLEAR SPANS ℓ_n[a]			
	Simple Span (ℓ_n/20)	One End Continuous (ℓ_n/24)	Both Ends Continuous (ℓ_n/28)	Cantilever Span (ℓ_n/10)
4	6′–8″	8′–0″	9′–4″	3′–4″
4½	7′–6″	9′–0″	10′–6″	3′–9″
5	8′–4″	10′–0″	11′–8″	4′–2″
5½	9′–2″	11′–0″	12′–10″	4′–7″
6	10′–0″	12′–0″	14′–0″	5′–0″
6½	10′–10″	13′–0″	15′–2″	5′–5″
7	11′–8″	14′–0″	16′–4″	5′–10″
7½	12′–6″	15′–0″	17′–6″	6′–3″
8	13′–4″	16′–0″	18′–8″	6′–8″
8½	14′–2″	17′–0″	19′–10″	7′–1″
9	15′–0″	18′–0″	21′–0″	7′–6″
9½	15′–10″	19′–0″	22′–2″	7′–11″
10	16′–8″	20′–0″	23′–4″	8′–4″

[a] (Sections 9.5.2.1, 9.5.3.4, and Table 9.5-a).

ANALYSIS

One-way slabs may be analyzed by the approximate method (coefficients for maximum shears and moments at critical sectons) or by the frame analysis method (Sections 8.4.2 and 8.5, respectively) (see Figs. 2-1, 2-2, 2-3). As noted in Chapter 2, the authors strongly recommend the use of approximate analysis where conditions of design permit, for maximum efficiency of design time and use of material.

EXAMPLE. ANALYSIS AND DESIGN. Design a one-way slab system for a nominally flat roof (where ponding is prevented by positive drainage). All spans are 16′ –0″ from center-to-center of 12 in. wide integral supports. Clear spans are 15′ –0″. Superimposed loads consist of 30 psf live and 16 psf dead. Use Grade 60 bars and a concrete strength, $fc' = 4{,}000$ psi.

Step 1: Thickness

See Table 3-3 for $l_n = 15' -0''$.

For end span, min $h = 7\text{-}1/2$ in., and for interior span, min $h = 6\text{-}1/2''$, unless computed deflections are within allowable limits.

Since the superimposed loads are very light and the maximum immediate deflection allowable under live load, L, is $l_n/180 = 1$ in., a 6 in. thickness will be assumed. Immediate deflection must be computed to verify that it is less than 1 in. (Section 9.5.2.1 and Table 9.5-b).

Step 2: Analysis by Coefficients (Section 8.4)

Design load, $U = 1.4\ D + 1.7\ L$ (Section 9.3, Eq. 9-1).

$$w_d = 1.4(75 + 16) = 127 \text{ psf};\ w_l = 1.7 \times 30 = 51 \text{ psf};\ w = 178 \text{ psf}$$

Shear at the first interior support, $V_u = (1/2)(15)(0.178)(1.15) = 1.53$ k.

Shear at all other supports, $V_u = 1/2(15)(0.178) = 1.33$ k.

$$wl_n^2 = (0.178)(15)^2 = 40.0 \text{ k-ft per 12 in. width of slab}$$

(Critical section for negative moment is at the face of the supports; for shear, at a distance, d, from the face of supports.)

Step 3: Determination of Steel Area Required and Bar Sizes for Use

The minimum steel (Section 7.13.1) permitted, $A_s = 0.0018$ bh.

$$\text{Min } A_s = (0.0018)(6 \times 12) = 0.130 \text{ in.}^2/\text{ft (See Table 3-1.)}$$

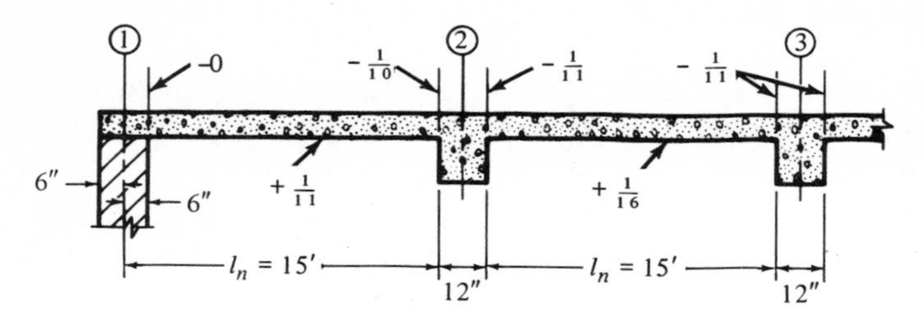

Fig. 3–1 Moment coefficients of factor, wl^2.

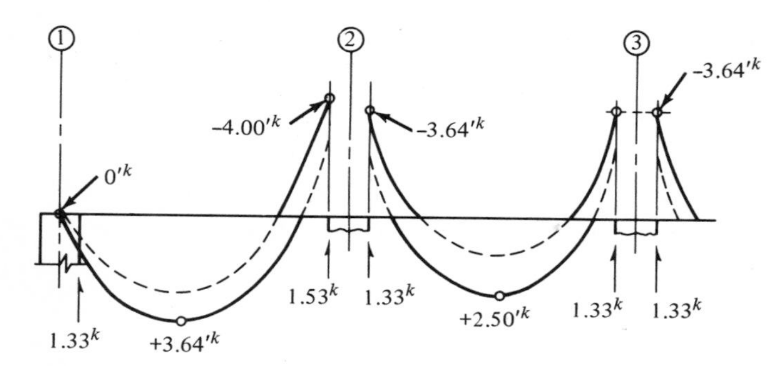

Fig. 3–2 Design moments and shears at critical sections.

Cover required = 3/4 in. (Section 7.14.1.1). For #4 bars, the effective depth, d = 6 in. − 0.75 in. − 1/2(0.5 in.) = 5 in. Required steel areas,

$$A_s = \frac{M_u}{\phi f_y(d - 1/2\ a)}$$

where $a = \dfrac{A_s f_y}{0.85\ f_c' b}$.* $a = 1.47\ A_s$. Assume $A_s = 0.2$ in.^{2}; $a = 0.3$ in.

---

* A direct solution for area of tension steel required is possible, but not feasibly for manual design. For use in computer design programs, it may be convenient to use the following direct solution derived by substituting the value for a into the equation for A_s:

$$A_s = 1.7\ f_c'\ bd\frac{A_s}{f_y} - 1/2\sqrt{\frac{2.89\ f_c'\ b^2d^2}{f_y^2} - \frac{6.8\ f_c'\ bM_u}{\phi\ f_y^2}}$$

For a rapid manual solution see Figs. 9-25(a) and (b).

Exterior span, bottom bars, using moment from Fig. 3-2.

$$A_s = \frac{(3.64) \times 12}{(0.90)(60)(5 - 0.15)} = 0.17 \text{ in.}^2/\text{ft. Use \#4 @ 14}$$

First interior support, top bars, using moment from Fig. 3-2,

$$A_s = \left(\frac{4.00}{3.64}\right)(0.17) = 0.19 \text{ in.}^2/\text{ft. Use \#4 @ 12}$$

Other interior supports, top bars: the moment is the same as for the exterior span bottom bars, use #4 @ 14. Interior spans, bottom bars:

$$A_s = \frac{2.50}{3.64}(0.17) = 0.12 \text{ in.}^2/\text{ft} < 0.130 \text{ in.}^2/\text{ft min}$$

The maximum spacing for bars (Section 7.4.3) is $3t = 3 \times 6$ in. $= 18$ in., but not more than 18 in. Since #4 bars have been selected for use elsewhere, use #4 @ 18 for the bottom bars in the interior spans and also for the temperature steel in the other direction. The requirement for minimum temperature reinforcement is the same as for minimum flexural reinforcement (Sections 10.5.2 and 7.13.1).

Step 4: Check Required Development Lengths for the #4 Bar Size

Since all of the bottom bars are #4, development length need only be checked at the most critcal point, which is the discontinuous end (Commentary, Section 12.2.3). The end embedment available is 12 in. − 2 in. (cover) = 10 in. for bottom bars at the discontinuous end; at interior supports an end embedment of 6 in. (Section 12.2.1) for all bars should be ample. Using only one length of bottom bar will simplify detailing, placing, and inspection, as well as design (see Fig. 3-3).

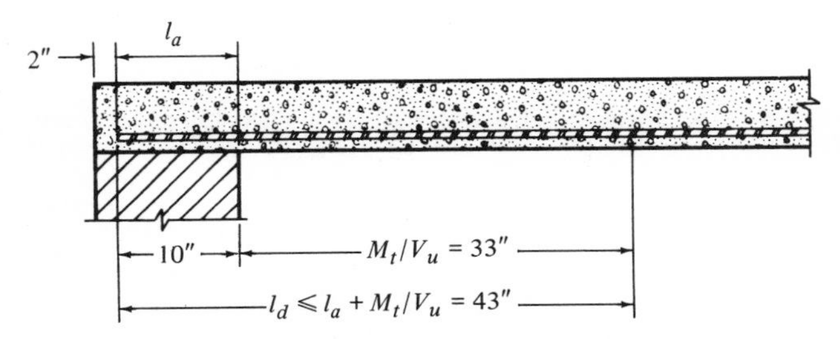

Fig. 3–3 Development length at the discontinuous end.

$$l_a = 12 \text{ in.} - 2 \text{ in.} = 10 \text{ in.}$$

$$\frac{M_t}{V_u} = \frac{3.64 \times 12}{1.33} = 33 \text{ in.}$$

$$l_d \leqq l_a + M_t/V_u = 43 \text{ in. (Section 12.2.3)}$$

$$\text{Required } l_d = 0.0004\ d_b f_y \text{ (Section 12.5.1-a)}$$

$$\text{For a \#4 bars, } l_d = 12 \text{ in. (See Table 13-4)}$$

The Code will permit two-thirds of the bottom bars at discontinuous ends and three-quarters at continuous ends to be cut short of the support (Section 12.2.1). Similarly, several lengths of top bars could be employed for a small saving in material. For this example, assume the saving in material is less than the added expenses for design, detailing, fabricating, placing, and inspection. In this case, then, no further check of development lengths is necessary.

Step 5: Check Shear

In any one-way flexural member where shear may be critical, the final investigation of shear cannot be made until moments have been determined and steel selected and detailed, including cutoff and bend-up points. The more exact formula for allowable shear assumed to be carried by concrete, $v_c = 1.9\sqrt{f_c'} + 2{,}500\ \rho_w\ V_u d/M_u$ . . . (Eq. 11-4), involves the tensile reinforcing steel ratio, ρ_w, at any section considered for shear, (Section 11.4.2). It may save time where shear is expected to be critical to make a preliminary shear investigation using the approximate formula for allowable shear, $v_c = 2\sqrt{f_c'}$, (Section 11.4.1).

In the example here, a one-way solid slab even less than minimum thickness/span ratio, the shear check is a waste of design time. The

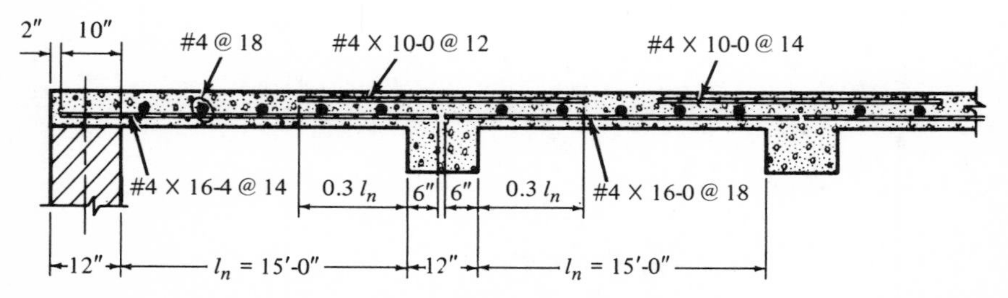

Fig. 3–4 Details of reinforcement.

Code requires, however, that shear be investigated. To demonstrate for completion of this example: the maximum shear at the face of the first interior support is 1.53 k per 12 in. width. The critical section, $d = 5$ in. from the face of the support, must resist a shear,

$$V_u = 1.53 - (0.178)(5/12) = 1.46\ k \quad \text{(Section 11.2.2)}$$

$$v_u = \frac{V_u}{\phi bd} \text{ (Section 11.2.1, Eq. 11-3)}$$

$$v_c = \frac{(1.46)(1000)}{(0.85)(12)(5)} = 29 \text{ psi} < 2\sqrt{f_c'} = 126 \text{ psi}$$

Step 6: Deflection Check

The coefficients of the approximate frame analysis give maximum moments for loadings which result in the following maximum deflections:

$$\text{End span } \Delta = \frac{3.25}{384}\frac{wl_n^4}{EI} \text{ (= 65 percent of simple span deflection)}$$

$$\text{Interior span } \Delta = \frac{2}{384}\frac{wl_n^4}{EI} \text{ (= 40 percent of simple span deflection)}$$

In this case the critical deflection occurs in the end span.

For deflection computations, the 1971 Code prescribes an effective moment of inertia

$$I_e = \left(\frac{M_{cr}}{M_a}\right)^3 I_g + \left[1 - \left(\frac{M_{cr}}{M_a}\right)^3\right] I_{cr}$$

where

$$M_{cr} = \frac{f_r I_g}{y_t} \text{ and } f_r = 7.5\sqrt{f_c'} = 475 \text{ psi for } f_c' = 4{,}000 \text{ psi}$$

(Section 9.5.2.2.)

$$M_{cr} = \frac{(475)(216)}{3} = 34{,}200 \text{ lb-in. } I_g = \frac{(12)(6)^3}{12} = 216 \text{ in.}$$

$y_t = 3$ in. (See Fig. 3-6.)

Service loads, w_s, $= 75 + 16 + 30 = 121$ psf

$$+M_a = (121/178)\ 3.64 = 2.47 \text{ kip-ft/12 in. width}$$

$$-M_a = (121/178)\ 2.94 = 2.00 \text{ kip-ft/12 in. width}$$

$$\text{Average } M_a = 2.23 \times 12 = 26.8 \text{ in.-kips/12 in. width}$$

$$\text{Average } I_{cr}\text{: average } \rho = \frac{0.18}{5 \times 12} = 0.003.$$

$$k = \sqrt{2\ n\rho + (n\rho)^2} - n\rho$$

$$k = 0.22$$

Average I_{cr} for negative and positive moment sections, neglecting compression steel at the supports:

$$I_{cr} = d^3[4\ k^3 + 12\ n\rho\ (1 - 2k + k^2)] = 27 \text{ in.}^4 \text{ per 12 in. strip}$$

$$I_e \left(\frac{34.2}{26.8}\right)^3 (216) + \left[1 - \left(\frac{34.2}{26.8}\right)^3\right] (27) > I_g\text{; use } I_g = 216 \text{ in.}^4$$

Instantaneous live load, $\Delta = \dfrac{(3.25)}{384} \dfrac{(30)(15)^4(1728)}{(3{,}600{,}000)(216)} = 0.029$ in. $<$ 1.00 in. allowable (Table 9.5-b).

Conservatively neglecting the compression steel at the support sections, the additional long-time deflection factor is equal to 2.0 (Section 9.5.2.3).

$$\text{Instantaneous dead load, } \Delta = (0.029)\left(\frac{91}{30}\right) = 0.086 \text{ in.}$$

$$\text{Additional long-time deflection, } \Delta = (0.086)(2.0) = 0.172 \text{ in.}$$

$$\text{For camber, total long-time deflection} = 0.258 \text{ in.}$$

If this roof is not supporting nor attached to nonstructural elements likely

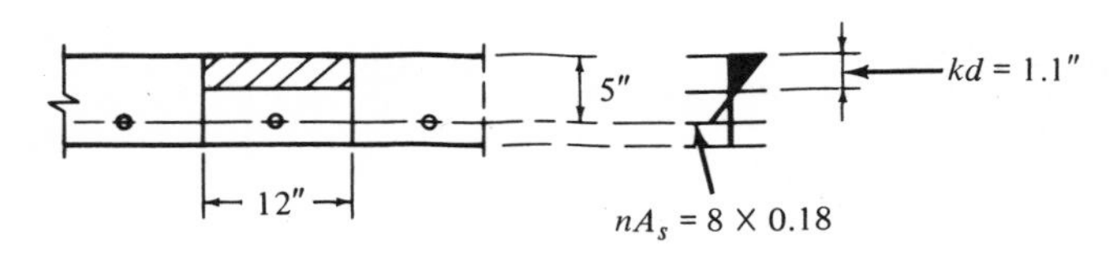

Fig. 3–5 Cracked transformed section.

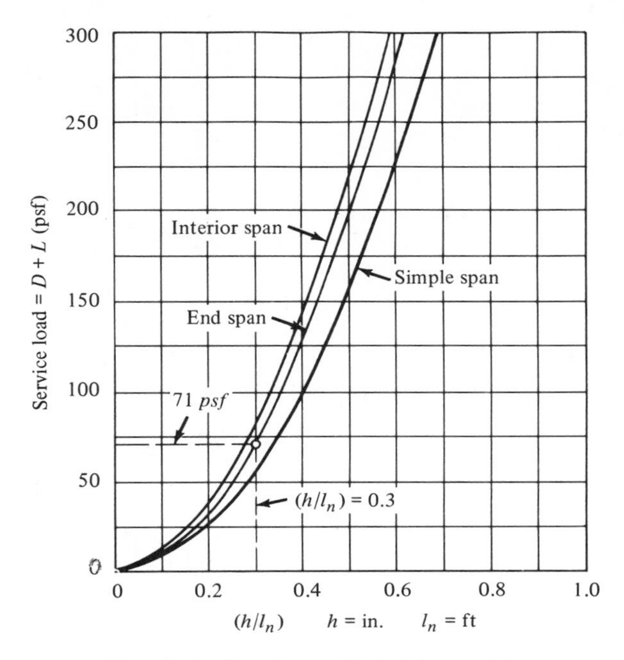

Fig. 3–6 Load at which $M_{cr} = M_a$.

to be damaged by large deflections, these calculations would satisfy both Code and practical requirements (Section 9.5, Table 9.5-b). If the roof *were* attached to or supporting nonstructural elements likely to be damaged by large deflections, the sum of the live load, plus additional long-time load deflections computed as above, must be no greater than $l_n/480$ (Section 9.5, Table 9.5-b). Additional long-time, plus live load $\Delta = 0.172 + 0.029 = 0.201$ in. $l_n/480 = \dfrac{180}{480} = 0.375$ in. $<$ 0.201 in. O.K. For a more complete discussion of deflection computations and limits thereon, see Chapter 4.

Note that whenever the cracking moment, M_{cr}, is larger than the positive or negative actual service load moments, the gross moment of inertia, I_g, can be used as the effective moment of inertia, I_e, saving computations for the cracked section moment (See Fig. 3-6).

Step 7: Bar Spacings for Crack Control (Section 10.6).

When the specified yield strength, f_y, for design exceeds 40,000 psi, bars must be spaced such that the quantity, $z = f_s\sqrt[3]{d_cA}$ . . . (Section 10.6.2,

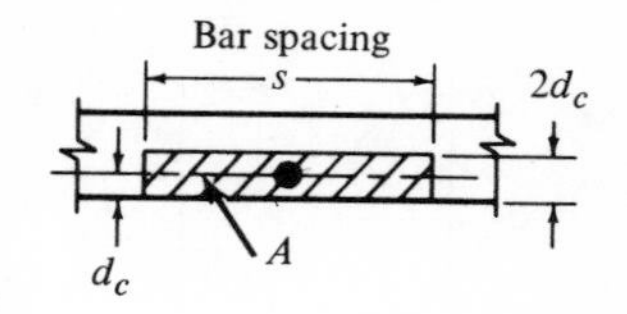

Fig. 3–7 Tributary tensile area per bar = *A*.

Eq. 10-2) does not exceed 175 for interior exposures and 145 for exterior exposures for beams and (1.2/1.35) times these values for one-way slabs (Section 10.6.3 and Commentary).

$$f_s = \text{tensile steel stress at service load; use } 0.60\, f_y$$

$$= 36 \text{ ksi unless calculated.*}$$

$$d_c = \text{cover from nearest surface to the center of the bar}$$

$$= \text{clear cover} + 1/2 \text{ bar diameter}$$

$$A = (2)(d_c)(s), \text{ where } s = \text{bar spacing}$$

For #4 bars with standard cover = 3/4 in., d_c = 1 in. Maximum $z = 175 \times \dfrac{1.2}{1.35} = 156$ for interior exposures. Solving for the maximum allowable spacing of bars, s:

$$\sqrt[3]{d_c A} = \frac{z}{f_s} = \frac{156}{36} = 4.33.$$

When $d_c = 1$, $\sqrt[3]{d_c A} = \sqrt[3]{2s}$ solving, maximum spacing, $s = 40$ in. > limit $3h = 18$ in. (Section 7.13.1).

DESIGN AIDS

In routine design the last three steps and part of Step 3 in the preceding example can be avoided by the use of design aids to ensure automatic conformance with all applicable code requirements. Such design aids are published in handbooks, including those by CRSI and ACI, and any special aids required for particular job conditions can be easily prepared.

---

* $f_s = \dfrac{\text{Area of steel required}}{\text{Area of steel furnished}} \times \dfrac{D + L}{w} \times 0.90\, f_y$ (Section 12.5-d)

where w = (1.4D + 1.7L)

For Step 3, selection of bar sizes, the minimum and maximum (singly reinforced) steel areas permitted for each common slab thickness can be tabulated for the standard Grade 60 reinforcement and concrete strengths generally used (see Table 3-1).

For Step 4, development lengths for each size of bar can be calculated in a matter of minutes for any concrete strength and grade of steel. Thereafter, the design step will consist merely of comparing the development length available, from a sketch of design dimensions, to that required for development of the various bar sizes. Ordinarily, a designer then simply selects the largest bar which can be developed. See Table 3-4 following and also complete splice and development tables in Chapter 13.

The deflection computation as noted in the text can be avoided entirely by conformance within the minimum thickness/span ratios prescribed (see Table 3-3). Where it is desirable to use lesser thicknesses of slab, for light loads and/or areas where immediate deflection is not likely to be harmful, design aids consisting of tabulated values for M_{cr} and I_{cr} for the various slab thicknesses and steel percentages, are helpful (see sample in Table 3-5). Also see Fig. 3-8 for an example of load capacities limited by strength requirements only versus loads limited by the immediate service live load deflection requirement, $\Delta < l_n/360$. (Fig. 3-8 was prepared from load tables in the *CRSI Handbook,* 1972.)

Finally, automatic conformance to the bar spacing limits (for crack control) can be assured with no computations at all in design for routine elements subject to the standard criteria for ordinary interior or exterior exposures. In selecting the bar sizes, the designer merely has to keep spacings within the tabulated limits for that bar size and exposure, as in sample Table 3-2. Special structures such as sewage treatment plants usually are provided with larger cover (commonly 1-1/2 in. minimum

TABLE 3-4 Development Lengths for Grade 60 Bars in Slabs*

(where thickness, $h < 13$ in.; normal weight concrete with $f_c' \geqq 4{,}000$ psi)

Bar Size	Compression Development Length, ℓ_d[a]	Compression Lap Splice Length[b]	Tension Development Length, ℓ_d[c]	Tension Embedment Length, *E*, with Standard End Hook[d]
#3	8 in.	12 in.	12 in.	7 in.
#4	10 in.	15 in.	12 in.	7 in.
#5	12 in.	19 in.	15 in.	7 in.
#6	14 in.	23 in.	18 in.	8 in.

[a] Section 12.6.
[b] Section 7.7.1.1.
[c] Section 12.5.
[d] See Table 13–6 for sketch showing E.
* From *Reinforcing Bar Splices,* 2nd Edition, 1971, by CRSI.

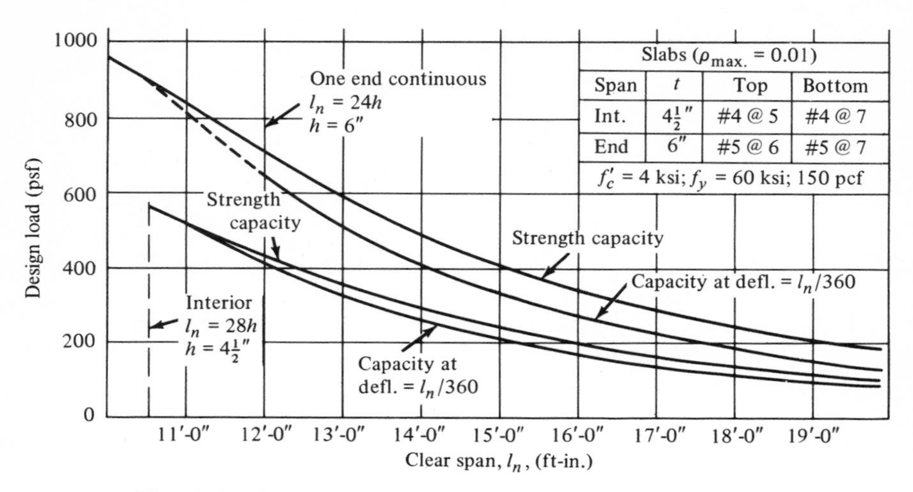

Fig. 3–8 One-way slabs, strength and deflection capacity.

clear cover) and the resulting deeper tributary areas of concrete in tension per bar will require closer spacings. Similarly, any structure intended to be watertight may utilize a lower service load stress or a lower limiting z-value or both. Special tables similar to Table 3-2 can be prepared as a preliminary design step to avoid any further crack control computations in the design of separate elements.

LONGER SPANS

For spans longer than those of Table 3-3, complete deflection calculations can often be avoided if the slab be lightly loaded. There is an inherent discontinuity in the Code-prescribed procedure for computing deflections. The expression for effective moment of inertia,

$$I_e = \left(\frac{M_{cr}}{M_a}\right)^3 I_g + \left[1 - \left(\frac{M_{cr}}{M_a}\right)^3 I_{cr}\right]$$

reduces to $I_e = I_g$ when the cracking moment, M_{cr}, is greater than the applied moment for which deflection is being computed, M_a. Immediate deflection under service live load is limited to $l_n/360$ for all floor construction (Table 9.5-b) and other deflection conditions can be computed proportionately. For immediate deflection M_a is the moment of the service dead and live loads.

$$M_{cr} = \frac{f_r I_g}{(1/2)(h)} \qquad \text{where } f_r = 7.5\sqrt{f_c'} \text{ (Section 9.5.2.2.)}$$

$$M_{cr} = 15\sqrt{f_c'}(h)^2 \text{ lb-in.} = 1.25\sqrt{f_c'}(h)^2 \text{ lb-ft.}$$

Equate the cracking moment, M_{cr}, to the maximum applied service load moment, M_a, and solve for the maximum applied service load when $M_{cr} = M_a$: Use h in inches: l_n in feet: for M in lb-ft.

Simple spans:

$$\frac{w(l_n)^2}{8} = 1.25\sqrt{f_c'}(h)^2 \tag{3-1}$$

$$w_{cr} = 10\sqrt{f_c'}(h/l_n)^2$$

Similarly,

End spans:

$$w_{cr} = 12.5\sqrt{f_c'}(h/l_n)^2 \tag{3-2}$$

Interior spans:

$$w_{cr} = 13.75\sqrt{f_c'}(h/l_n)^2 \tag{3-3}$$

At loads equal or less than the maximum from Eq. (3-1), (3-2), or (3-3), $M_{cr} \geq M_a$ and so I_g is used in the computation for deflection. When I_g is used in the equations for deflection (see Example Step 6), and equated to a total service load deflection $\Delta = l_n/360$, the following equations result:

Simple span: $w_{max} = 5{,}250\ (h/l_n)^3$

End span: $w_{max} = 8{,}050\ (h/l_n)^3$

Interior span: $w_{max} = 13{,}050\ (h/l_n)^3$

EXAMPLE. For a 6 in. thick slab, end span = 20′–0″, $f_c' = 4{,}000$ psi, find the maximum load where $M_{cr} = M_a$ and check if deflection exceeds $l_n/360$. Use h in inches; l_n in feet. $(h/l_n) = 0.30$ (see Fig. 3-6).

$$w_{cr} = (12.5)(63.2)(6/20)^2 = 71 \text{ psf for } M_{cr} = M_a$$

Compare to load causing $\Delta = l_n/360$ computed with I_g:

$$w_{max} = (8{,}050)(0.3)^3 = 217 \text{ psf} > 71 \text{ psf, and so } \Delta < l_n/360$$

Note that 71 psf is the total actual service load = $D + L$. Note also that the last step above will seldom be necessary. When load is limited to less than the cracking moment, deflections will seldom be large. See Fig 3-8 for the effect of the deflection limitations on load capacity of more heavily loaded (and more heavily reinforced) slabs. See Table 3-5 for a convenient tabulation of values of I_g, M_{cr}, and I_{cr} for common slab thicknesses and steel ratios.

WELDED WIRE FABRIC

Welded wire fabric (WWF) is commonly used for reinforcing short one-way slabs. Two types of wire are used to form WWF—plain and deformed (plain Sections 3.5.5 and 3.5.6, and deformed, Sections 3.5.7 and 3.5.8). Under the latest ASTM specifications, plain wire sizes are designated by the letter, *W-*, followed by the cross-sectional area in hundredths of an inch units from W-31 (5/8 in. in diameter) to W-0.5 (0.08 in. in diameter); in half sizes to W-6; and multiples of two (hundredths) for sizes above W-8. Deformed wire sizes are designated similarly in sizes from D-31 to D-1. Welded intersections comprise the bond value for plain WWF; the Code limits the maximum spacing of welded plain wire intersections to 12 in. (Section 3.5.6). For deformed WWF welded intersections are limited to a maximum spacing of 16 in. (Section 3.5.8).

Since a great variety of wire size and spacing combinations are possible, it is best to determine those readily available in the area for proposed use, and to use as few combinations as possible for practical applications. Even combinations of plain and deformed wire fabric can be obtained as special orders.

For one-way slabs with spans less than 10–0″ (center-to-center) a single layer of draped fabric may be employed for both top and bottom reinforcement (Section 7.3.3). (Usually, wire sizes for draped reinforcement should be kept to about 1/4 in. diameter or less to permit draping in short distances.) The Code also permits the use of equal top and bottom moments for design in slabs with spans 10′ –0″ or less, $\pm M_u = 1/12\ wl_n^2$ (Section 8.4.2). For one-way spans within these limits, WWF is capable of being placed in a minimum of time. Cross-wires are designed to satisfy the requirements for minimum temperature and shrinkage steel (Section 7.13) (see Fig. 3-9).

For use in one-way spans 10′0″ or less, the design can be quite simple, In addition to moment, shear, and minimum steel areas only the lap splice and end embedment requirements need to be computed (Sections 12.1.1

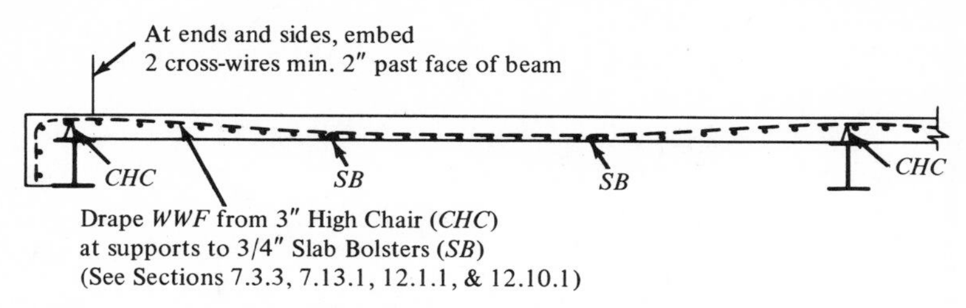

Fig. 3–9 Welded wire fabric—end spans (interior spans similar).

and 7.13). The splice arrangements and end embedments desired can usually be shown most simply as typical details. End embedment for plain WWF consists of two cross wires with the nearest one embedded at least 2 in. past the point of maximum moment (Section 12.10.1). For deformed wire, the required embedment is slightly less, but the formulas for computing development length include the number of cross-wires, n, which is unknown until the development length is compared to the cross-wire spacing (Section 12.10.2). For quick, safe design, the authors recommend simply specifying the same end anchorage for both. Similarly, splices for plain WWF are very simple—two cross-wires overlapped a minimum of 2 in. for stresses in excess of $1/2\ f_y$ and one overlapped a minimum of 2 in. for stress of $1/2\ f_y$ or less (Sections 7.8.1 and 7.8.2). Two formulas, one containing n, are required to compute minimum lap splice lengths, whichever is greater, for deformed WWF (Section 7.9). Again, to avoid field confusion and a waste of design and inspection time, the authors recommend use of the same typical details for both plain and deformed WWF lap splices. See Figs. 13-20 through 13-23 for wire and WWF splice and embedment typical details.

Flat sheets of heavier wire fabric usually with the same wire sizes and same spacings in each direction ("two-way" fabric) are used in two-way construction for somewhat longer spans.

EXAMPLE. Use welded wire fabric as the top and bottom reinforcement in a floor slab to span 10′ –0″ from center-to-center of supporting steel beams. The live load specified is 80 psf. Use concrete with $f_c' = 4{,}000$ psi weighing 150 pcf (see Fig. 3-8).

Try thickness, $h = 4$ in. $d = 4.00 - 0.75$ (cover) $- 0.15$ in. (1/2 dia.)

$$= 3.1 \text{ in.} \qquad D = \frac{4}{12}(150) = 50 \text{ psf.}$$

TABLE 3-5 Moments of Inertia and Cracking Moments for Normal Weight Concrete One-Way Slabs

h for I_g and M_{cr} / d for I_{cr}	I_g in.^{4} per 12 in. strip	M_{cr}[a] in.-k per 12 in. strip	I_{cr} (IN.^{4} PER 12 IN. WIDTH)[a] Tensile Reinforcement Ratio, $\rho = A_s/bd$ (for continuous members, use average top and bot. ρ)									
			0.0018	0.002	0.003	0.004	0.005	0.006	0.008	0.010	0.012	0.014
3″	27	8.54	3.7	4.1	5.8	7.5	9.0	10.4	13.1	15.5	17.8	19.9
3½″	43	11.65	5.9	6.5	9.3	11.8	14.3	16.5	20.8	24.6	28.2	31.6
4″	64	15.2	8.8	9.7	13.8	17.7	21.3	24.7	31.0	36.8	42.1	47.1
4½″	91	19.2	12.6	13.8	19.7	25.2	30.3	35.1	44.1	52.4	60.0	67.1
5″	125	23.7	17.3	19.0	27.0	34.5	41.6	48.2	60.5	71.8	82.3	92.0
5½″	166	28.8	23.0	25.3	36.0	46.0	55.3	64.2	80.6	95.6	110	122
6″	216	34.2	29.9	32.8	46.7	60.0	71.8	83.3	105	124	142	159
6½″	275	40.0	38.0	41.7	59.4	75.8	91.3	106	133	158	181	202
7″	343	46.5	47.4	52.1	74.2	94.7	114	132	166	197	226	252
7½″	422	53.2	58.3	64.1	91.2	117	140	163	204	242	278	310
8″	512	60.8	70.8	77.7	111	141	170	197	248	294	337	377
8½″	614	68.7	84.9	93.2	133	170	204	237	297	353	404	452
9″	729	77.0	101	111	158	201	242	281	353	419	480	536
9½″	857	85.5	119	130	185	237	285	331	415	493	564	631
10″	1000	95.0	138	152	216	276	332	386	484	575	658	756

[a] For $f'_c = 4{,}000$ psi; $f_r = 7.5\sqrt{f'_c} = 475$ psi; $n = 8$

$M_{cr} = 2 f_r h^2$; $k = \sqrt{2pn + (pn)^2} - pn$

$I_{cr} = d^3 [4k^3 + 3pn(1 - 2k + k^2)]$

TABLE 3-6 Standard Wire Sizes for Reinforcement

STANDARD STEEL WIRE GAUGES				DEFORMED WIRE (A496)	PLAIN WIRE (A82)	NOMINAL	
Diameter (in.)	AS&W Gauge	Diameter (in.)	Area (sq. in.)	Size	Size	Diameter (in.)	Area (sq. in.)
1/2	—	.5000	.19635	D-31	W-31	0.628	0.310
—	7/0	.4900	.18857	D-30	W-30	0.618	0.300
15/32	—	.46875	.17257	D-29	—	0.608	0.29
—	6/0	.4615	.16728	D-28	W-28	0.597	0.280
7/16	—	.4375	.15033	D-27	—	0.586	0.27
—	5/0	.4305	.14556	D-26	W-26	0.575	0.260
13/32	—	.40625	.12962	D-25	—	0.564	0.25
—	4/0	.3938	.12180	D-24	W-24	0.553	0.240
3/8	—	.3750	.11045	D-23	—	0.541	0.23
—	3/0	.3625	.10321	D-22	W-22	0.529	0.220
11/32	—	.34375	.092806	D-21	—	0.517	0.21
—	2/0	.3310	.086049	D-20	W-20	0.504	0.200
5/16	—	.3125	.076699	D-19	—	0.491	0.19
—	0	.3065	.073782	D-18	W-18	0.478	0.180
—	1	.2830	.062902	D-17	—	0.465	0.17
9/32	—	.28125	.062126	D-16	W-16	0.451	0.160
—	2	.2625	.054119	D-15	—	0.437	0.15
1/4	—	.2500	.049087	D-14	W-14	0.422	0.140
—	3	.2437	.046645	D-13	—	0.406	0.13
—	4	.2253	.039867	D-12	W-12	0.390	0.120
7/32	—	.21875	.037583	D-11	—	0.374	0.11
—	5	.2070	.033654	D-10	W-10	0.356	0.100
—	6	.1920	.028953	D-9	—	0.338	0.09
3/16	—	.1875	.027612	D-8	W-8	0.319	0.080
—	7	.1770	.024606	D-7	W-7	0.298	0.070
—	8	.1620	.020612	D-6	W-6	0.276	0.060
5/32	—	.15625	.019175	—	W-5.5	0.264	0.055
—	9	.1483	.017273	D-5	W-5	0.252	0.050
—	10	.1350	.014314	—	W-4.5	0.240	0.045
1/8	—	.1250	.012272	D-4	W-4	0.225	0.040
—	11	.1205	.011404	—	W-3.5	0.211	0.035
—	12	.1055	.0087147	D-3	W-3	0.195	0.030
3/32	—	.09375	.0069029	—	W-2.5	0.178	0.025
—	13	.0915	.0065755	D-2	W-2	0.159	0.020
—	14	.0800	.0050266	—	W-1.5	0.138	0.015
—	15	.0720	.0040715	D-1	W-1	0.113	0.010
—	16	.0625	.0030680	—	W-0.5	0.080	0.005

$$U = 1.4\ D + 1.7\ L \text{ (Section 9.3.1)}$$

$$U = (1.4)(50) + (1.7)(80) = 206 \text{ psf}$$

$$\pm M_u = 1/12\ wl_n^2 \text{ (Section 8.4.2)}$$

$$\pm M_u = (1/12)(0.206)(10)^2(12) = 20.6 \text{ in-kip/ft}$$

Resisting moment, $M_u = \phi\ A_s f_y(d - 1/2a)$, where $a = \dfrac{f_y A_s}{0.85 f_c{}'b}$ and $a =$ $1.47\ A_s$. Take $a \simeq 0.20$:

$$A_s = \frac{20.6}{(0.90)(60)(3.1 - 0.1)} = 0.127 \text{ in.}^2/\text{ft} = \text{area for moment}$$

Area of steel (minimum): $A_s = 0.0018\ bh$ (Section 7.13.1). Minimum $A_s = (0.0018)(12)(4) = 0.0864$ in.²/ft (see Table 3-6 for wire sizes). Use: 6″ × 6″ W-7/W-4.5 or 4″ × 4″ W-4.5/W-3 whichever fits the remainder of the floor system and availability best. Check shear.

$$V_u = 1.15\ w\ (1/2) \quad \text{(Section 8.4.2)}$$

$$V_u = (1.15)(206)(1/2)(10) = 1{,}185 \text{ lbs/ft}$$

Unit shear stress, $v_u = \dfrac{V_u}{\phi bd}$ at a critical section a distance d from the face of the support (Sections 11.2.1 and 11.2.2). Allowable unit shear on the concrete, $v_u = 2\sqrt{f_c'}$ (Section 11.4.1).

$$v_u = \frac{1{,}185}{(0.85)(12)(3.1)} = 37.5 \text{ psi} < v_c = 126 \text{ psi OK}$$

For end embedment and support details, see Fig. 3-9. The design would be completed with typical splice details selected from those illustrated in Chapter 13.

4 one-way joist systems

GENERAL

Concrete joist construction consists of a monolithic combination of regularly spaced ribs and a top slab. It may be constructed with permanent or removable fillers between ribs (joists) arranged to span in one direction or in two orthogonal directions (Section 8.8.1). Two-way joist construction which spans in two orthogonal directions, commonly referred to as waffle slab construction, must conform to the requirements of Chapter 13 of the 1971 Code for slab systems reinforced in flexure in more than one direction. Chapter 7 of this book gives a detailed explanation of these requirements. One-way joist construction with joists parallel to one another must conform to the requirements for analysis and design of Chapter 8 of the 1971 Code which will be discussed in this chapter.

FORMS

Removable form fillers can be made with ready-made steel "pans" of standard size or from hardboard, fiberboard, glass reinforced plastic, or corrugated cardboard fillers. Some firms that supply such removable fillers also subcontract to erect and remove them. Other firms only rent or sell such fillers.

Lightweight and normal weight concrete or clay tile filler blocks that remain in place as permanent fillers are also used in some proprietary designs.

Standard Reusable Forms

The U.S. Department of Commerce Voluntary Product Standard No. PS 16-69 standardized removable filler depths and widths to include 20 and 30 in. widths and 6, 8, 10, 12, 14, 16, and 20 in. depths. Tapered end fillers and 10 and 15 in. wide fillers are also available from industry as special items.

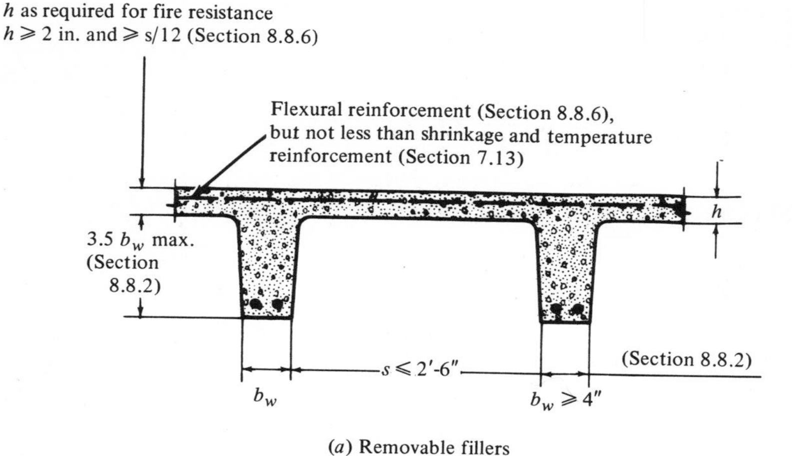

(a) Removable fillers

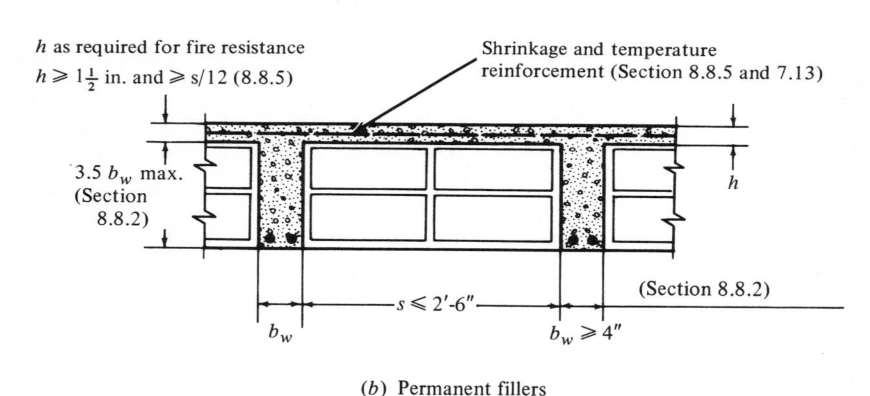

(b) Permanent fillers

Fig. 4–1 Limiting dimensions—concrete joist construction.

JOIST-SLAB CONSTRUCTION LIMITATIONS

Concrete joist construction that does not conform to the dimensional limitations of the Code (Sections 8.8.1 and 8.8.2) must be designed as slabs and beams (Section 8.8.3). Distribution (bridging) joists are not required by the Code.*

Limitations on the width and spacing of joist ribs, depth of fillers,

* Industry practice as recommended by CRSI is to furnish distributing ribs (bridging) with one #4 bar top and bottom for floor construction; one rib for spans 20-30 ft; two for spans 30-40 ft; and three for spans more than 40 ft.

and thickness of top slabs are shown in Fig. 4-1 for joists with removable and permanent fillers.

Top slabs that contain horizontal conduits or pipes that are allowed by the Code (Section 6.3) must meet the thickness limitations shown in Figs. 4-1 and in addition must have a thickness not less than 1 in. plus the total overall depth of the conduit or pipe (Section 8.8.7).

MINIMUM AND MAXIMUM REINFORCEMENT

The positive moment flexural reinforcement ratio, ρ, for one-way joist construction, must be equal to or greater than $200/f_y$, unless the area of both the positive and negative reinforcement at every section is equal to one-third greater than the amount required by analysis. For reinforcement in the ribs, this minimum ratio for positive moment is based on the concrete area of the joist rib, $b_w d$, where b_w is the width of the stem and does not include the flange area (Section 10.5.1).

The positive and negative moment reinforcement ratio, ρ, must be equal to or less than 0.75 of the ratio which produces balanced conditions in pure flexure (Section 10.3.2). In determining the maximum reinforcement ratio for positive moment, the top slab or compression flange is included in the concrete area bd. For negative moment the maximum reinforcement ratio is based on the area of the joist rib $b_w d$. There is no minimum limitation on the area of negative moment reinforcement.

Negative moment reinforcement should be distributed in zones of maximum concrete tension and a part of the main tension reinforcement must be placed in the effective flange width of the joist (Section 10.6.2). The entire top slab will be within the effective flange width (Section 8.7.2) for joist construction ($s \leq 16h$ in Fig. 4-1), and the authors recommend that the negative moment bars should be uniformly spaced throughout the entire top flange for effective crack control.

MAXIMUM SPACING OF TOP BARS (CRACK CONTROL)

The maximum spacing of negative moment reinforcing bars in the top slab is limited by $z = f_s \sqrt[3]{d_c A}$ (Section 10.6.3, Eq. 10-2). This equation can be expressed in terms of the bar spacing, $s_b = (z/f_s)^3/2d_c^2$, as shown in Fig. 4-2.

Table 4-1 gives maximum bar spacings, s_b, in one-way slabs for crack control with $f_s = 0.6\ f_y$, Grade 60 bar sizes, the usual range of concrete cover, and for interior and exterior exposures. In one-way slabs for interior exposure, z (Section 10.6.3) is (175) (1.2/135) = 156 ksi; and for exterior exposure, (145) (1.2/135) = 129 ksi (Commentary 10.6).

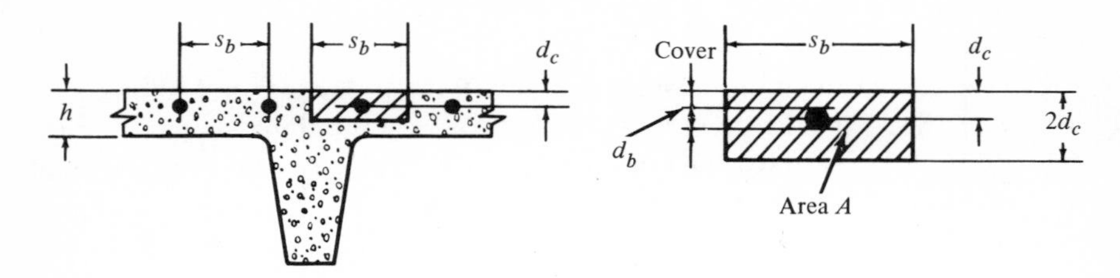

Fig. 4–2 Crack control.

For exterior exposure, Code Eq. (10-2) does not apply to structures subjected to very aggressive exposure or to structures which are to be watertight (Section 10.6.3). Special precautions are required for each exposure.

The values of $z = 175$ and 145 kips per inch for interior and exterior exposure, respectively, were selected to limit crack widths in beams to 0.016 and 0.013 in. and assure many fine hair-cracks rather than a few wide ones (Commentary 10.6).

It is obvious from Table 4-1 that the Code (Section 10.6) does not limit the maximum spacing of joist construction negative moment top bars for interior conditions of exposure except for reinforcing bar sizes #9, #10, #11 with 1 in. concrete cover unless, of course, the spacing is further limited by a maximum of five times the thickness of the top slab. For exterior conditions of exposure the Code (Section 10.6) does limit the maximum spacing of bar sizes #3 through #11 with 1-1/4, 1-1/2, and 2 in. of concrete cover. #14 and #18 bars are not practical for use as negative moment bars in top slabs of joist construction and were, therefore, omitted from Table 4-1.

TABLE 4-1 Joist Construction—Maximum Spacing of Negative Moment Reinforcing Bars in Top Slab for Crack Control*

Bar Size	INTERIOR EXPOSURE COVER IN INCHES			EXTERIOR EXPOSURE COVER IN INCHES		
	5/8″	3/4″	1″	1-1/4″	1-1/2″	2″
#3	18	18	18	11	8	—
#4	18	18	18	10	7½	—
#5	18	18	18	9½	7	—
#6	18	18	18	—	6½	4
#7	18	18	18	—	6	3½
#8	18	18	18	—	5½	3½
#9	18	18	16½	—	5½	3½
#10	18	18	15	—	5	3
#11	18	18	14	—	4½	3

* Spacing must be equal to or less than $5h$ and 18 in. (Sections 10.5.2, 7.13, and 8.7.5).

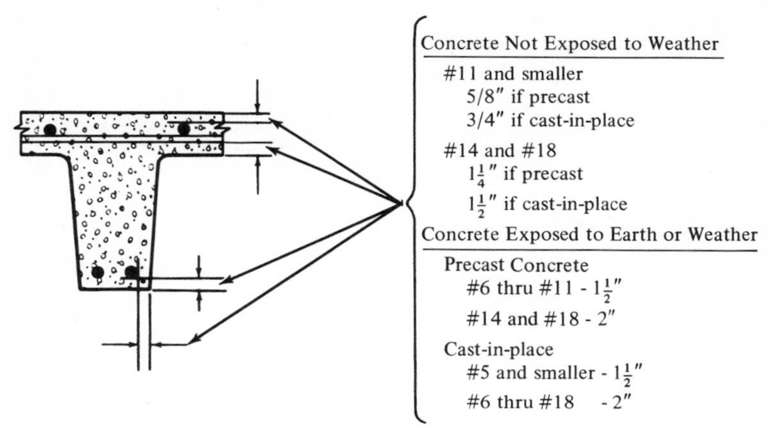

Fig. 4–3 Concrete cover for protection of reinforcement.

MINIMUM CONCRETE COVER

Required concrete cover for protection of joist reinforcement against exposure to weather (Section 7.14.1 and 7.14.2) is shown in Fig. 4-3. When the "general code" requires greater concrete protection for more severe conditions of exposure or for fire resistive ratings, it should be provided (Section 7.14.5).

Minimum Concrete Cover and Top Slab Thickness for Fire Resistance

The required thickness of concrete top slabs and the required concrete cover of reinforcement for different fire resistive ratings depends on the "general code" that is being followed. Table 4-2 lists the top slab thickness and the thickness of concrete cover for protection of reinforcement required for fire resistance by the *Uniform Building Code,* 1970 Edition, Vol. I, for joist construction made from Class A normal weight and structural lightweight concrete with aggregates such as limestone, calcareous gravel, trap rock, slag, expanded clay, shale, slate or any other aggregates possessing equivalent fire properties.

In addition to the fire resistive requirement, the thickness of the top slab of concrete joist construction must be sufficient to meet the dimensional requirements shown in Fig. 4-1 and to provide adequate cover for protection of reinforcement as shown in Fig. 4-3. Table 4-3 shows the minimum concrete top slab thickness necessary for cover of negative moment reinforcement with #3 shrinkage and temperature bars perpendicular to the joists.

TABLE 4-2 Joist Construction—Minimum Top Slab Thickness and Reinforcement Cover for Various Fire Resistive Ratings*

Type of Concrete	FIRE RESISTIVE RATING (HOURS)		
	1 Hour	2 Hours	3 Hours
Normal Weight Concrete			
Reinforcement cover[b]	¾″	1″	1¼″
Top slab thickness[c]	3½″[a]	4½″	5½″
Structural Lightweight Concrete			
Reinforcement cover[b]	¾″	1″	1¼″
Top slab thickness[c]	3″	4″	4½″

[a] The thickness can be reduced to three inches (3″) where limestone aggregate is used.
[b] Table 43A item 37 footnote 9.
[c] Table 43C items 1 and 2 footnote 2.
* From *Uniform Building Code,* 1970 Edition, Vol. 1.

The Code does not limit the minimum amount of negative moment reinforcement. For one-way joist construction, it is the authors' opinion that, if the positive moment steel is not lapped at the supports, sufficient negative moment reinforcement should be provided at the supports to equal 0.0018 times the total area of the joist (rib and top slab) to resist temperature and shrinkage effects.

Shrinkage and temperature steel must be provided for one-way joist construction (Section 7.13). Such reinforcement, placed perpendicular to the ribs, must not be spaced further apart than five times the slab thickness nor more than 18 in. The amount of such reinforcement, expressed as a ratio of reinforcement area to concrete area, varies with the yield strength of the steel.

The required shrinkage and temperature steel of Grade 60 bars and

TABLE 4-3 One-Way Joist Construction—Minimum Top Slab Thickness for Various Thicknesses of Concrete Cover of Reinforcement with #3 Temperature Reinforcement Perpendicular to Joists

Bar Size	CONCRETE COVER IN INCHES					
	5/8″	3/4″	1″	1-1/4″	1-1/2″	3″
#3	2″	2½″	3″	3½″	4″	5″
#4	2½″	2½″	3″	3½″	4″	5″
#5	2½″	2½″	3″	3½″	4″	5″
#6	2½″	3″	3½″	4″	4½″	5½″
#7	2½″	3″	3½″	4″	4½″	5½″
#8	3″	3″	3½″	4″	4½″	5½″
#9	3″	3″	3½″	4″	4½″	5½″
#10	3″	3½″	4″	4½″	5″	6″
#11	3½″	3½″	4″	4½″	5″	6″

TABLE 4-4 Deformed Grade 60 Shrinkage and Temperature Steel Per Foot of Slab Width for One-Way Joist Construction

Thickness of Top Slab (Inches)	SHRINKAGE AND TEMPERATURE REINFORCEMENT	
	Required Area (square inches)	Size and Spacing
2	0.043	4 x 12–W1.5/W1
2½	0.054	4 x 12–W2/W1
3	0.065	4 x 12–W2.5/W1
3½	0.076	#3 @ 17½″
4	0.086	#3 @ 15″
4½	0.097	#3 @ 13½″
5	0.108	#3 @ 12½″
5½	0.119	#3 @ 11
6	0.130	#3 @ 10

WWF for the top slab of one-way joist construction, placed perpendicular to the joists, is shown in Table 4-4.

SHEAR

The average design shear stress, v_u, used as a measure of diagonal tension stress is computed by the familiar formula $v_u = V_u/\phi b_w d$ (Section 11.2.1, Eq. 11-3) in which b_w is equal to the width of the joist rib. When permanent burned clay or concrete fillers with a compressive strength equal to that in the concrete joists used as permanent fillers, the vertical shells of the fillers in contact with the joists can be included in the width b_w (Section 8.8.4).

The average permissible shear stress, v_c, carried by the concrete for one-way and two-way concrete joist construction can be taken as 10 percent greater than the values used for other structural elements (Section 8.8.8).

The shear that can be carried by the concrete of joist construction not subjected to axial load or torsion must be less than the following amounts unless a detailed analysis is made that considers the tension reinforcement ratio, the design moment, and shear at the section of the joist considered.

Normal Weight Concrete (Section 11.4.1)

$$v_c = 2.2\sqrt{f_c'}$$

Lightweight Aggregate Concrete—f_{ct} average splitting tensile strength specified (Section 11.3.1)

$$v_c = 2.2\ f_{ct}/6.7 \text{ where } f_{ct}/6.7 \leq \sqrt{f_c'}$$

Sand-Lightweight Concrete—f_{ct} not specified (Section 11.3.2)

$$v_c = 1.87\sqrt{f_c'}$$

All-Lightweight Concrete—f_{ct} not specified (Section 11.3.2)

$$v_c = 1.65\sqrt{f_c'}$$

When a detailed analysis is made that considers the tension reinforcement ratio and the design moment and shear, then v_c can be increased to the following amounts, providing $V_u d/M_u$ is taken not greater than 1.0.

Normal Weight Concrete (Section 11.4.2)

$$v_c = 2.09\sqrt{f_c'} + 2750\ \rho_w V_u d/M_u \leqq 3.85\sqrt{f_c'}$$

Lightweight Aggregate Concrete—f_{ct} specified

$$v_c = 2.09\ f_{ct}/6.7 + 2750\ \rho_w V_u d/M_u \leqq 3.85\sqrt{f_c'}$$

Sand-Lightweight Concrete—f_{ct} not specified

$$v_c = 1.78\sqrt{f_c'} + 2338\ \rho_w V_u d/M_u \leqq 3.27\sqrt{f_c'}$$

All-Lightweight Concrete—f_{ct} not specified

$$v_c = 1.57\ f_c' + 2062\ \rho_w V_u d/M_u \leqq 2.89\sqrt{f_c'}$$

When shears exceed the values allowed for v_c, stirrups are required and should be provided (Section 11.6).

DESIGN

Code requirements for joist construction will be demonstrated by use of a design example of one-way joist construction as shown in Fig. 4-4 for an interior clear span of 30′ –0″, constructed from normal weight concrete, with $f_c' = 4{,}000$ psi and Grade 60 reinforcement. A 3-1/2 inch top slab for a 1 hr fire rating will be used with 6 in. wide joist ribs and 14 in. deep by 30 in. wide removable filiers with 3 ft. long tapers at each end of the span. The tapered end fillers increase the shear capacity by increasing the width of the joist from 6 in. to 11 in. at the support.

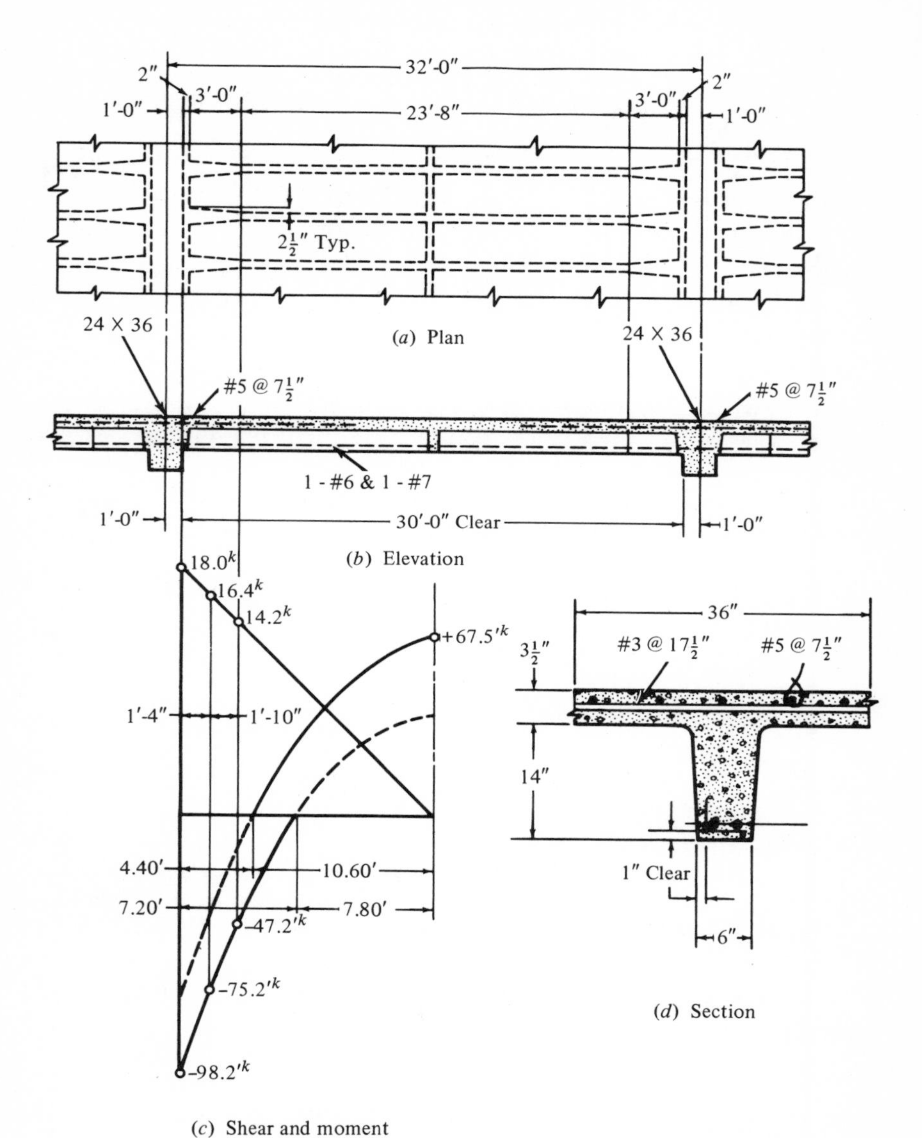

(c) Shear and moment
$w = 1.20$ k/ft

Fig. 4–4 One-way joist construction—example.

EXAMPLE

Loads. The design loads selected for the example shown in Fig. 4-4 are as follows:

Joist and bridging	77 psf	
Ceiling	10 psf	
Piping	17 psf	
	(104)(1.4) =	145 psf
Live load	(150)(1.7) =	255 psf
$w_d + w_l$		400 psf

$w_d + w_l$ per lineal foot of joist rib = (.40)(3) = 1.2 klf

Analysis

Design shears and moments were calculated using the simplified assumptions of frame analysis (Section 8.4.2).

Negative moment at the face of supports:

$$-M_u = -(1/11)(1.2)(30)^2 = -98.2 \text{ ft-kips per rib}$$

Positive moment at center of span:

$$M_u = +(1/16)(1.2)(30)^2 = 67.5 \text{ ft-kips per rib}$$

Flexural Reinforcement Required for Moments

Positive and negative reinforcement areas were calculated using assumed values for j. (Any error in this assumption must be adjusted after reinforcement has been selected.) See Figs. 9-25(a) and (b).

Negative moment reinforcement at face of supports ($j = 0.92$):

$$-A_s = M_u/\phi f_y jd = (98.2)(12)/(.9)(60)(.92)(16) = 1.48 \text{ in.}^2$$

Use #5 @ $7\frac{1}{2}$ Top bars. $A_s = (0.31)(36)/7\frac{1}{2} = 1.49 \text{ in.}^2$

Positive moment reinforcement at center of span ($j = 0.98$)

$$+A_s = M_u/\phi f_y jd = (67.5)(12)/(.9)(60)(.98)(16) = 0.96 \text{ in.}^2$$

Use 1-#6 and 1-#7 Bot. Bars. $A_s = 1.04 \text{ in.}^2$

Check Required Steel for Minimum and Maximum Limits

The negative moment reinforcement ratio, $A_s/b_w d = 1.49/(11)(16) = 0.0084$, is less than $0.75\ \rho_b = 0.0213$ (Section 10.3.2) and is satisfactory for strength. The ratio of negative moment reinforcement area to gross area of concrete joist, $1.49/[(7.17)(14) + (36)(3\text{-}1/2)] = 0.0066$, is greater than 0.0018 required for shrinkage and temperature, thus eliminating the need to lap bottom bars at the supports.

The positive moment reinforcement ratio, $A_s/bd = 1.04/(36)(16) = 0.0018$, is sufficient for shrinkage and temperature steel, and also less than $.75\rho_b$ and satisfactory for strength. The positive moment ratio, $A_s/b_w d = 1.04/(6)(16) = 0.0108$, is greater than the minimum required of $200/f_y = 200/60{,}000 = 0.0033$ (Section 10.5.1).

Shear

Check the shear at a distance d from the support where the minimum joist width is $6 + (5)(22)/(36) = 9.05$ in. and at the end of the taper where the minimum joist width is 6 in. An average joist width is used in the shear calculations and is based on an assumed slope on the sides of the removable fillers of 1 in 12.

Shear at d from support (See Fig. 4-4)

$$v_u = \frac{V_u}{\phi b_w d} = \frac{16{,}400}{(0.85)(9.05 + 1.17)(16)}$$

$$= \frac{16{,}400}{(0.85)(10.22)(16)}$$

$$= 118 \text{ psi OK} < 2.2\sqrt{f_c'} = 139 \text{ psi}$$

Shear at end of 3′-0″ taper (See Fig. 4-4)

$$v_u = \frac{V_u}{\phi b_w d} = \frac{14{,}200}{(0.85)(6 + 1.17)(16)}$$

$$= \frac{14{,}200}{(0.85)(7.17)(16)}$$

$$= 146 \text{ psi} > 2.2\sqrt{f_c'}$$

Because the shear at the end of the taper is greater than $2.2\sqrt{f_c'} = 139$ psi a more detailed calculation of the shear carried by the concrete will be made as follows:

$$v_c = 2.09\sqrt{f_c'} + 2750\ \rho_w V_u d/M_u$$

$$= 2.09\sqrt{4000} + \frac{(2750)(1.49)(14{,}200)(16)}{(7.17)(16)(47{,}200)(12)}$$

$$= 132 + (2750)(0.0130)(0.402)$$

$$= 132 + 14$$

$$= 146 \text{ psi which equals the shear at end of 3'-0'' taper}$$

A graphical presentation of the shearing stresses due to design loads and those allowed to be carried by the concrete are shown in Fig. 4-5.

Fig. 4-5 shows that for one-way joist construction without stirrups the use of tapered fillers increases the clear span range from 26′ –4″ to 30′ –0″ or approximately 15 percent. The critical section for shear occurs at the end of the tapered filler. Between the end of the tapered filler and a distance *d* from the support, the calculated shearing stress, v_u, decreases at a faster rate than the permissible shearing stress, v_c, calculated in accordance with Code Eq. (11-4) for one-way joist construction (Sections 11.4.2 and 8.8.8).

The complicated relationships shown in Fig. 4-5, which result from the tapered stem width, b_w, and allowable shear by the long formula, including the tensile steel ratio, become much more complicated with truss bent

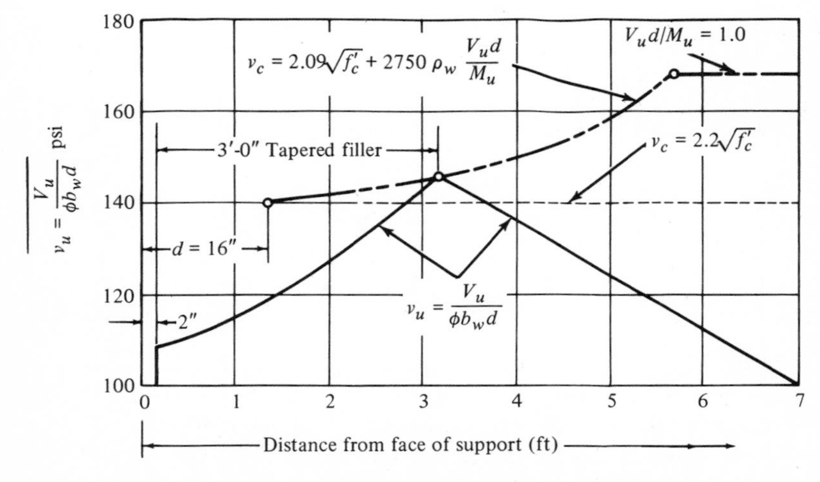

Fig. 4–5 Shear stress example—tapered end joist.

bars. The authors find that it is not possible to indicate a single critical section in a single drawing when using truss bent bars, following standard practice. It has become standard practice for designers using truss bent bars to show the bend-up point at a fixed fraction of the span, usually 0.25 l_n, on "typical bending details."* When spans are 12′ –0″ or less, the bend-up point is inside the 3′ –2″ distance for the tapered end; for longer spans, it may be bent up through the section at the end of the taper or completely outside it. When the resulting variable amount of tensile steel is included with a tapered width, the steel ratio, ρ_w, changes and allowable shear becomes unpredictable. The authors recommend use of straight bars only. If truss bent bars are used with tapered ends, the authors recommend use of the short formula for allowable shear. Even with a computer program, use of the long formula with the term, ρ_w, will become difficult and will probably require a separate shear investigation at short intervals (about 3 in.) beginning near the face of the support and continuing nearly to midspan.

The depth of the joist construction shown in Fig. 4-4 barely meets the minimum thickness limitation of $l_n/21$ for interior span ribbed one-way slabs in the Code (Table 9.5a):

$$17\text{-}1/2 > (30)(12)/(21) = 17.10''$$

and computations for deflection are not required (Section 9.5.2.1). The deflection is calculated here to demonstrate the application of deflection criteria in the Code. In this computation l_n is equal to the clear span (Section 8.5.2).

SECTION PROPERTIES FOR DEFLECTION CALCULATIONS

Determine the gross moment of inertia of the concrete section for positive moment ($+I_g$) (Section 9.0).

$$\begin{aligned} (30)(3.5) &= 105 \times 1.75 = 184 \\ (17.5)(6) &= \underline{105} \times 8.75 = \underline{919} \\ & \quad 210 \qquad\qquad\quad 1093 \end{aligned}$$

$$y_t(\text{top}) = 1093/210 = 5.20''$$

* ACI *Manual of Standard Practice for Detailing Reinforced Concrete Structures;* CRSI *Manual of Standard Practice.*

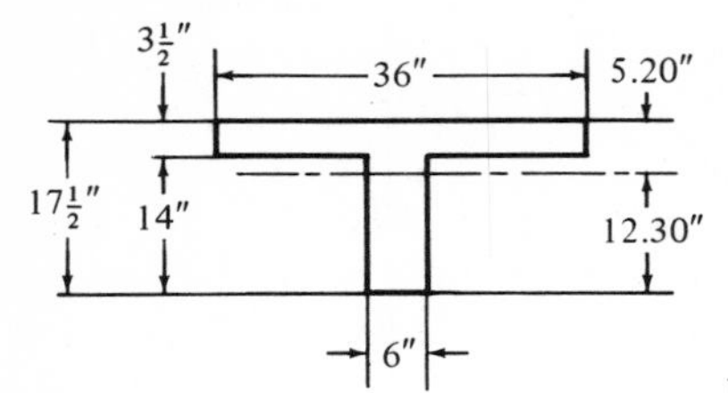

Fig. 4–6 I_g for + moment.

$$y_b(\text{bottom}) = 17.5 - 5.20 = 12.3''$$

$$\begin{aligned}
(12.30)^3(6)/(3) &= 3721 \\
(5.20)^3(36)/(3) &= 1687 \\
& \quad 5408 \\
(-1.70)^3(30)/(3) &= -49 \\
+I_g &= 5359 \text{ in.}^4
\end{aligned}$$

Determine the gross moment of inertia of the concrete section for negative moment ($-I_g$) (Section 9.0).

$$\begin{aligned}
(25)(3.5) &= 87.5 \times 1.75 = 153 \\
(17.5)(11) &= 192.5 \times 8.75 = 1684 \\
& \quad 280.0 \qquad\qquad 1837
\end{aligned}$$

$$y_t(\text{top}) = 1837/280 = 6.56''$$

$$y_b(\text{bottom}) = 17.5 - 6.56 = 10.94''$$

$$\begin{aligned}
(10.94)^3(11)/(3) &= 4820 \\
(6.56)^3(36)/(3) &= 3390 \\
& \quad 8210 \\
(-3.06)^3(25)/(3) &= -239 \\
-I_g &= 7971 \text{ in.}^4
\end{aligned}$$

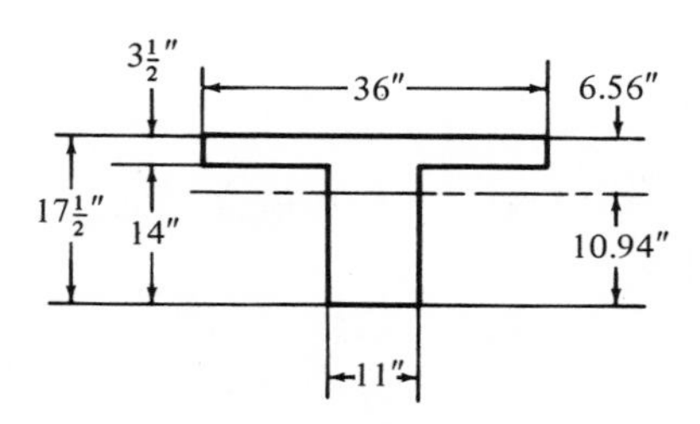

Fig. 4–7 I_g for − moment.

Determine the cracked moment of inertia for positive moment ($+I_{cr}$) (Section 9.0).

$$(36)(kd)^2/(2) = (8.32)(16 - kd)$$

$$kd = 2.50''$$

$$(36)(2.5)^3/(3) = 188$$
$$(8.32)(13.5)^2 = 1516$$
$$+I_{cr} = 1704 \text{ in.}^4$$

Determine the cracked moment of inertia for negative moment ($-I_{cr}$) (Section 9.0)

$$(11)(kd)^2/(2) + (16.6)(kd - 1.5) = (11.9(16 - kd)$$

$$kd = 4.17''$$

$$(11)(4.17)^3/(3) = 265$$
$$(11.9)(11.83)^2 = 1658$$
$$(16.6)(2.67)^2 = 118$$
$$-I_{cr} = 2041 \text{ in.}^4$$

Note that $2n$ is used for the modular ratio of the compression steel (Section 8.10.1.4).

Determine the modulus of rupture of the concrete (f_r) (Section 9.5.2.2).

$$f_r = 7.5\sqrt{f_c'} = (7.5)(\sqrt{4000}) = 474 \text{ psi}$$

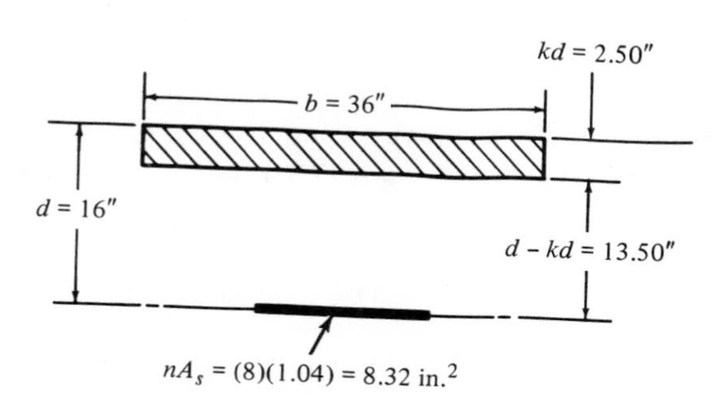

Fig. 4–8 I_{cr} for + moment.

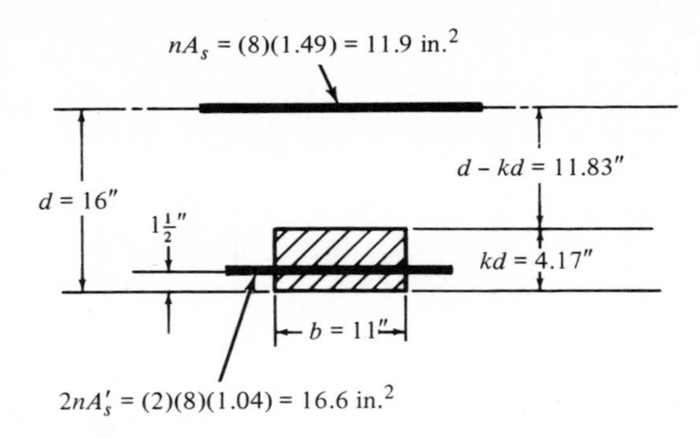

Fig. 4–9 I_{cr} for − moment.

Determine cracking moments (M_{cr}) (Section 9.5.2.2).

$$-M_{cr} = f_r I_g / y_t$$

$$= (474)(7971)/(6.56) = 576{,}000 \text{ in lbs.}$$

$$= 31.5 \text{ ft-kips}$$

$$+M_{cr} = (474)(5359)/(12.30) = 206{,}000 \text{ in. lbs.}$$

$$= 17.2 \text{ ft-kips}$$

The cracking load based on equal spans and a loading pattern that results in a maximum positive moment of $wl_n^2/16$ is

$$w_{cr} = (17{,}200)(16)/(3)(30)^2$$

$$= 102 \text{ psf}$$

Determine the maximum moment for service dead and service live load (M_a) for the stage of loading that produces maximum deflection (Section 9.0)

$$M_a = -M = +M = wl_n^2/16$$

$$= (.254)(3)(30)^2/(16)$$

$$= 42.9 \text{ ft-kips}$$

Determine the effective moment of inertia for positive moment ($+I_{eff}$) (Section 9.5.2.2).

$$+I_{eff} = (M_{cr}/M_a)^3 I_g + [1 - (M_{cr}/M_a)^3] I_{cr} \leq I_g$$

$$= (17.2/42.9)^3(5359) + [1 - (17.2/42.9)^3](1704)$$

$$= 344 + 1594$$

$$= 1938 \text{ in.}^4$$

Determine the effective moment of inertia for negative moment ($-I_{eff}$) (Section 9.5.2.2).

$$-I_{eff} = (31.5/42.9)^3(7971) + [1 - (31.5/42.9)^3](2041)$$

$$= 3160 + 1232$$

$$= 4392 \text{ in.}^4$$

The necessary parameters essential to calculate the deflection of the interior span joist are now available (Section 9.5.2).

DEFLECTION COMPUTATIONS

The instantaneous deflection under service dead load and service live load is based on the clear span, l_n, as allowed for two-way construction, because the negative gross moment of inertia of the joist increases about 20 times in the width of the beams. The loading condition for deflection under service dead and live load is that which produces maximum positive moment ($wl_n^2/16$) and maximum deflection (2) $wl_n^4/384\ E_cI$. The effective moment of inertia is used in deflection calculations and is taken equal to the average of the values for the critical positive and negative moment sections (Sections 9.5.2.2) as recommended by Subcommittee 1 of ACI Committee 453. Note that $I_{cr} \leq I_{eff} \leq I_g$.

$$\Delta_{(D+L)} = \frac{(2)(D + L)l_n^4}{384\ E_c I_{eff}}$$

$$= \frac{(2)(254)(3)(30)(360)^3}{(384)(3{,}600{,}000)(1/2)(1938 + 4392)}$$

$$= \frac{(2)(254)(3)(30)(360)^3}{(384)(3{,}600{,}000)(3115)}$$

$$= 0.496 \text{ in.}$$

Computed Immediate Live Load Deflection

The instantaneous deflection for service live load, based on the effective moment of inertia for service dead and live load, is

$$\Delta_L = (0.496)(105)/(254) = 0.293 \text{ in.}$$

Computed Long-Time Deflections

It is recommended, however, that calculations for the effect of the sustained service load ($D = 104$ psf) be based on the moment of inertia of the cracked section in lieu of I_{eff} because "tests seem to suggest that the tension in the concrete diminishes or even disappears with time for moments above the cracking moment."* Actually, the cracking moment per se becomes unimportant once cracking has occurred. Under *any* sustained load thereafter the tension carried by the concrete between cracks diminishes and additional deflection occurs with time.

The instantaneous deflection for equal spans with all spans loaded with a service dead load of 104 psf using the average cracked moment of inertia is

$$\Delta_{(D\,=\,104)} = \frac{Dl_n^4}{384\,E_cI_{\text{cr}}} = \frac{(104)(3)(30)(360)^3}{(384)(3{,}600{,}000)(1/2)(1704 + 2041)}$$

$$= \frac{(104)(3)(30)(360)^3}{(384)(3{,}600{,}000)(1872)}$$

$$= 0.1682 \text{ in.}$$

The multiplier to be applied to the instantaneous deflection of 0.1682 in. for a dead load of 104 psf to calculate the additional long-time deflection (Section 9.5.2.3) is

$$2 - 1.2(A_s'/A_s) \geqq 0.6 = 2 \text{ (no compression steel with positive moment)}$$

The additional long-time deflection for a sustained service load of 104 psf is

$$\Delta_{(\text{long time})} = (0.1682)(2) = 0.337 \text{ in.}$$

---

* "Allowable Deflections," *Journal of the American Concrete Institute,* June 1968, Chap. 2.

Deflection After Installation of Non-Structural Items

The floor of the example shown in Fig. 4-4 is attached to a plaster ceiling and piping that is likely to be damaged by large deflections. For such a floor the Code recommends that the sum of the additional long-time deflection due to all sustained loads and the immediate deflection due to any additional live load be not more than 1/480 of the span unless adequate measures are taken to prevent damage to these attached elements (Code Table 9.5b). The sum of additional long-time deflection due to sustained service load ($D = 104$ psf) and immediate live load ($L = 150$ psf is equal to

$$\Delta_{(\text{additional long-time plus instantaneous live load})} = 0.337 + 0.293$$

$$= 0.630 \text{ in.}$$

This deflection is equal to 1/572 of the clear span and is less than the maximum value of 1/480 of the span allowed by Table 9.5b of the Code. A joist depth of 17-1/2 in. is therefore, satisfactory because it meets Code deflection criteria (Section 9.5.2.1).

SUMMARY DEFLECTION COMPUTATIONS

A graphical presentation of the deflection of the interior span joist of Fig. 4-4 is shown in Fig. 4-10. The deflection for a cracking load of 102 psf

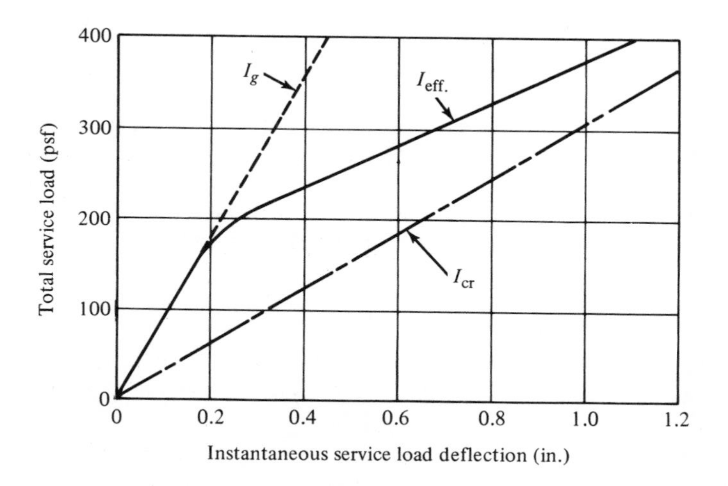

Fig. 4–10 Deflection of interior span joist.

was calculated for equal spans with a maximum positive moment of $wl_n^2/16$ and is

$$\Delta_{cr} = \frac{2wl_n^4}{384\ E_cI_g} = \frac{(2)(102)(3)(30)(360)^3}{(384)(3{,}600{,}000)(5359)}$$

$$= 0.1155 \text{ in.}$$

In order to study the shape of the load-deflection curve from cracking load to design load two additional instantaneous deflections were calculated, one for D + L = 200 psf and another for $1.4D + 1.7L = 400$ psf. Both of these calculations were based on equal spans with alternate panels loaded for an assumed maximum positive moment equal to $wl_n^2/16$ and an assumed maximum deflection equal to (2) $wl_n^4/384E_cI$. The average of the effective moment of inertia (I_{eff}) of the critical positive and negative moment sections was used in these calculations, which are as follows:

$$\Delta_{(D+L=200\text{ ps})} = \frac{2wl_n^4}{384\ E_cI_{eff}}$$

$$= \frac{(2)(22)(3)(30)(360)^3}{(384)(3{,}600{,}000)(1/2)(2189 + 6849)}$$

$$= \frac{(2)(200)(3)(30)(360)^3}{(384)(3{,}600{,}000)(4519)}$$

$$= 0.269 \text{ in.}$$

$$\Delta_{(1.4D+1.7L=400\text{ psf})} = \frac{2wl_n^4}{384\ E_cI_{eff}}$$

$$= \frac{(2)(145 + 155)(3)(30)(360)^3}{(384)(3{,}600{,}000)(1/2)(1764 + 2649)}$$

$$= \frac{(2)(400)(3)(30)(360)^3}{(384)(3{,}600{,}000)(2206)}$$

$$= 1.10 \text{ in.}$$

There is a sudden change in the slope of the load deflection curve for loads greater than the cracking load. An effective moment of inertia (I_{eff}) for deflection calculations is prescribed between the gross moment of inertia (I_g) and the cracked moment of inertia (I_{cr}) (Section 9.5.2.2, Eq. 9-4). Note the transition in Fig. 4-10 for the joist shown in Fig. 4-4.

11-60 Food Service Equipment-Contd.

11-64 Food Preparation Machines
11-65 Food Preparation Tables
11-66 Food Serving Units
11-67 Refrigerated Cases
11-68 Sinks & Drainboards
11-69 Soda Fountains
11-70 Ice Makers
11-71 Garbage Grinders

11-80 GYMNASIUM EQUIPMENT

11-81 Basket-Ball Backstops
11-82 Gymnasts' Equipment
11-83 Gymstands
11-84 Score Indicators
11-85 Indoor Sports Equipment
11-86 Indoor Rifle Range

11-90 INDUSTRIAL EQUIPMENT

11-100 LABORATORY EQUIPMENT

121'4

121'6½

25-0

12

3-0

23-

37

25-4

11-101 Laboratory Casework
11-102 Sinks & Countertops
11-103 Laboratory Equipment
11-104 Fume Hoods

11-110 LAUNDRY EQUIPMENT

11-111 Washers
11-112 Tumblers
11-113 Extractors
11-114 Pressers
11-115 Ironers
11-116 Linen Trucks

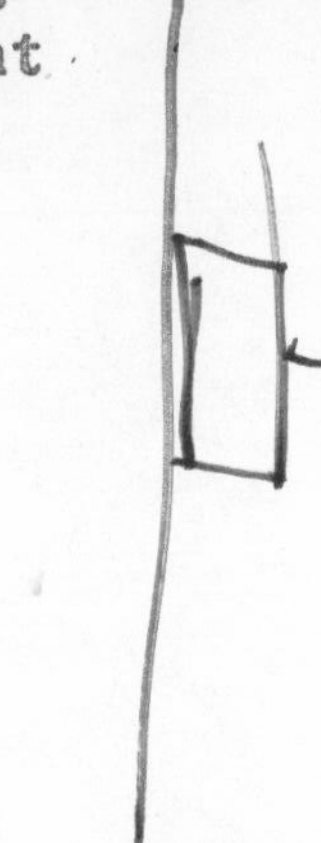

11-120 LIBRARY EQUIPMENT

11-121 Bookshelving
11-122 Bookstacks
11-123 Charging Counters
11-124 Card File System
11-125 Carrels

5 two-way solid flat plate design

GENERAL

Following accepted terminology, a flat plate is considered here as a two-way reinforced flat slab of uniform depth without interior beams, drop panels, brackets, or capitals that is supported directly by columns spaced to form a square or rectangular grid. Spandrel beams at edges and large openings will usually be required for heavy loads, except where large columns are available, or where special provisions are made for shear transfer of loads to edge columns. Cantilever edges can be structurally advantageous particularly where spandrel beams are undesirable for architectural reasons.

The 1971 ACI Building Code is intended to become part of general statutory building codes. Actual live load, as distinguished from design live load (Section 2.1), is defined as the live load specified by the general code. Most building codes allow a reduction of the specified live load based on the area supported by the member, the magnitude of the live load, and the ratio of the dead load plus live load to the live load. The Uniform Code, the BOCA Code, and others allow such live load reductions for flat slabs considering the area of the panel as the area supported by the member.* (Refer to Minimum Design Loads in Buildings and other Structures A58.1–1955 from American Standard Building Code Requirements as sponsored by the National Bureau of Standards.) For 20 × 20 ft panels, a specified live load of 60 psf and a dead load of 100 psf, this code would allow a 32 percent reduction in live load which would reduce the total design load from (1.4 × 100 + 1.7 × 60) or 242 psf to (1.4 × 100 + 0.68 × 1.7 × 60) or 209 psf.

In the 1971 Code, requirements for flat slabs and "two-way slabs" have been combined and completely revised. Format, terminology, analysis, and design for both moment and shear, and even detail lengths for reinforcement are all new. Perhaps the first major-appearing change to those familiar with previous codes is the coefficient 0.125 wl_n^2 in the formula

* See Fig. 2-1.

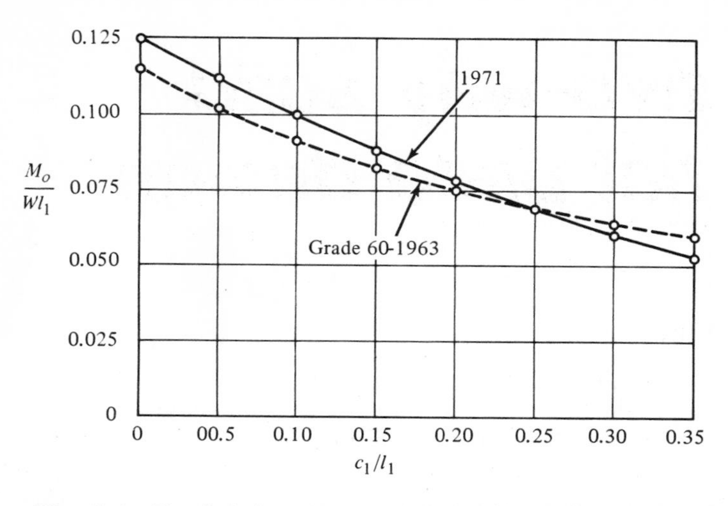

Fig. 5–1 Total design moment, M_o—1963 vs 1971 Codes.

for the total panel moment, M_o. The effect of this change upon the final design is trivial compared to other new provisions, however, and so the practical user is advised merely to accept it and proceed to more significant revisions.* The big changes in analysis and design procedures are the use of the concept of an *equivalent column* in which torsional stiffnesses are considered, and replacing the former "empirical" method with the "direct design" method.

The new direct design method is a direct single cycle solution of a moment distribution frame analysis using very refined stiffness calculations. The previous simple Type A, B, C edge support conditions for external moment have been replaced by a stiffness analysis that considers the column gross moment of inertia between slabs, infinite column moment of inertia within the depth of the slab, and the torsional stiffness of the slab and, if any, beam at the column transverse to the direction in which flexural moments are to be determined. If no beam is used, torsional stiffness of a piece of the slab as wide as the column depth is considered. For the slab flexural stiffness, the gross moment of inertia of the slab is used for the portion of the span between the columns (Section 13.4.1.3). For the portion of the span within the depth of the column the moment of inertia must be taken as the gross moment of inertia of the slab divided by the quantity $(1 - c_2/l_2)^2$ (Section 13.4.1.4). The unbalanced negative slab

* Ample documentation for theorists can be found in the Code Commentary, Chapter 13, and the selected references therein. See Fig. 5-1.

design moments at the columns must be transferred to the columns through flexure and shear in accordance with a formula involving column width, depth, and effective depth of slab (Section 11.13.2). Shear due to transfer of unbalanced moment is combined with uniform vertical shear due to gravity load and becomes a critical value, particularly at edge columns without spandrels, beams, or cantilevers.

The same refined stiffness calculations are also employed in the alternate equivalent frame method. The equivalent frame method must be used when the structure does not fall within the limitations on the use of the direct design method. These limitations are changed from those on empirical design in previous codes. The length-to-width ratio limit has been increased from 1.33 to 2.0. A limit on live-to-dead load ratio of 3.0 has been added. The previous limitation on story and structure height for combination of lateral force moments has been dropped.

Those engineers frequently undertaking flat slab design will find use of a design handbook, a computer design program, or at least design aids for determining various element stiffnesses essential to profitable operations. In the remainder of this chapter, step-by-step procedures with applicable code section references will be given. Some typical design aids for manual calculation of stiffnesses are presented. This material should aid the user in developing a computer design program or a full set of large-scale design aids for precise manual calculation. Finally, some simple approximations for quick longhand calculation are offered. These approximations should give acceptable preliminary designs for estimates, and may often serve as final designs, since the Code permits a 10 percent adjustment of all moments within the required total panel moment (Section 13.3.4.6).

DIRECT DESIGN METHOD (Section 13.3)

In addition to the principal reference section, the new user will find it necessary to refer to new definitions (Section 13.0), new provisions for transfer of moment and shear applicable to either method of design (Sections 13.2.4 and 13.2.5), as well as various expressions under Section 13.4, "Equivalent Frame Method," to solve for terms as defined in Section 13.0. Sections 13.2.4 and 13.2.5 refer the user to Chapter 11, "Shear and Torsion." Finally, a number of the formulas necessary for shear computations are located in the Code Commentary Chapter 11. The best way to illustrate the sequence of references applicable to direct design is by an example. (All of the operations illustrated here apply also to flat slabs, waffle flat slabs, and two-way slabs. In subsequent chapters only the additional operations required for these designs are shown, with references to the simpler example here.)

Sample Problem—Direct Design Method

The most straightforward example is a flat plate without spandrels, with the outer edge flush with the outer face of the exterior columns. As a simple case which might serve for a low-rise motel, dormitory, etc., consider a superimposed ultimate load of 100 psf on square panels with spans of 20 ft. between column centerlines. The story height is 12 ft. The same square column cross section will be maintained from footings to roof for proven overall economy. Lateral forces are to be resisted principally by other stiffer elements. Assume normal weight concrete, $f_c' = 4{,}000$ psi, is an economical choice in the area. Use Grade 60 bars for the maximum economy in reinforcement.

SOLUTION. (Note that the direct design method is a review, requiring the designer to first select trial design dimensions and then check same for strength and serviceability.) A step-by-step procedure follows:

Step 1. Plate thickness (h) Even if the designer intends to compute deflection and to reduce the first trial thickness to the absolute minimum without exceeding the maximum allowable computed deflection (Section 9.5.3.4), he must first complete a trial design through selection of reinforcement, since both the cracking moment and the maximum moment capacity are required for the deflection computations (Section 9.5.2.2).

If, as usual, the designer wishes to avoid deflection calculations entirely, the minimum thickness must satisfy the three limits of Section 9.5.3.1, Eqs. (9-6), (9-7), and (9-8). Note for first trials that Eq. (9-8) sets an upper limit. For flat plates without beams, the controlling limit on thickness of interior panels is Eq. (9-8). See Fig. 5-2 for the solutions of minimum thickness for flat plate edge panels. Select 8 in. as the first trial. Compute the various floor loads: design live load, $w_l = 100$ psf; dead load $= 8/12 \times 150$ psf; design dead load, $w_d = 100 \times 1.4 = 140$ psf; total design load $= 240$ psf. (Note that the term "design load" means specified load times the appropriate load factor in the new code.) Ratio of the live to-dead loads $= \dfrac{100/1.7}{140/1.4} = \dfrac{59}{100} = 0.59$. This ratio applies whenever the term "design" is omitted from Code provisions.

Step 2. Trial column dimensions. For square columns, $c_1 = c_2$. There are no limits, minimum or maximum, on column stiffness. For the usual reinforced concrete column monolithic with the slab, however, an additional step of computing additional bottom reinforcement can be avoided if the required ratio of column-to-slab stiffnesses, α_{min}, is provided (Section

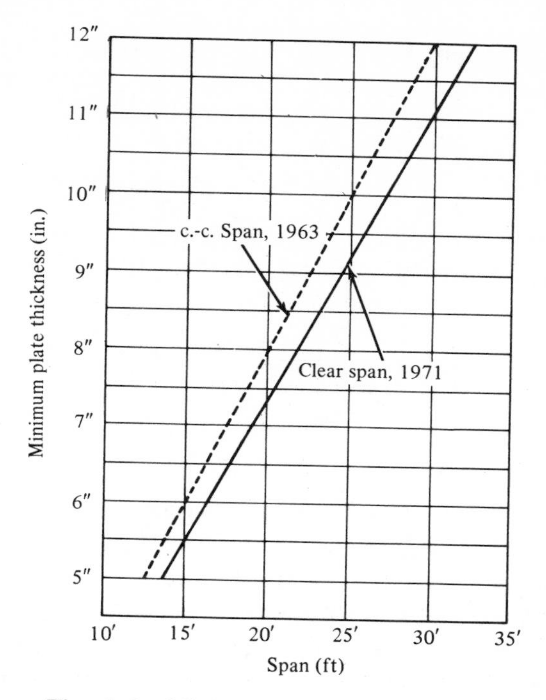

Fig. 5–2 Minimum flat plate thickness.

13.3.6.1a). Note that design for absolute minimum size very flexible columns is permitted (Section 13.3.6.1-b) but will probably involve design of special shear heads (Section 11.1), increased positive moments (Section 13.3.6.1-b), and "slender" column capacity reduction calculations (Section 10.11).

See Fig. 5-3 for required values of α_{min}. For $L/D = 0.59$, $l_1 = l_2$ (square panel), and $\alpha = 0$ (no beams), required $\alpha_{min} = 0.126$. To select the square column dimensions, $c_1 = c_2$, for a ratio $\alpha_c \geqq \alpha_{min}$, a trial size must be assumed. Try $c_1 = c_2 = 16$ in. (see guide to selection of minimum column size, Fig. 5-37). Figure 5-37 shows minimum size for square exterior columns (17 in.) with minimum size square interior columns (12 in.). When both are to be made the same size, the exterior minimum may control, but the increased stiffness of the larger-than-minimum interior column will permit some reduction in the exterior column size shown.

By definition, $\alpha_c = \Sigma K_c/\Sigma K_s$ (Section 13.0) $\Sigma K_c = 2\ K_c$ for columns above and below the plate floor. $\Sigma K_s = 1 \times K_s$ for an exterior column-plate joint. (Note that for roofs, $\Sigma K_c = K_c$, and at interior columns $\Sigma K_s =$

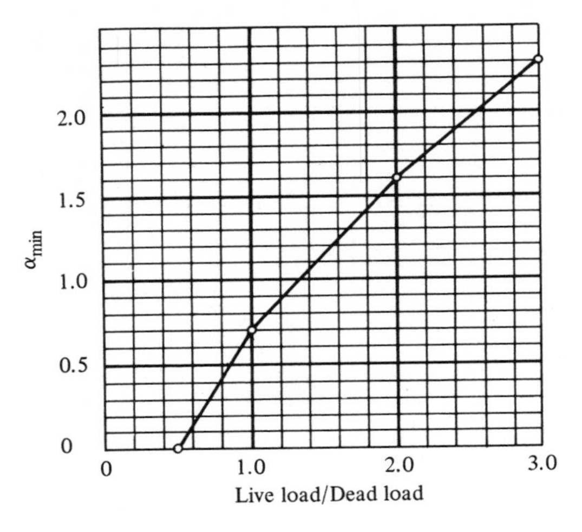

Fig. 5–3 Required α_{min} for square flat plates.

2 K_s.) Flexural stiffness of the column, K_c, must be computed considering the portion within the plate as infinitely stiff (Section 13.4.1.5) and based on the gross cross section elsewhere (Section 13.4.1.3). Slab stiffness, K_s, is based on the gross section between faces of the columns (Section 13.4.1.3) and, for the length from column face to column centerline, the same value multiplied by the factor $\dfrac{1}{(1 - c_2/l_2)^2}$. The subscript 2 indicates direction transverse to that of the moments being determined (Section 13.4.1.4).

Small-scale sample charts illustrate results of these complicated operations for the determination of K_c and K_s (see Figs. 5-4 and 5-5). A large-scale set of similar charts or computer analysis will be required to avoid very time-consuming manual solutions for precise results. The authors find that the following simple approximations for manual solution give reasonably accurate results in the ordinary ranges of span, load, size, etc. for flat plates:

$$K_c = \frac{4\ EI_c}{l_c - 2\ h},$$ where I_c = gross section moment of inertia, column

l_c = story height; and h = plate thickness (all in inches)(see Fig. 5-4)

$$K_s = \frac{4\ EI_s}{l_1 - (0.5)c_1},$$ where I_s = gross section moment of inertia, slab;

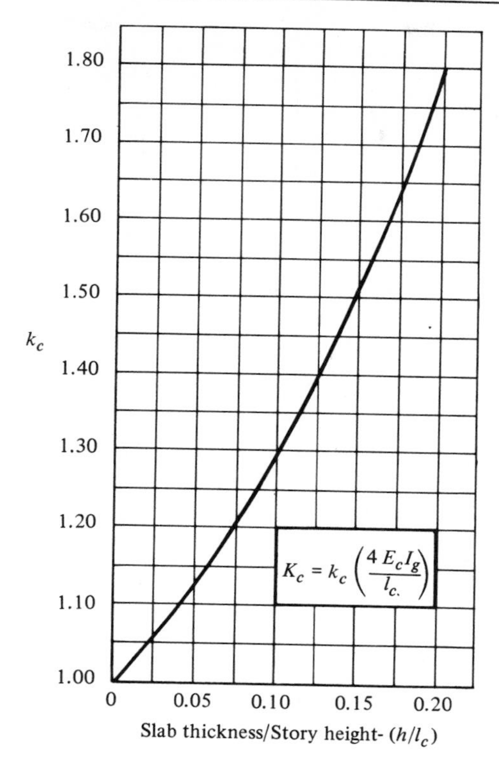

Fig. 5–4 Column stiffness, K_c.

l_1 = center-to-center span in the direction of flexure; and c_1 = column side dimension in the direction of flexure and at the joint considered if columns at each end differ in size; see Fig. 5-5 (*a*) and (*b*).

Using the numerical values of the example,

$$I_c = 5460 \text{ in.} \qquad I_s = 10{,}240 \text{ in.}$$

$$K_c/E = \frac{4 \times 5460}{144 - (2 \times 8)} = 171 \qquad K_s/E = \frac{4 \times 10{,}240}{240 - (1/2 \times 16)} = 176$$

$$\alpha_c = \frac{2 \times 171}{176} = 1.945 > \alpha_{\min} = 0.126$$

Step 3. Ratio of the equivalent column-to-slab stiffness factors for the determination of exterior panel moments, α_{ec}

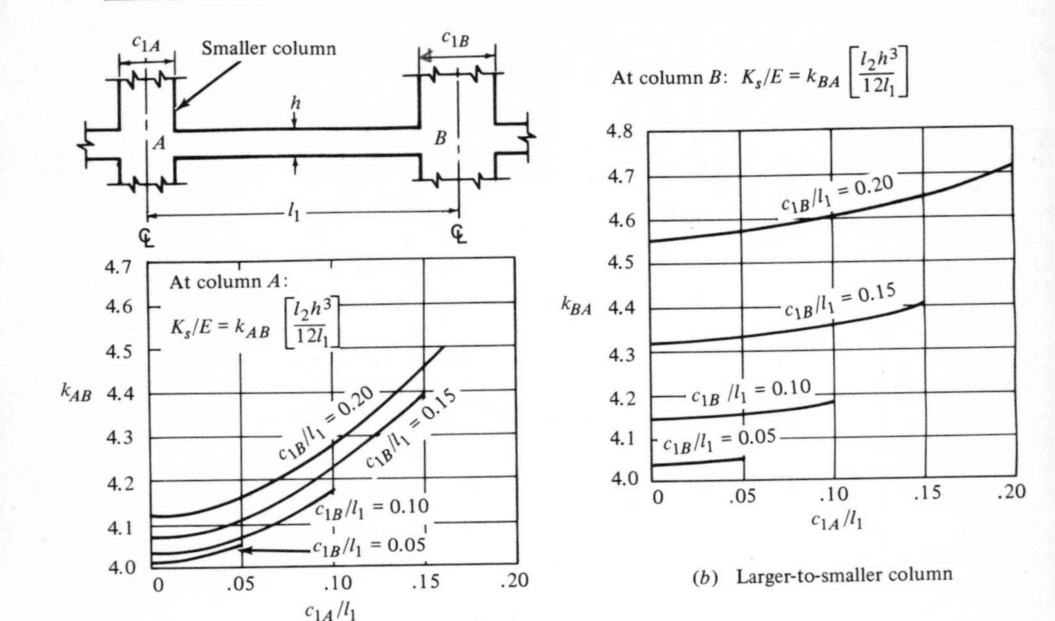

(a) Smaller-to-larger column

(b) Larger-to-smaller column

Fig. 5–5 Flexural stiffness, K_s, for flat plates.

$$\alpha_{ec} = \frac{K_{ec}}{\Sigma K_s} \text{ (Section 13.0).} \qquad \frac{1}{K_{ec}} = \frac{1}{\Sigma K_c} + \frac{1}{K_t} \text{ (Section 13.4.1.5).}$$

$$K_t = \frac{\Sigma 9EC}{l_2(1 - c_2/l_2)^3}$$ (Section 13.4.1.5, Eq. 13-6), the torsional stiffness of the plate at the side of the column.

For an edge column, double the quantity for identical plate sections on each side of the column, $2 \times 9 = 18$. For a corner column $\Sigma = 1.0$.

$$C = \Sigma(1 - 0.63\ x/y)\ \frac{x^3y}{3} \text{ (Section 13.4.1.5, Eq. 13-7)}$$

For a flat plate without spandrels, x and y are the dimensions of a transverse strip of the plate with width equal to the column side, c_1, and a depth equal to the plate thickness, h. x is the smaller of the two dimensions. In this example, $x = 8$ in. and $y = 16$ in. The summation sign here is for spandrel beam areas; $\Sigma = 1$ in this case. $C = (1)[(1 - 0.63 \times 8/16)]$ $\frac{(8)^3(16)}{3} = 1870$. $K_t/E = \frac{2 \times 9 \times 1870}{(240)(1 - 0.0667)^3} = 172$. See sample chart

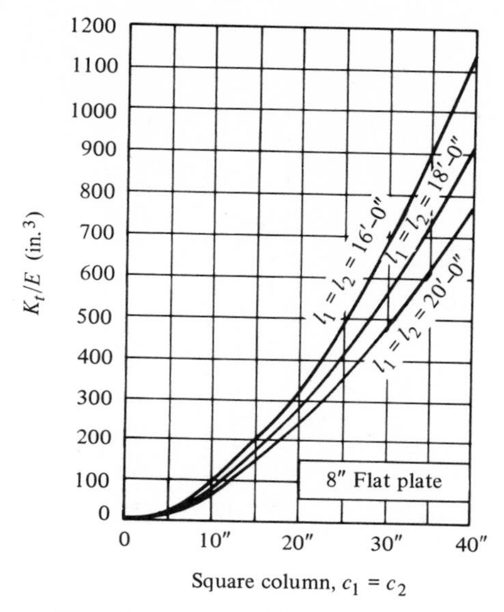

Fig. 5–6 Torsional stiffness, K_t.

for the solution of K_t for $h = 8$ in. (Fig. 5-6). The authors recommend preparation of similar charts for the various plate thicknesses required in routine practice for manual design.

$$E/K_{ec} = E/\Sigma K_c + E/K_t$$

$$\frac{E}{K_{ec}} = \frac{E}{2 \times 171} + \frac{E}{172} = 0.00874\ E$$

$$\frac{K_{ec}}{E} = 114 \qquad \alpha_{ec} = \frac{114}{176} = 0.650$$

See the small scale curves, Fig. 5-7, which show the entire range of solutions of K_{ec} for practicable flat plate applications. Note the very sensitive relationship at the lower end of the range. In this area accurate solutions are more important. For routine manual design, large scale charts in this range will save much time if many column sizes or trials are encountered.

Step 4. Exterior panel moment distribution

$$M_o = \frac{wl_2l_n^2}{8} \text{ (Section 13.3.2.1, Eq. 13-2)}$$

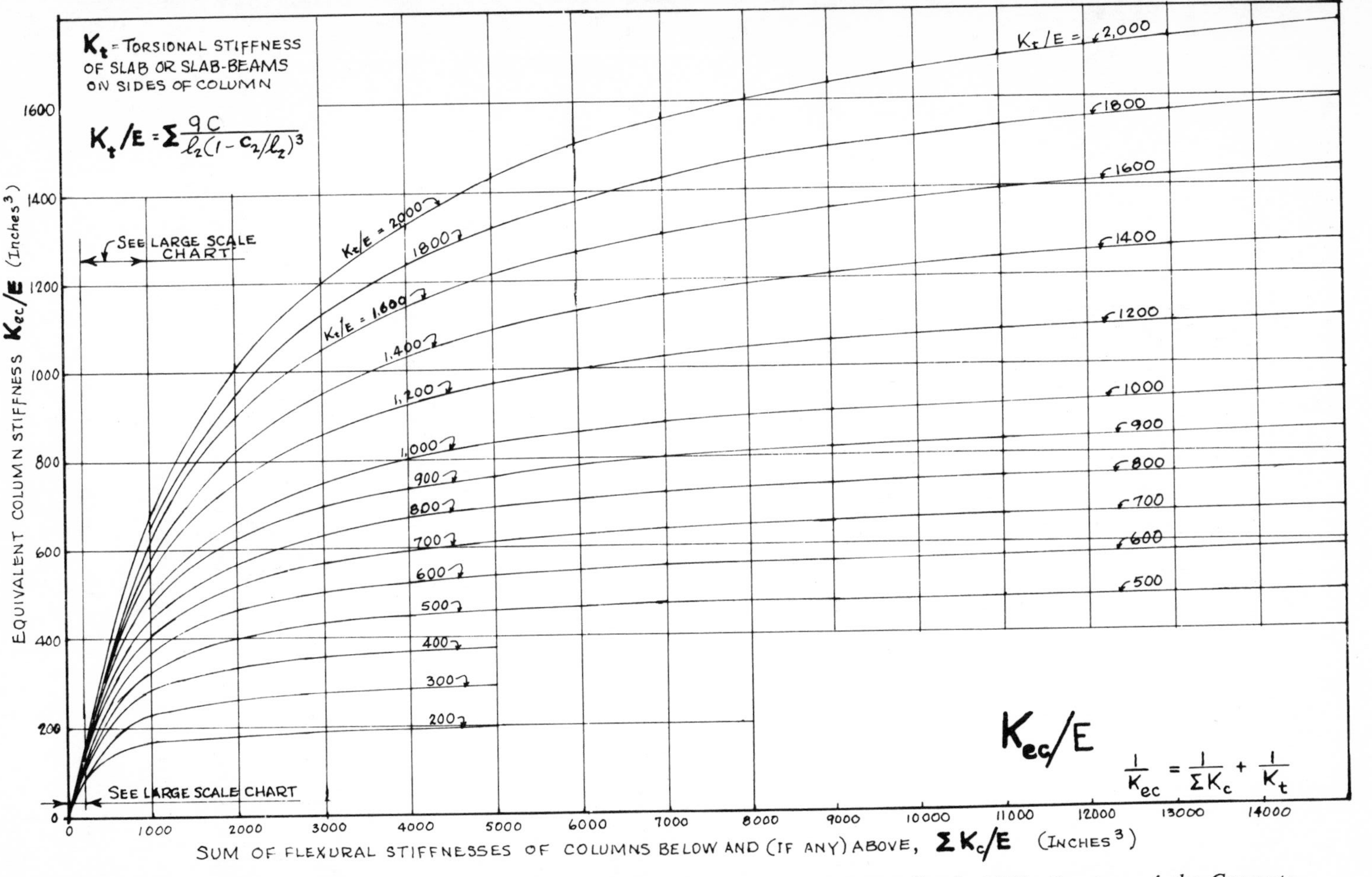

Fig. 5–7 Flexural stiffness, K_{ec}, of equivalent columns. (*From the CRSI Handbook, 1972. Courtesy of the Concrete Reinforcing Steel Institute*)

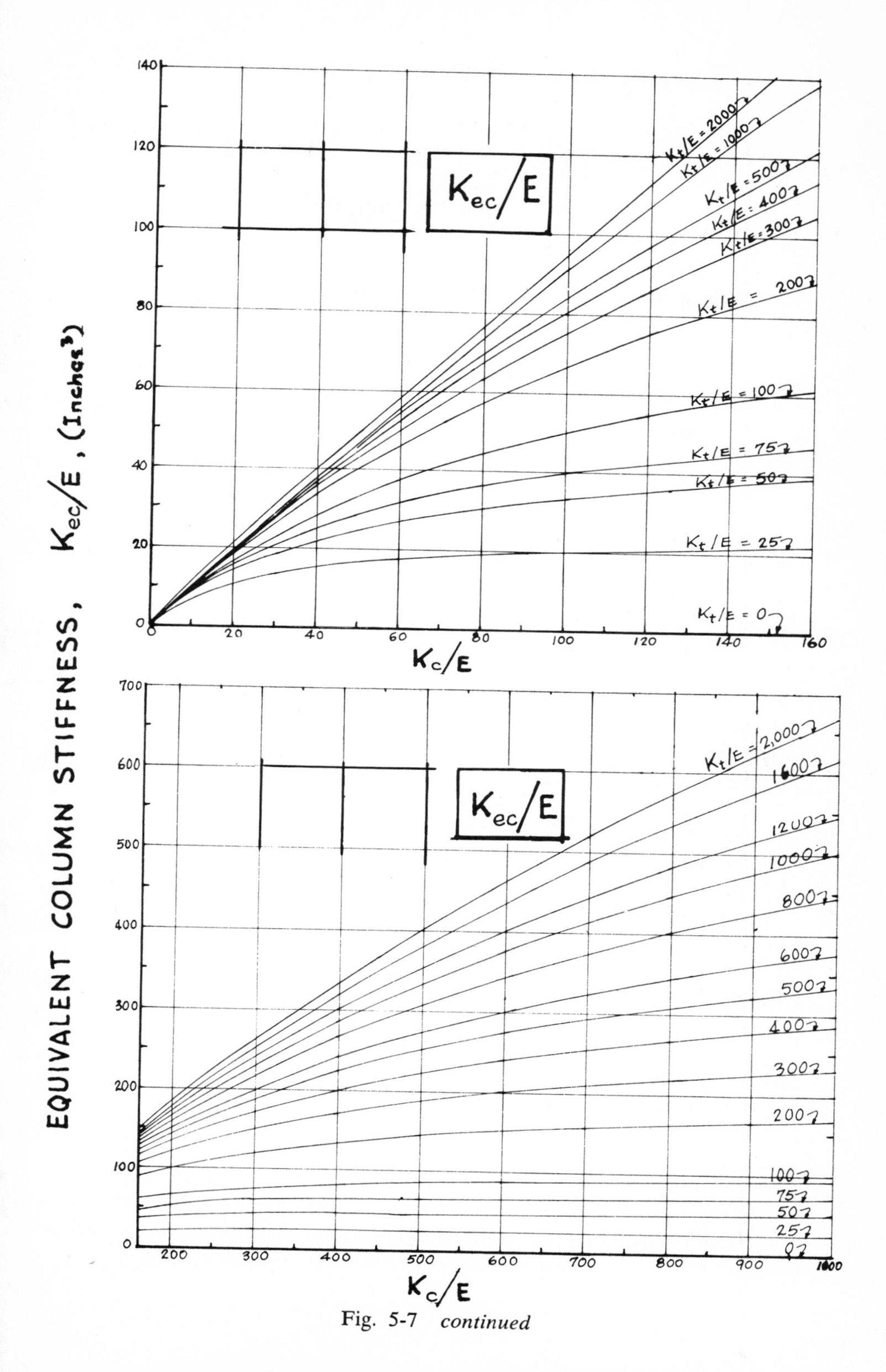

Fig. 5-7 *continued*

$l_n = 20 \text{ ft} - 16/12 = 18.67$ ft, assuming 16 in. square first interior column

$$M_o = \frac{(0.240)(20)(18.67)^2}{8} = 209 \text{ ft-kips total panel moment}$$

Distribution (Section 13.3.3.3). Compute the quantity $(1 + 1/\alpha_{ec})$, $1 + 1/0.650 = 2.54$.

1st interior column: $-M_u = -(0.75 - 0.10/2.54)M_o = -148$ ft-kips

Positive moment: $+M_u = +(0.63 - 0.28/2.54)M_o = +108$ ft-kips

Exterior column: $-M_u = -(0.65/2.54)M_o = -54$ ft-kips

(Note that these are panel moments to be apportioned to column and middle strips; see Fig. 5-8). Note the critical effect of the exterior column stiffness on the exterior panel moments until columns become very large, such as those in lower floors of very tall buildings, etc.

Considering the 10 percent redistribution factor permitted (Section 13.3.4.6), the experienced designer will immediately recognize several short cuts within any given or usual range of column sizes that he may encounter. For practical placing, congestion of the top slab bars passing through the column verticals can be a problem. With no edge beam, all the top bars must be anchored at the edge of the slab, and must be in the column strip. For $\beta_t = 0$, use 100 percent of panel moment for the column strip (Section 13.3.4.2). Furthermore, the code encourages placing up to 60 percent of these bars within the column width plus half the plate thickness on each side, $c_2 + h$. When these requirements get too restrictive for practical placing, decrease the top bar area at the edge by 10 percent, and add approximately 5 percent to the area of the bottom bars. For the example, one could reduce $-M_{ext}$, $0.10 \times 54 = 5.4$ ft-kips and add half, 2.7 ft-kips at the center of the span which equals 2.5 percent times $+ M_u$. Negative moment change at the first interior column can be disregarded as negligible; see Fig. 5-8 (*a*).

Step 5. Shear.

Unfortunately, the designer must proceed through the first four steps before a close trial of the assumed column size for shear can be attempted. Under the 1971 Code, investigation for shear at an exterior column has become quite complicated.

It will be convenient to subdivide this step into three parts: computation of shear constants for the sections in the joint, adjustment of the exterior

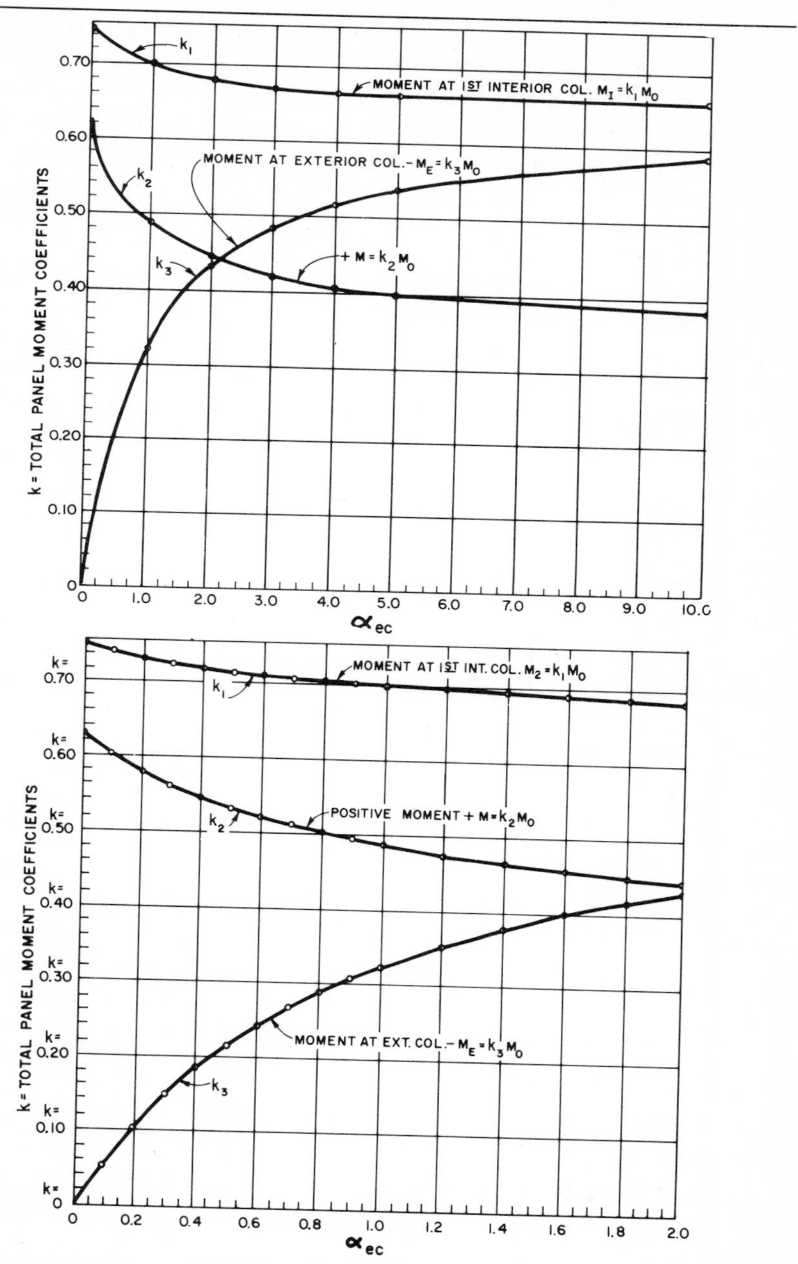

Fig. 5–8 Total panel moment coefficients for edge panels. (*From the CRSI Handbook, 1972. Courtesy of the Concrete Reinforcement Steel Institute*)

moment to the "center of shear" (c.s.), and calculations of the maximum unit shear combining a uniform vertical load shear with moment-induced shear in an assumed elastic distribution.

It will be helpful to avoid numerous trials if the designer can select a reasonably close first trial size column. For square edge columns flush with the edge, the authors find moment-induced shear about half the total allowable value. For such columns, try a column size for which the gravity load shear uniformly distributed around the 3-sided critical section is 50 percent of the allowable shear ($2\sqrt{f_c'}$). For rectangular columns with long side parallel to the edge, try 55 percent ($2.2\sqrt{f_c'}$); with long side at right angles to the edge, 45 percent ($1.8\sqrt{f_c'}$). For square interior columns, neglect moment-induced shear in selecting the first trial size.

Constants. (See Code Section 11.13.2 and for formulas the Commentary,

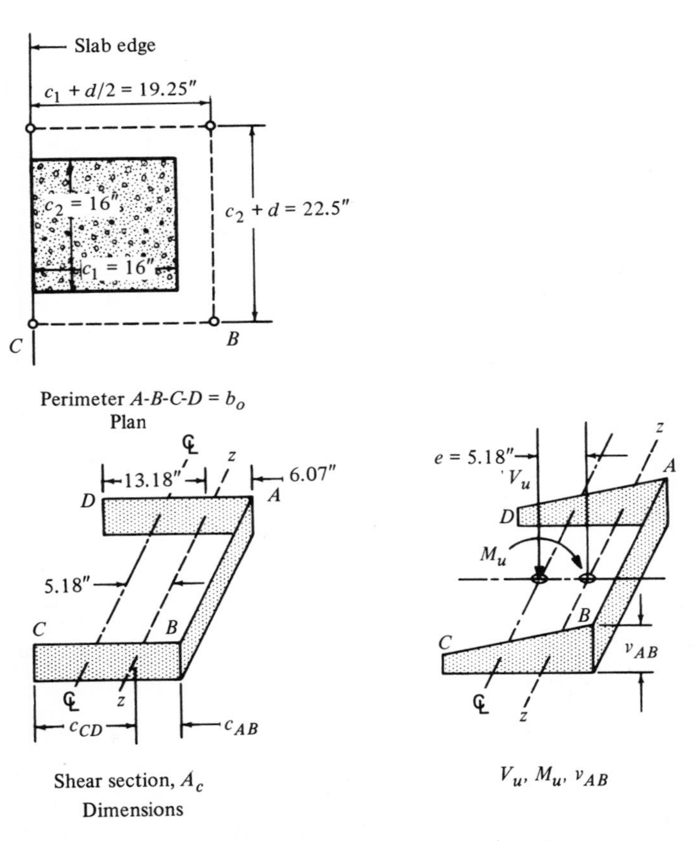

Fig. 5–9 Shear constants, example.

same section.) First consider the exterior column of our example as most likely to be critical. Average effective depth for two-way bars assuming maximum size #6 bars and 3/4 in. cover, $d = 8 - 3/4 - 3/4 = 6\ 1/2$ in. See Fig. 5-9 for designations identifying dimensions. Perimeter of the critical shear section $= b_o$ (Section 11.10.2). For derivation of shear equations, see Fig. 5-33.

$$b_0 = 2(c_1 + d/2 + (d/2 + c_2 + d/2) = 61 \text{ in.}$$

Area of critical shear section, $A_c = b_o d = 61 \times 6.5 = 396$ in. Axis Z–Z passes through the center of gravity of the shear area; point c.s. on the Z-axis is the centroid of shear areas.

$$c_{AB} = \frac{(2)(6.5)(19.25)(1/2)(19.25)}{(396)} = 6.07 \text{ in.}$$

$$c_{CD} = 19.25 - 6.07 = 13.18 \text{ in.}$$

Center of column to the c.s. = 13.18 − 8.00 = 5.18 in.
Face of column to the c.s. = 8.00 − 5.18 = 2.82 in.

$$J_c = \frac{(19.25)(6.5)^3}{6} + \frac{(6.5)(19.25)^3}{6}$$

$$+ 2(6.5)(19.25)(19.25/2 - 5.18)^2 + (6.5)(22.5)(6.07)^2$$

$$J_c = 16{,}980 \text{ in.}^4$$

Let the fraction of the unbalanced moment to be transferred to the column by shear about the center of shear $= K$ (Section 11.13.2):

$$K = 1 - \frac{1}{1 + \frac{2}{3}\sqrt{\frac{c_1 + d/2}{c_2 + d}}} = 0.381$$

The reduction in shear at the face of the exterior column due to the difference in end moments at the exterior and first interior columns is:

$$\frac{(M_{\text{ext}} - M_{int})}{\text{clear span}} = \frac{(54 - 148)}{18.67} = 5.0 \text{ kips}$$

Shear at the face of the column: $V_f = (0.5)(0.240)(20)(18.67) - 5 = 39.8\ k$. See Fig. 5-10.

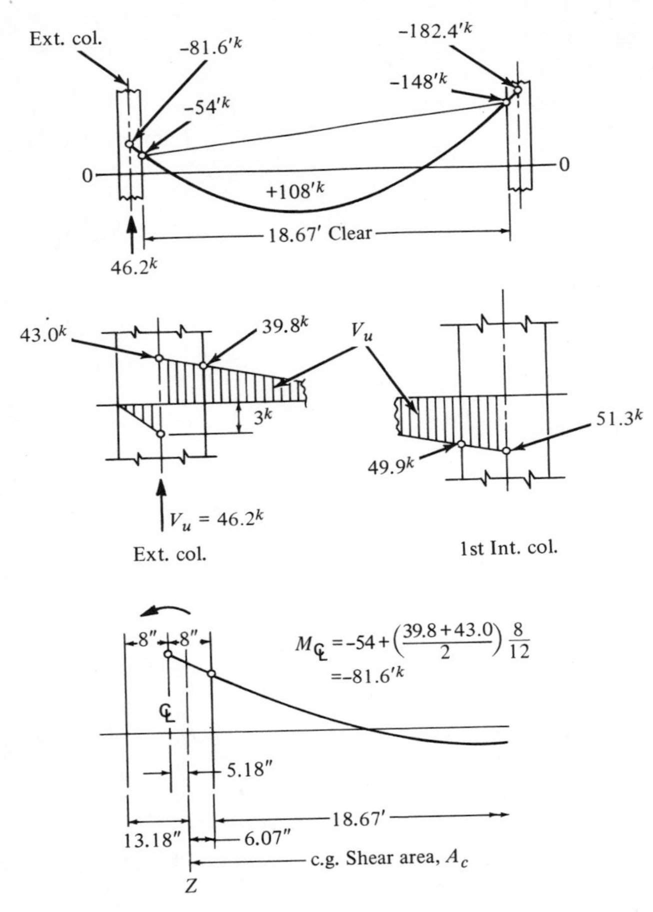

Fig. 5–10 Shear and moment at column centerlines.

Reaction at exterior column: $V_{\text{c.s.}} = 39.8 + (0.240)(20.00)(16/12) =$ 46.2 kips.

Moment at the centerline of the column:

$$M \quad = -54.0 - \frac{(39.8 + 43.0)}{2} \frac{(8)}{(12)} = -81.6 \text{ ft-kips}$$

Moment at the centroid of shear area, A_c:

$$M_{\text{c.s.}} = -M_{\text{un}} + V_{\text{c.s.}}(e) = -[81.6 - 1.1] + (46.2)\frac{(5.18)}{12} = -60.6 \text{ ft-kips}$$

Combined shear due to gravity loads and to moment transfer:

$$v_{AB} = \frac{V_{c.s.}}{\phi A_c} + \frac{KM_{c.s.}c_{AB}}{\phi J_c} \text{ (Commentary 11.13.2)}$$

$$v_{AB} = \frac{(46{,}200}{(0.85)(396)} + \frac{(0.381)(60{,}600)(12)(6.07)}{(0.85)(16{,}980)}$$

$$= 137 \text{ psi} + 117 \text{ psi}$$

$$= 254 \text{ psi}$$

Allowable shear when shear reinforcement is not provided, v_c:

$$v_c = 4\sqrt{f_c'} = 253 \text{ psi}$$

Consider the column size selected satisfactory for shear provided that it is not necessary to use bars larger than #6 both ways, and conservative if the average effective depth is more than that assumed, 6.5 in., due to the use of smaller bars. No shear reinforcement will be required (Section 11.10.3).

Step 6. Shear at the first interior column. Most of the calculations required are the same as in Step 5, and will not be repeated in detail here. Even with the same size of column, the shear constants will change because the four sides of the interior column form a critical shear symmetrical about both axes. The column centerline becomes the center of the shear area. Also, since only the prescribed unbalanced moment need be transferred to the column (Section 13.3.5.2) the combined shear calculation is much less critical. See Fig. 5-11 for dimensions of the critical shear

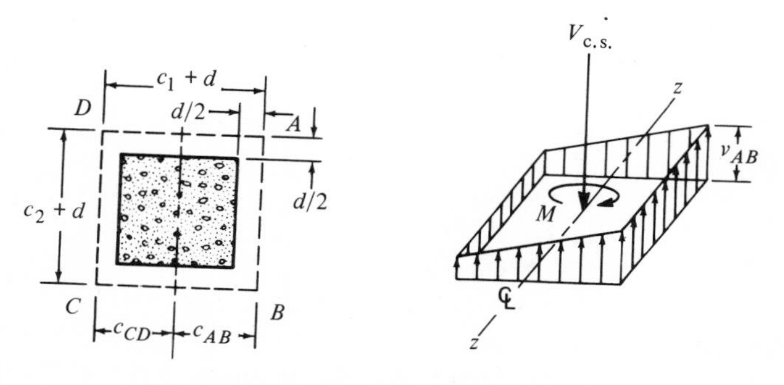

Fig. 5–11 Shear at interior column.

section. For our example, try a 16 in. square column here also. $c_1 = c_2 =$ 16 in. $A_c = 585$ in.² (four sides). $J_c = 50{,}400$ in.⁴ The moment transfer fraction, $K = 0.4$, since $(c_1 + d)/(c_2 + d) = 1$. $c_{AB} = 0.5(16 + 6.5) =$ 11.25 in. The unbalanced moment to be transferred, half to the column above and half to the column below, M_{un} for square panels, l_1 and l_2, where

$$\alpha_{ec} = \frac{K_{ec}}{\Sigma K_s} = \frac{114}{2 \times 176} = 0.325 \text{ reduces to}$$

$$M_{un} = \frac{(0.08)(0.5\ w_l)(l_2)(l_n)^2}{1 + 1/\alpha_{ec}}$$

$$M_{un} = 2{,}370 \text{ ft-lb}$$

Shear at interior columns other than first line = panel load. Panel load = 0.240 (20)² = 96 kips. Also, for the first interior panel, add the shear deducted from exterior column due to difference in negative moments in the exterior panel. Shear at interior column, $V_{c.s.} = 96.0 + 5.0 = 101.0$ kips, all panels loaded. $v_u = \dfrac{101{,}000}{0.85 \times 585} = 203$ psi < 253 allowable. Combined shear plus moment transfer for panel with one-half live load on one side only $V_u = 5.0 + (140)(20)^2 + 0.5(100)(20)^2 = 81{,}000$ lb.

$$v_{AB} = \frac{81{,}000}{(0.85)(585)} + \frac{(0.4)(2{,}370)(12)(11.25)}{(0.85)(50{,}400)} = 163 + 3 = 166 \text{ psi}$$

$$= 166 \text{ psi} < 253 \text{ psi allowable}$$

Beam shear. An investigation of beam shear is required (Section 11.10.1-a). Although it might be critical for long rectangular panels, it is unnecessary in square flat panels. In this example, $b_o = 240$ in.; $A_c = b_o d = 240 \times 6.5 = 1560$ in.² Shear is a maximum at a distance d from the face of the first interior column, $V_{max} = 1/2$ exterior panel load + the increase from the exterior column moment adjustment (See Figs. 5-11 and 5-12).

$$V_{max} = 49{,}900 - (0.240)(20)(6.5/12) = 47{,}300 \text{ lb}$$

$$\text{Beam shear, } v_u = \frac{47{,}300}{0.85 \times 1560} = 35 \text{ psi} < v_c = 2\sqrt{f_c'} = 126 \text{ psi}$$

Step 7. Distribution of panel moments to column and middle strips

The distribution factors used in the previous codes have been revised in the 1971 Code. The new factors are based upon elastic conditions of distribution measured in extensive load tests. With no beams, $\alpha_1 = 0$

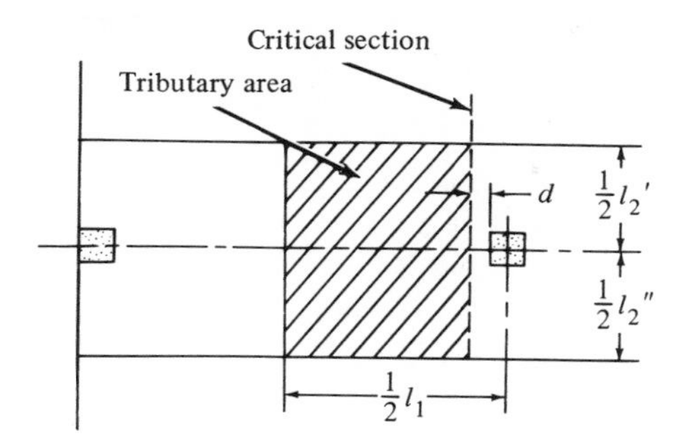

Fig. 5–12 Beam shear.

(Section 13.0), and $\beta_t = 0$ if the attached torsional member is a small strip of slab as in the example. For square panels, $c_1 = c_2$, the moments are distributed in the following percentages for the example used:

Location	Panel Moment	Column Strip	Middle Strip	Reference Sections
Face of ext. column	−54	100%	0%	13.3.4.2
+ Moment ext. panel	+108	60%	40%	13.3.4.3
Face of 1st int. col.	−148	75%	25%	13.3.4.1
+ Moment int. panels	$0.35M_o$	60%	40%	13.3.4.3
Face other int. cols.	$0.65M_o$	75%	25%	13.3.4.1

Step 8. Selection of bars (Section 13.5).

Computation of the steel areas required follows principles familiar to all users of the previous codes. This final step in design is identical whether the design moments have been determined by direct design or by equivalent frame analysis. Where there are two approximately identical crossing layers of bars required, as in square symmetrical panels at centers of the spans or at interior columns, use of the average depth is suggested for computations. Even if the two spans, l_1 and l_2, differ by a small amount, a conservative design can be based upon the longer span and both layers of bars can be made identical. This device saves design computation time, time in specification of the steel placing sequence, and the engineer's time for field inspection. It will also simplify fabrication and reduce field placing time (and avoid possible costly errors) for cost savings to all concerned.

The experienced code user will note that revisions to the typical detail drawing showing minimum extensions, bending, and cutoff points have been made. For greater ease in detailing, fabricating, and placing, simpler combinations are shown (see (Code) Fig. 13.5.6). Unfortunately, one new design requirement shown could increase time for design, detailing, and placing—the specification of all extensions into the span, cutoff points, bend points, etc. on the basis of the clear span. With columns of variable size and shape, some offset from the column centerlines is permitted, and perhaps being reduced in size for upper floors, this requirement literally followed could become a nightmare for all concerned. The authors would advise users to show a typical bending diagram on their structural drawings using all straight bars (end hooks excepted) with dimensions converted conservatively to percentages of center-to-center spans and conservatively referenced to same. See Fig. 5-13 for an example based on our sample problem. Conversion of the code coefficients for bar extensions based on clear span and measured from the face of supports, to the basis of centerline span, l_1, measured from centerline is easily accomplished. Let code coefficient $= k'$, new coefficient $= k$, and the average depth of column at each end of span $= c$. $kl_1 = k'l_n + c/2 = k'(l_1 - c) + c/2$. $0.20 \leq k' \leq 0.33$, and $0.17\,c \leq (-k' + 1/2)c \leq 0.30\,c$. For 16 in. square columns, $c = 0.067\,l_1$ and the correction varies from a minimum of 0.006 to 0.011 times l_1. Use $0.01\,l_1$. $k = k' + 0.01$.

In computing required steel areas and selecting bar sizes, the following controls will ensure conformance to the code and a practical design:

(1) Minimum $\rho = 0.0018\,bh$ for Grade 60 bars (Section 7.13) for either top or bottom steel, or where both are fully developed, the sum of the two. (Section 13.5.3).

(2) Maximum bar spacing is $2h$ (Section 13.5.1), but not to exceed 18 in. (Section 7.4.3). Note that crack control requirements for the various exposure conditions anticipated may control the maximum spacing of the bars (Section 10.6.2). See the explanation, "Crack Control," in this chapter under "Special Conditions" for a practical way to ensure automatic conformance without separate calculations for each panel.

(3) Maximum top bar spacing at all interior locations subject to construction traffic (no code limit). The CRSI recommends not less than #4 @ 12 to provide adequate rigidity and to avoid displacement of top bars with standard bar support layouts under ordinary foot traffic.*

(4) Maximum ρ is limited to $0.75\,\rho b$ (Section 10.3.2); however, the authors recommend that $\rho_{max} \leqq 0.50\,\rho_b$ to permit redistribution (Section 8.6.1), to provide ductility, to avoid too flexible systems subject to objectionable vibration or deflection, and for a practical balance to achieve overall economy of materials, construction, and design time.

* *CRSI Manual of Standard Practice.*

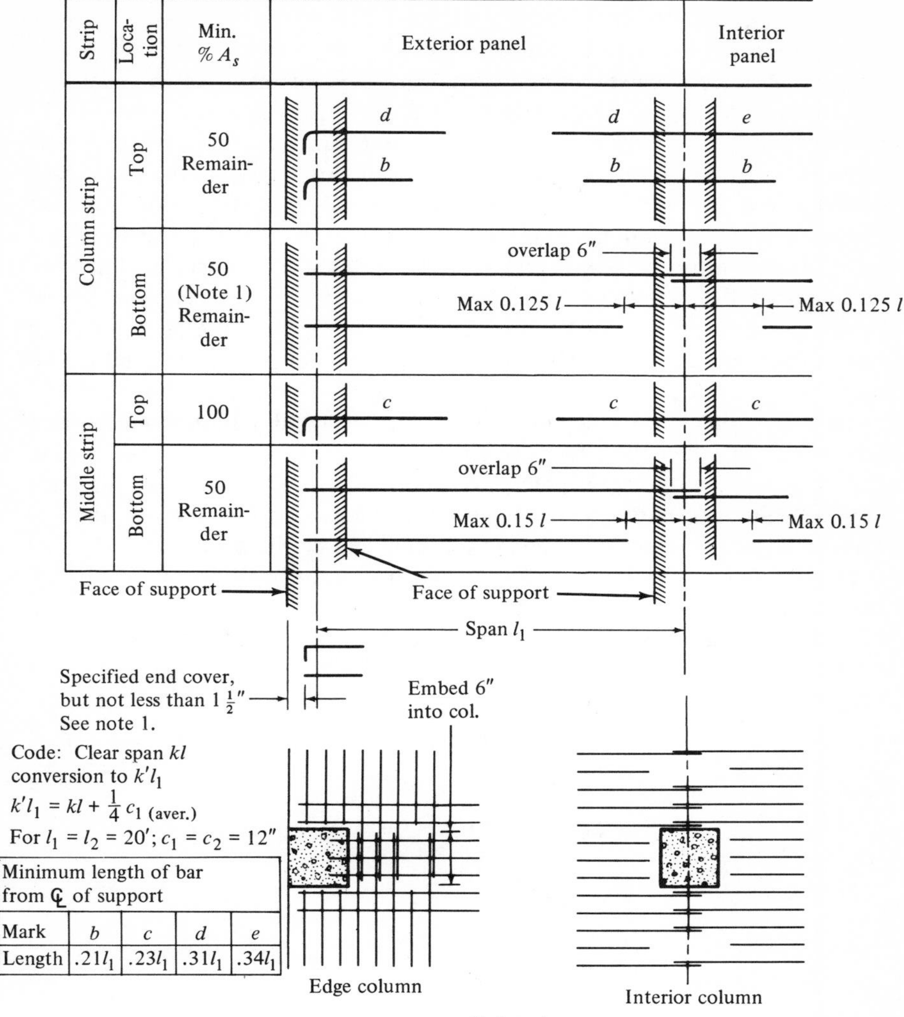

Minimum length of bar from ℄ of support

Mark	b	c	d	e
Length	$.21l_1$	$.23l_1$	$.31l_1$	$.34l_1$

Notes:

1. Place alternate lengths of column strip bottom bars except within the width of the column verticals. Use the shorter length bars in this width to clear the column verticals as shown, except at edge columns, embed 6″ into edge columns.
2. Place half of the column strip top bars within the middle third of the column strip.
3. Place alternate lengths of middle strip bottom bars.

Fig. 5–13 Flat plate reinforcement—recommended details.

(5) Size of bars. Generally, the largest size of bars which will satisfy the maximum limits on spacing will provide overall economy. Critical dimensions which limit size are the thickness of plate available for hooks and the distance from the critical section to edge of plate. In the example, this distance for embedment of top bars at the edge is equal to the column side dimension, *c*, minus the end cover. For full efficiency, the top bars must be fully developed at the critical section for moment, the inner face of exterior columns (Section 13.3.3.1). When excess steel area must be provided to satisfy minimum area or maximum spacing requirements, the required development length based on full f_y may be reduced in the ratio of A_s req'd/A_s used (Section 12.5.d). See Chapter 13, tabulation of end hook embedments, for full f_y. Cantilevered edges afford the designer great flexibility here in selecting the bar size and providing space for full development.

For use with either the all straight bar or combination straight-truss bar detail, note that the new code shows separate top bars with 90° hooks for the negative reinforcement at discontinuous edges. In the exterior panels, half-truss bars are used (straight extension in bottom at discontinuous edge). The separate hooked top bars can be tilted to fit into shallow depth flat plates. Full truss bars with end hooks that will not fit within the slab depth have either caused job delays or resulted in loss of depth assumed for design when tilted. The practice by some designers of specifying hooks bent sideways for such conditions increases fabricating costs disproportionately to the benefit obtained.

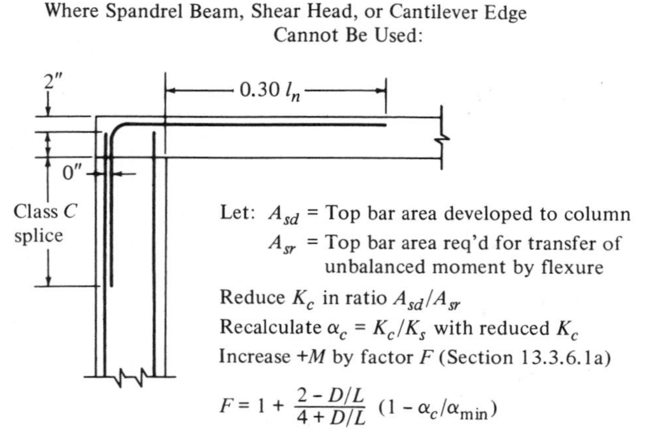

Fig. 5–14 Roof plate design adjustment.

Roof Construction.

The principal change in design from floor slabs to roof slabs results from the loss of the column above the flat plate. This change is similar to the effect of doubling the story height for a column above and below a floor slab. It will require revisions of the reinforcement in the exterior panel. The principal change will be a reduction in the exterior panel negative moment at the edge and a corresponding increase in the positive moment. See the relationship of design moment versus α_{ec} (Fig. 5-8). Another problem is the connection of column to slab for transmission of moment by flexure at the exterior column particularly (see suggestions, Fig. 5-14). Note that a spandrel beam may be used to advantage in the roof slab even if not required on floors below, particularly with flush edges (no cantilever), or that a cantilever at the roof level has especial structural utility.

EQUIVALENT FRAME METHOD

References

In addition to the principal reference (Section 13.4), the new user will find it necessary to refer to new notations (Section 13.0), new definitions (Section 13.1), and new provisions for transfer of unbalanced bending moment and shear between slab and column (Sections 13.2.4 and 13.2.5). All of these provisions are applicable to both direct design and equivalent frame methods. Sections 13.2.4 and 13.2.5 refer the user to Chapter 11, "Shear and Torsion." Formulas for shear computations are also located in the Code Commentary on Chapter 11.

Equivalent Frame Concept

The equivalent frame method of a flat plate design is an elastic analysis of a structural frame consisting of a row of equivalent column members and of horizontal slab members that are one panel in width and extend transversely each side of the centerline of the row of equivalent columns halfway to the next adjacent row of columns. The equivalent column concept is new and is necessary because of the fact that moments "leak" around the column which is not recognized by the 1963 Code. This new concept can best be explained as follows.

If a beam 2 ft wide framed into an exterior column also 2 ft wide and of infinite flexural stiffness in the plane of the beam, there would be no rotation of the joint within the column. The negative moment in the beam at the face of the column would equal the fixed end moment. Under

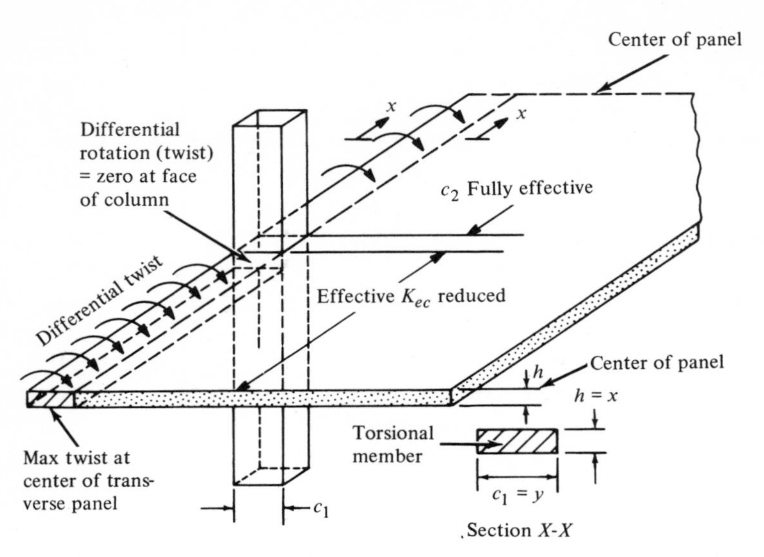

Fig. 5–15 Equivalent column concept.

the new concept an equivalent column is defined as the column and any attached transverse members which contribute torsional resistance to the joint. In this case there are no torsional members and the joint is complete within the width of the column. For this case then, the stiffness of the equivalent column is equal to the stiffness of the column alone.

If a two-way slab 20 ft wide framed into the same column, however, the condition of no rotation of the joint would exist for only the central portion of the slab 2 ft wide at the column (see Fig. 5-15). For the remaining two portions of slab extending 9 ft transversely to the column on each side, rotation of the slab would occur and increase from zero at the side face of the column to a maximum at the end of each 9 ft portion of slab. Because of his rotation, the negative moment at the face of the column would be less than the full fixed end moment. Obviously, the equivalent (or effective) stiffness of the exterior column is decreased by the rotation of the attached transverse members.

The equivalent column stiffness, K_{ec}, is, therefore, always less at a joint with a two-way slab than the stiffness of the column itself, K_c, since the differential torsional rotation between slab and column will occur unless the torsional members possess infinite torsional stiffness. The flexibility (reciprocal of stiffness) of the equivalent column is increased by the flexibility of the torsional member (see Fig. 5-7 (*a*). Note that the ratio, K_{ec}/K_c approaches unity only when the torsional stiffness, K_t, is much greater than the column stiffness, K_c; when $K_t \geq 50\ K_c$, the chart shows that $K_{ec} \approx K_c$.

Scope of the Equivalent Frame Method

There are no limitations on the maximum ratio of live load to dead load, of longitudinal to transverse span, or of the amount that successive span lengths can differ one from the other. The analysis of the equivalent frame is made for full design live load on all spans when the (actual) live load is variable, but does not exceed 0.75 times the (actual) dead load (Section 13.4.1.8). A pattern loading sequence with three-fourths of the design live load must be used for determining maximum moments and shears when the live load specified (in the General Code) exceeds 0.75 times the dead load. "Design live load" (w_l) is the specified live load multiplied by the live load factor. "Live load" and "dead load" mean specified or actual loads (see Chapter 2 and Fig. 2-1).

Shear reinforcement consisting of bars, wires, or of steel I—or channel shapes can be used with flat plate construction to minimize column sizes (Section 11.11). For usual flat plate loads and spans, shear reinforcement is not required. If column loads are small, the minimum column size is determined by limiting the shear stress on the critical shear section, located a distance $d/2$ from the column, caused by gravity loads and unbalanced gravity load moments to the value, $4\sqrt{f_c'}$ allowed on concrete (Section 11.10).

The sequence of references applicable to the equivalent frame analysis is shown by use of the same example used with the direct design method—which makes possible a comparison of the two methods. (All of the operations illustrated for the equivalent frame method also apply to two-way slabs supported on beams, flat slabs with drop panels and/or capitals, which utilize two-way joists (see Chapters 6, 7, and 8). For these designs, only the additional operations required by the modifications therein are shown.)

SAMPLE PROBLEM. Superimposed design live load, w_l, of 100 psf on square panels with spans of 20 ft between column centerlines. The story height is 12 ft. The same square column cross section will be maintained from footings to roof for proved economy. Lateral forces are to be resisted principally by other stiffer elements. Assume that normal weight concrete $f_c' = 4{,}000$ psi is an economical choice in the area. Use Grade 60 bars for maximum economy in reinforcement. The outer face of edge columns will be flush with the edge of the slab.

SOLUTION. (Note that as with the direct design method, the equivalent frame method is a review. The designer must first select trial concrete dimensions and then make an equivalent frame analysis to check the assumed sizes for structural adequacy.)

Step 1. Plate thickness (h)

Same procedure as for direct design method (See Fig. 5-2, $h = 8'' \geq 6.7$).

Step 2. Trial column dimensions

For square columns $c_1 = c_2$. There are no limits on column size other than that the column be large enough to sustain the load with its eccentricity and provide a peripheral shear area that can transfer the load and a part of the unbalanced moment from the slab to the column.

If shear reinforcement is not used, a trial size exterior column can be obtained from Fig. 5-37 or by determining the area of the critical section shown in Fig. 5-16 that will limit the shearing stress due to gravity load only to a conservative value, $2\sqrt{f_c'}$.

Determine trial size exterior column size.

From Fig. 5-37, for minimum interior *and* exterior columns, $c_1 = 17$ in. or required

$$A_c = \frac{(w_d + w_l)(l_2)(l_1/2)}{\phi(2)\sqrt{f_c'}} = \frac{[(1.4)(100) + 100]\ (20)(20)/2}{(0.85)(2)(\sqrt{4000})}$$

$$= 446 \text{ in.}^2$$

For square exterior columns with exterior face flush with the exterior

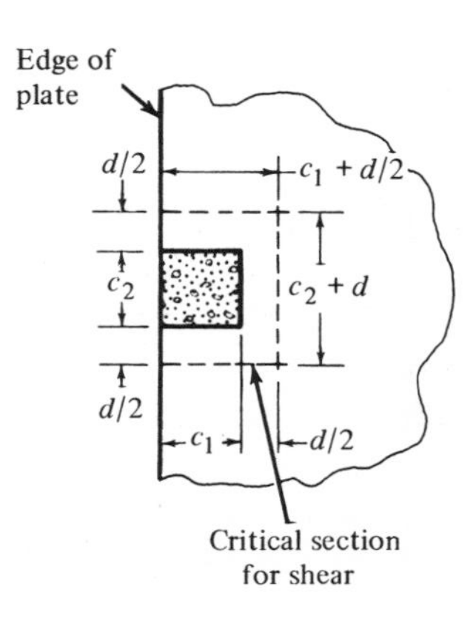

Fig. 5–16 Plan at exterior column.

edge of the slab, the minimum size exterior square column can be calculated as follows:

$$d = h - 1.5 = 8 - 1.5 = 6.5 \text{ in.}$$

$$(3c_1 + 2d)d = 446 \text{ in.}^2$$

$$[3c_1 + (2)(6.5)]\,(6.5) = 446 \text{ in.}^2$$

$$c_1 = 18.6 \text{ in.}$$

If both interior and exterior columns are made to the same size, the trial column size selected should lie between the two sizes shown for minimum exterior and interior columns shown in Figs. 5-34 through 5-39. Figure 5-37 shows minimum exterior square column, 17 in., minimum interior column. 12 in. Try 16 in. square columns for both.

A trial size for the interior column can be made as for the exterior one, except limiting the shear stress for gravity load only, V_u/A_c (neglecting unbalanced gravity load moment), to $4\sqrt{f_c'}$ (see Fig. 5-17). A 16 in. square column should be conservative.

Step 3. Determine column stiffness, K_c

The stiffness of the exterior column or interior columns can be obtained by the simple approximations outlined in Step 2 of the direct design method. If greater accuracy is desired, see Fig. 5-4. The stiffness can be also calculated directly by use of column analogy as follows:

$$K_c = \frac{1}{A_{ac}} + \frac{Mc}{I_{ac}}$$

A_{ac} = Area of analogous column

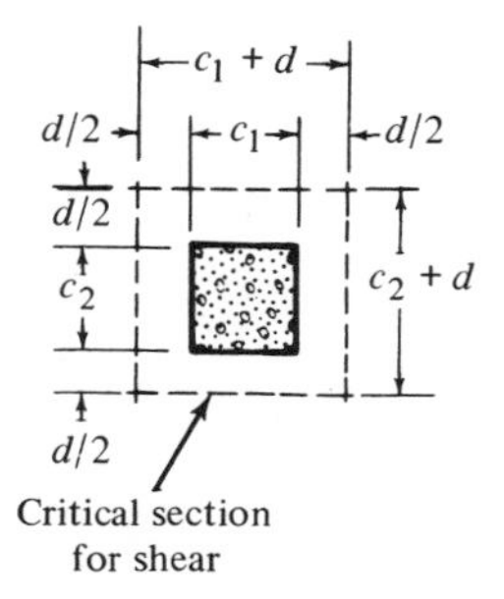

Fig. 5–17 Plan at interior column.

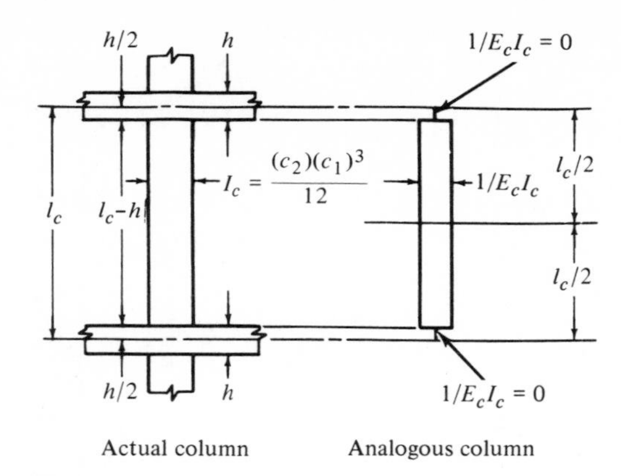

Fig. 5–18 Column stiffness, K_c, by column analogy.

I_{ac} = Moment of inertia of analogous column about neutral axis x–x

M = Moment due to a unit load at the extreme fiber of the analogous column which is located at the center of the slab = $l_c/2$.

$$A_{ac} = \frac{l_c - h}{E_cI_c}$$

$$I_{ac} = \frac{(l_c - h)^3}{12E_cI_c}$$

$c = l_c/2$ or the distance from the neutral axis to the extreme fiber of the analogous column.

$$K_c = \frac{E_cI_c}{l_c - h} + \frac{3E_cI_cl_c^2}{(l_c - h)^3}$$

$$K_c/E_c = \frac{(I_c)}{(l_c - h)}\left[1 + \frac{3l_c^2}{(l_c - h)^2}\right]$$

For the trial column size of 16 in. from Step 2, the column stiffness will be determined by the column analogy procedure and compared with the approximate method outlined in Step 2 of the direct design method.

$$K_c/E_c = \frac{5461}{(144 - 8)}\left[1 + \frac{(3)(144)^2}{(144 - 8)^2}\right]$$

$$= 175 \text{ in.}^3$$

A value of 171 in.^{3} was obtained for K_c/E_c by the authors' approximation which is less than 3 per cent in error.

Step 4. Slab stiffness, K_s

The flexural stiffness of the slab, like the flexural stiffnesses of columns, can be determined using the simple approximate methods outlined in Step 2 of the direct design method, by using Figs. 5-5 (*a*) and 5-5 (*b*) or by the column analogy method.

Numerical values for the flexural stiffness of the slab will be determined by the column analogy method and compared to the results obtained from Figs. 5-5(*a*) and 5-5(*b*) and those previously determined by the approximations shown with the direct design method.

$$K_s = \frac{1}{A_{ac}} + \frac{Mc}{I_{ac}}$$

$$I_g \text{ (slab between columns)} = \frac{(240)(8)^3}{12} = 10{,}240 \text{ in.}^4$$

$$I_g \text{ (slab at column)} = \frac{I_s}{\left[1 - \frac{c_2}{l_2}\right]^2} = \frac{10{,}240}{\left[1 - \frac{16}{240}\right]^2}$$

$$= 11{,}756 \text{ in.}^4$$

$$\begin{aligned} A_{ac} = (0.0000976)(18.67)(12) &= 0.02186 \\ (2)(0.0000850)(0.667)(12) &= \underline{0.00136} \\ &\ 0.02322 \end{aligned}$$

$$\begin{aligned} I_{ac} = (0.0000850)(240)^3/12 &= \ 97.9 \\ + (0.0000126)(224)^3 &= \ \underline{11.8} \\ &\ 109.7 \end{aligned}$$

$$K_s/E_c = \frac{1}{A_{ac}} + \frac{(1)(l_1/2)(l_1/2)}{I_{ac}}$$

$$= \frac{1}{0.02322} + \frac{(1)(120)(120)}{109.7}$$

$$= 43 + 131$$

$= 174$ in.^{3} (175 in.^{3} obtained by approximation shown with direct design method)

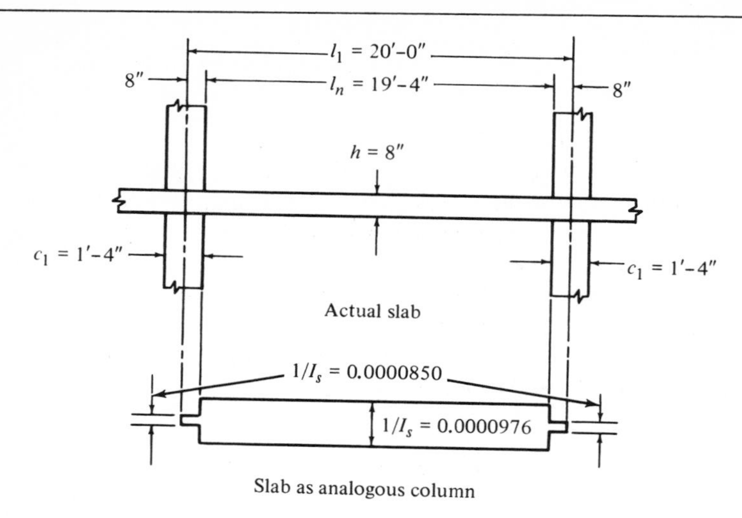

Fig. 5–19 Slab stiffness, K_s by the column analogy.

K_s using Figs. 5-5(a) and 5-5(b)

$$c_{1A} = c_{1B} = \frac{16}{240} = 0.0667$$

$$k_{AB} = k_{BA} = 4.09 \text{ (See Figs. 5-5a or 5-5b)}$$

$$I_g = \frac{l_2 h^3}{12} = \frac{(1)(20)(12)(8)^3}{12} = 10{,}240 \text{ in.}^4$$

$$\frac{K_s}{E_s} = \frac{k_{AB} I_s}{l_1} = \frac{k_{AB}(10{,}240)}{(20)(12)}$$

$$= (4.09)(42.7)$$

$$= 175 \text{ in.}^3 \text{ (176 in.}^3 \text{ obtained by approximation shown with direct design method)}$$

Step 5. Stiffness of exterior equivalent column, K_{ec}.

Use the same procedure shown in Step 3 of the direct design method for determining K_t/E_s.

$$\frac{K_c}{E_c} = 175 \text{ in.}^3 \text{ (Refer to step 3 of equivalent frame method)}$$

$$\frac{K_t}{E_s} = 172 \text{ in.}^3 \text{ (Refer to step 3 of direct design method)}$$

$$\frac{K_{ec}}{E_c} = \frac{\Sigma K_c}{1 + \Sigma \dfrac{K_c}{K_t}} = \frac{2 \times 175}{1 + \dfrac{(2)(175)}{172}}$$

$$= 115 \text{ in.}^3 \text{ (114 in.}^3 \text{ by approximation in direct design method example)}$$

Step 6. Moment distribution factors.

For distributing slab moments at exterior and interior columns. At the exterior column, the distribution factor is as follows:

$$\text{DF (Ext. Col.)} = \frac{K_s}{K_s + K_{ec}} = \frac{174}{174 + 115} = \frac{174}{289} = 0.602$$

$$\text{DF (Int. Col.)} = \frac{K_s}{K_s + K_{ec}} = \frac{174}{(2)(174) + 115} = \frac{174}{463} = 0.376$$

Step 7. Design loads for moment distribution.

The specified live load is equal to 100/1.7 or 59 psf. The dead load is computed: $8/12 \times 150$ or 100 psf. The live load does not exceed three-quarters of the dead load ($59 < 0.75 \times 100$). The equivalent frame can be analyzed one floor at a time with the far ends of the columns assumed fixed for full design gravity load on all spans and patterned loading neglected (Section 13.4.1.8).

Frame loads per foot of span (across 20 ft width l_2)

$$w_d(klf) = \frac{(1.4)(100)l_2}{1000} = \frac{(1.4)(100)(20)}{1000}$$

$$= 2.80 \text{ klf}$$

$$w_l(klf) = \frac{(59)(1.7)l_2}{1000} = \frac{(100)(20)}{1000}$$

$$= 2.00 \text{ klf}$$

Step 8. Carry-over factors.

COF_{AB} and COF_{BA} for the equivalent frame elastic analysis will be obtained by the column analogy method (see Fig. 5-19).

$COF_{AB} = COF_{BA}$ (by column analogy method)

$$= \frac{\dfrac{1}{A_{ac}} - \dfrac{(1)(l_1/2)(l_1/2)}{I_{ac}}}{\dfrac{1}{A_{ac}} + \dfrac{(1)(l_1/2)(l_1/2)}{I_{ac}}} = \frac{43 - 131}{43 + 131}$$

$$= \frac{-88}{174} = -0.506$$

$COF_{AB} = COF_{BA}$ (Commentary Table 13 C–1, for

$$\frac{c_{1A}}{l_1} = \frac{c_{1B}}{l_1} = \frac{16}{240} = 0.0667$$

$$= 0.505$$

Step 9. Fixed end moment factors.

M_{AB} and M_{BA} can be obtained by the column analogy method (see Fig. 5-20); compare with the Commentary, Table 13-1, if desired.

area M/I diagram (P_{ac})

1. $\frac{(2)}{3}(18.67)(12)(0.245) \quad = 36.59$

2. $(18.67)(12)(0.36) \quad = 8.06$

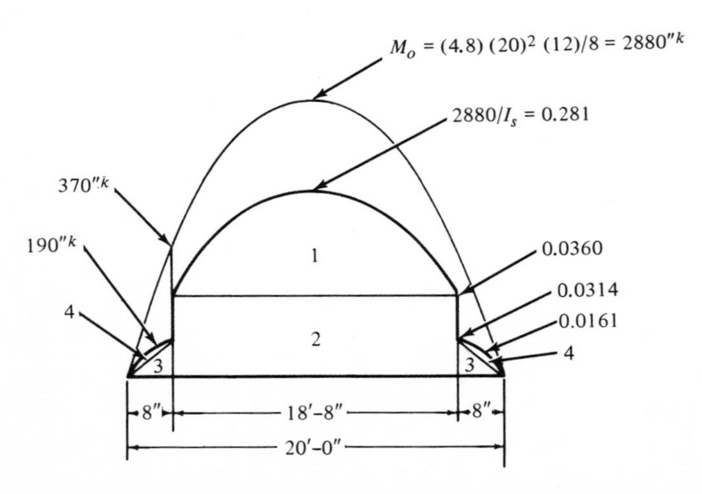

Fig. 5–20 Slab fixed end moments by column analogy.

3. $(2)(\underline{1})(0.667)(12)(0.0314) = \quad 0.25$
$\quad\quad 2$

4. $(2)(\underline{2})(0.667)(12)(0.0004) = \quad \underline{\quad —\quad}$
$\quad\quad 3 \quad\quad\quad\quad\quad\quad\quad\quad\quad\quad\quad\quad 44.90$

$$FEM_{AB} = FEM_{BA} \text{ (symmetrical moment diagram)} = \frac{P_{ac}}{A_{ac}} = \frac{44.9}{0.0232}$$

$$= 1935 \text{ in.-kips} = 161 \text{ ft-kips}$$

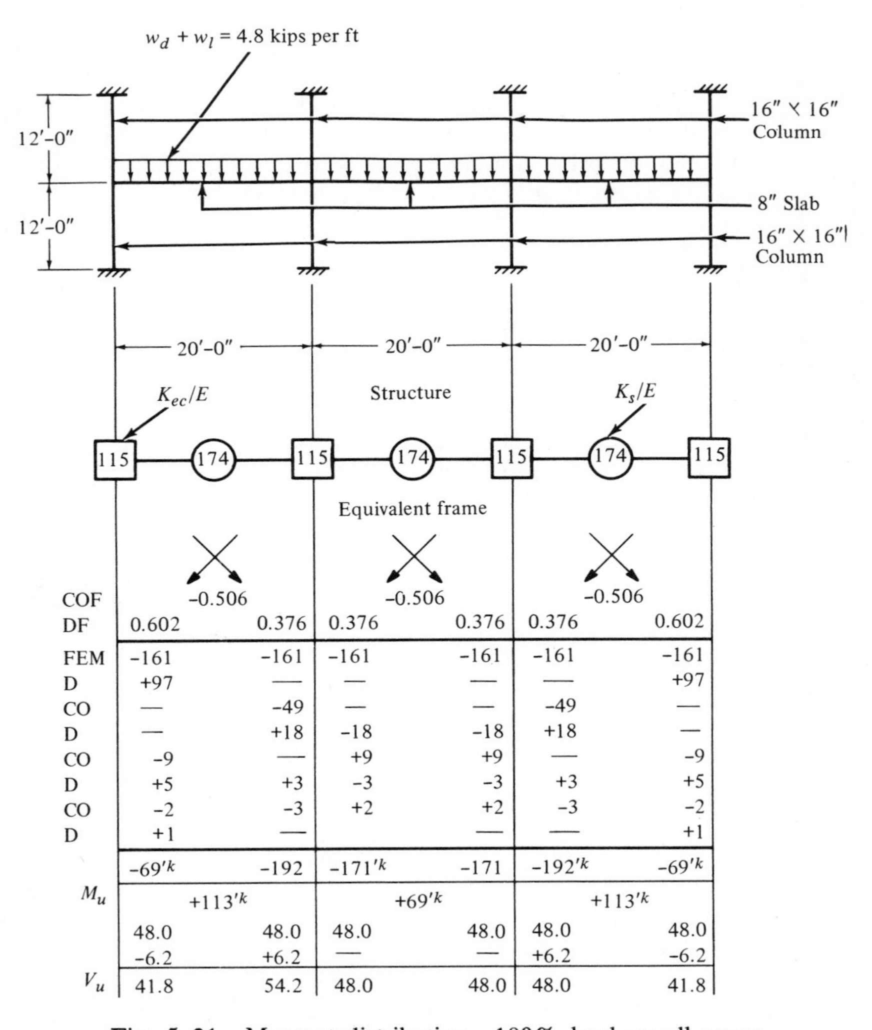

COF	-0.506		-0.506		-0.506	
DF	0.602	0.376	0.376	0.376	0.376	0.602
FEM	-161	-161	-161	-161	-161	-161
D	+97	—	—	—	—	+97
CO	—	-49	—	—	-49	—
D	—	+18	-18	-18	+18	—
CO	-9	—	+9	+9	—	-9
D	+5	+3	-3	-3	+3	+5
CO	-2	-3	+2	+2	-3	-2
D	+1	—		—	—	+1
	-69′ᵏ	-192	-171′ᵏ	-171	-192′ᵏ	-69′ᵏ
M_u	+113′ᵏ		+69′ᵏ		+113′ᵏ	
	48.0	48.0	48.0	48.0	48.0	48.0
	-6.2	+6.2	—	—	+6.2	-6.2
V_u	41.8	54.2	48.0	48.0	48.0	41.8

Fig. 5–21 Moment distribution—100% load on all spans.

Step 10. Moment distribution analyses.

1. Moment distribution for full load on all spans (see Fig. 5-21).
2. Compare results to those by the direct design method: see Fig. 5-10.

Step 11.

Compute slab shear at exterior trial column. The same procedure shown in Step 5 of the direct design method is used in the equivalent frame method (see Fig. 5-10). The shear (V_{cs}) is equal to the exterior column reaction which includes the slab load between the column centerline and the exterior face of the slab at the exterior face of the column. The moment at the centroid of the critical section is equal to the unbalanced moment at the column centerline minus (V_{cs}) (e) where e is equal to the distance between the column centerline and the centroid of the critical section. For shears and moments, see Fig. 5-22.

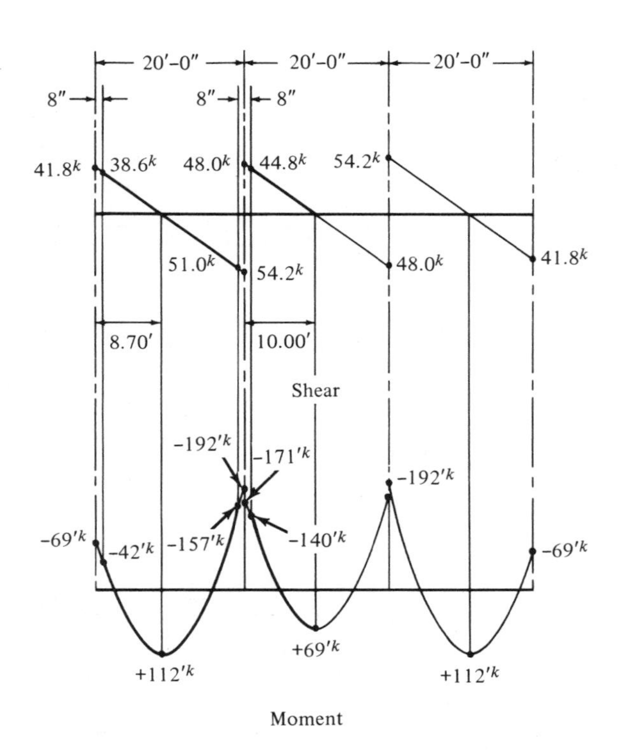

Fig. 5–22 Moment distribution results—100% load on all spans.

$$V_{cs} = 41.8 + (4.8)\left(\frac{8}{12}\right)$$

$$= 41.8 + 3.2 = 45.0 \text{ kips}$$

$$M_{cs} = 69 - (4.8)\left(\frac{8}{12}\right)\left(\frac{4}{12}\right) - (V_{cs})(e)$$

$$= 69 - 1 - (45.0)\left(\frac{5.18}{12}\right) = 69 - 1 - 19$$

$$= 49 \text{ ft-kips}$$

Using the geometrical properties of the critical section as determined in Steps 5 and 6 of the direct design method on page 82 (see Fig. 5-9), the shearing stresses at the exterior column based on the equivalent frame analysis method are calculated as follows:

$$v_{AB} = \frac{V_{cs}}{\phi A_c} + \frac{KM_{cs}c_{AB}}{\phi J_c}$$

$$= \frac{45{,}000}{(0.85)(396)} + \frac{(0.381)(49{,}000)(12)(6.07)}{(0.85)(17{,}160)}$$

$$= 134 + 93$$

$$= 227 \text{ psi} < 4\sqrt{f_c'} = 254 \text{ allowed.}$$

$$v_{CD} = \frac{V_{cs}}{\phi A_c} - \frac{KM_{cs}c_{cd}}{\phi J_c}$$

$$= 134 - \frac{(0.381)(49{,}000)(12)(13.18)}{(0.85)(17{,}160)}$$

$$= 134 - 202$$

$$= -68 \text{ psi} < 4\sqrt{f_c'}$$

Step 12. Slab shears based on interior trial column.

Where the same size columns are used, the shearing stress is not usually critical for the interior columns. The maximum shearing stress at the critical section of the first interior column with total load on all spans and shears and moments as shown in Fig. 5-22 is as follows:

$$V_{cs} = 48.0 + 54.2 = 102.2 \text{ kips}$$

$$M_{cs} = 192 - 171 = 21 \text{ ft-kips}$$

$$v_u = \frac{V_{cs}}{\phi A_c} + \frac{KM_{cs}c}{\phi J_c}$$

$$v_u = \frac{102{,}200}{0.85 \times 585} + \frac{(0.40)(21{,}000)(12)(11.25)}{(0.85)(50{,}389)}$$

$$= 206 + 26 = 226 \text{ psi} < 4\sqrt{f_c'} = 254 \text{ psi}$$

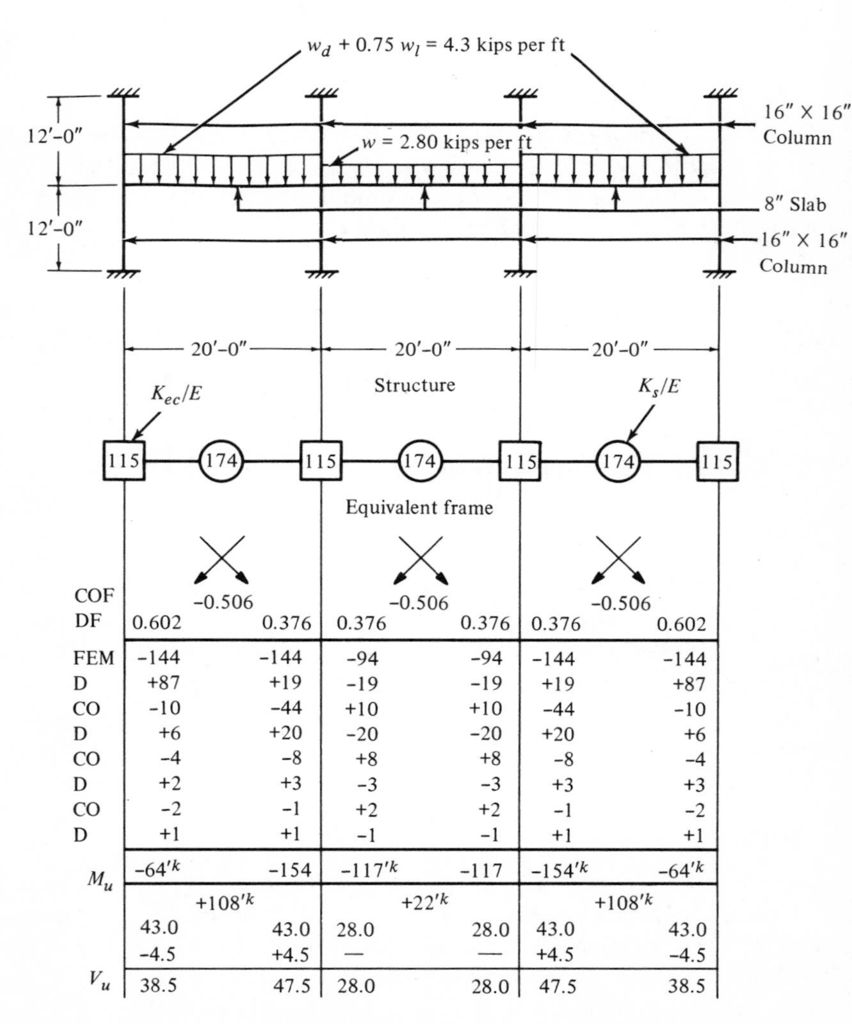

COF	-0.506		-0.506		-0.506	
DF	0.602	0.376	0.376	0.376	0.376	0.602
FEM	-144	-144	-94	-94	-144	-144
D	+87	+19	-19	-19	+19	+87
CO	-10	-44	+10	+10	-44	-10
D	+6	+20	-20	-20	+20	+6
CO	-4	-8	+8	+8	-8	-4
D	+2	+3	-3	-3	+3	+3
CO	-2	-1	+2	+2	-1	-2
D	+1	+1	-1	-1	+1	+1
M_u	-64′k	-154	-117′k	-117	-154′k	-64′k
	+108′k		+22′k		+108′k	
	43.0	43.0	28.0	28.0	43.0	43.0
	-4.5	+4.5	—	—	+4.5	-4.5
V_u	38.5	47.5	28.0	28.0	47.5	38.5

Fig. 5–23 Moment distribution—75% W_l on alternate spans.

The maximum shearing stress based on a patterned loading with three-fourths of w_l on alternate spans as shown in Fig. 5-23 is calculated as follows:

$$V_{cs} = 47.5 + 28.0 = 75.5^k$$

$$M_{cs} = 154 - 117 = 37 \text{ ft-kips}$$

$$v_u = \frac{V_{cs}}{\phi A_c} + \frac{K M_{cs} c}{\phi J_c}$$

$$= \frac{75{,}500}{(0.85)(585)} + \frac{(40)(37{,}000)(12)(11.25)}{(0.85)(50{,}389)}$$

$$= 152 + 47 = 199 \text{ psi} < 4\sqrt{f_c'} = 254 \text{ psi}$$

Step 13. Beam shear. The maximum beam shear for total design load is as follows:

$$v_u = \frac{\left[51{,}000 - (4{,}800)\left(\frac{6.5}{12}\right)\right]}{(0.85)(240)(6.5)}$$

$$= \frac{48{,}400}{(0.85)(240)(6.5)}$$

$$= 37 \text{ psi} < 2\sqrt{f_c'} = 126$$

Both exterior and interior trial size columns have now proved to be adequate in shear at the critical slab sections. Assuming the columns are adequate for axial load and moment, the reinforcing steel for column and middle strips of the flat plate can now be determined by the same method as demonstrated in Steps 7 and 8 of the direct design method.

Total end panel moments, obtained by both the direct design method and the equivalent frame analysis method are shown in Table 5.1 and compared with results obtained using the empirical method of the 1963 ACI Code.

TABLE 5-1 End Panel Moments

Method of Design	Direct Design Method 1971	Equivalent Frame Method 1971	1963 Code Empirical Method
$-M_u$ at face of exterior column	$-54'^k$	$-42'^k$	$-102'^k$
$+M_u$	$+108'^k$	$+112'^k$	$+84'^k$
$-M_u$ at face of 1st interior column	$-148'^k$	$-157'^k$	$-139'^k$

DESIGN MODIFICATIONS FOR LONGER SPANS AND/OR HEAVIER LOADS

Reason for Modification

The example used in the explanations for the "Direct Design Method" and the "Equivalent Frame Method" was a straightforward flat plate. Load, span, and column size were all in the range for which many flat plate designs are employed. As loads or spans become greater, however, modifications of the simple flat plate become necessary in the exterior panels. In order to extend the range of the very economical flat plate concept, the transfer of vertical load and a part of the unbalanced moment by shear, particularly at exterior columns, becomes critical. Nearly as critical is the transfer of the remaining part of the unbalanced moment by flexure into the exterior column.

Types of Modification

Perhaps the simplest design modification to utilize flat plates for heavier loads is the cantilever edge. As the cantilever moment approaches that of an interior column (balanced layout) the problem of moment transfer to exterior column disappears and the cross-sectional area of the shear perimeter is almost evenly stressed by balanced vertical shear (see Fig. 5-32). Where the cantilevered edge layout cannot be employed, as for architectural reasons, and a flush edge is required, the new Code permits at least three design modifications without resort to excessively large exterior columns: (1) structural steel or reinforcing steel shearheads (Sections 11.11, 13.4.1.6). (2) brackets (Sections 13.0, 13.1.6, 13.3.2.2, 13.4.2), and (3) spandrel beams (Sections 13.0, 13.3.4.1, 13.3.4.4, 13.3.4.7, 13.3.4.8, 13.3.4.10, 13.4.1.5).

Effect of Load and Story Height on Column Size

As an indication of the size increase required for exterior columns without these modifications, consider the example used in "Direct Design." (Story height 12′ –0″, 8 in. plate floor on 20 ft square panels with design live load 100 PSF). Minimum column sizes for shear and stiffness were computed for a range of design live loads of 50 to 400 psf for floors with 8′ –0″, 12′ –0″, and 22′ –8″ story heights. (The 22′ –8″ story height column stiffness for a floor slab is about equivalent to a roof slab with 12′ –0″ story height.) See Table 5-2.

TABLE 5-2 Effect of Load and Story Height on Flush Edge Square Column Sizes Required for Shear

w_ℓ Design Live Load psf	ℓ_c = 8′–0″			ℓ_c = 12′–0″			ℓ_c = 22′–8″		
	EXT. COLUMN		Int. Column Size $C_1 = C_2$ in.	EXT. COLUMN		Int. Column Size $C_1 = C_2$ in.	EXT. COLUMN		Int. Column Size $C_1 = C_2$ in.
	size $C_1 = C_2$ in.	$-M_u$ ft kips		size $C_1 = C_2$ in.	$-M_u$ ft kips		size $C_1 = C_2$ in.	$-M_u$ ft kips	
50	10	21	10	10	17	10	10	11	10
100	18	67	12	17	59	12	10	14	12
150	24	99	15	23	94	16	22	83	17
200	28	123	19	28	121	19	27	114	19
250	33	147	22	32	144	22	32	141	22
300	36	166	26	36	165	26	36	162	26
400	44	204	32	44	203	32	44	201	32

Shearheads and Shear Reinforcement

The Code has extensive provisions for design of shear reinforcement in flat plates around interior columns (Section 11.11). The term "shearhead" is defined in the Code (Section 11.11.2.1) as a welded assembly of four steel "shapes" at right angles and continuous through the (interior) column section. Provision for bar or wire shear reinforcement is also made (Section 11.11.1). Critical section sketches for common styles are illustrated (Commentary Fig. 11-7). If bars or wires are used as shear reinforcement, the maximum shear stress permitted on critical section $d/2$ from the face of the column can be increased from $4\sqrt{f_c'}$ to $6\sqrt{f_c'}$ (Section 11.10.3). The shear stress, v_c, carried by the concrete shall not exceed $2\sqrt{f_c'}$ at the critical section or at any successive crtical sections where the shear reinforcement is reduced or ended. If steel I- or channel shapes are used as shearheads, the maximum shear stress can be further increased to $7\sqrt{f_c'}$ at the critical section $d/2$ from the column. Shear stress carried by the concrete at this section, v_c, must not exceed $4\sqrt{f_c'}$ (Section 11.10.3). These increases, of course, are allowed only if the shearhead is designed as prescribed (Section 11.11), and if the shear stress on the minimum critical slab section on the perimeter at three-quarters the effective length of the shearhead does not exceed $4\sqrt{f_c'}$ (Section 11.11.2.3). It is recommended that this value be reduced to $3.3\sqrt{f_c'}$ if the shearhead extends beyond the column face a distance equal to the column dimension and to $3\sqrt{f_c'}$ for long shearheads. Locations of critical section for various shapes of shearheads are shown (Commentary Fig. 11-7).

The design engineer enjoys great freedom to exercise ingenuity in shearhead design because of the variety of possible solutions, and the many proprietary devices with load test data, etc. that are available. Specific Code provisions for shearhead design using four identical arms do not apply to exterior columns. Where shear is transferred to a column at an edge or a corner of a slab, special shearhead designs are required (Section 11.11.2).

EXAMPLE. STEEL SHEARHEADS. The sequence of references applicable to the design of an interior column structural steel shearhead will be illustrated by an example of a low-rise building with a superimposed ultimate live load (design live load w_l) of 300 psf to be supported by an 8 in. flat plate floor with square panels, spans of 20′ –0″ between column centerlines and story heights above and below the floor of 12′ –0″. This example is based on the desirability of a maximum interior column size of 12 in. Table 5-2 shows shearheads are required for square columns smaller than 26 in.

Shear at first critical section (d/2 from columns)

The total design shear is

$$V_u = [(1.4)(0.100) + 0.300](20)^2$$

$$= 176 \text{ kips}$$

The shearing stress is first checked at a distance $d/2$ from the column face and compared with that permitted by Code (Section 11.10.3) when steel shearheads are used. The shearing stress at $d/2$ is

$$v_u = \frac{V_u}{\phi b_o d} = \frac{176{,}000}{(0.85)[(4)(12 + 6.5)](6.5)}$$

$$= 430 \text{ psi OK} < 7\sqrt{f_c'} = 7\sqrt{4000} = 442 \text{ psi}$$

Size of shearhead

The required minimum perimeter of the critical section of the shearhead can be determined by limiting the maximum shearing stress to $v_u \leq 3\sqrt{f_c'}$ as recommended for long shearheads. The required b_o is

$$b_o = \frac{V_u}{\phi 3\sqrt{f_c'} d} = \frac{176{,}000}{(0.85)(3)(\sqrt{4000})(6.5)} = 168 \text{ in.}$$

For square columns and a shearhead with each arm made from a structural shape, the perimeter can be defined by the following expression in which W_s equals the width of the compression flange of the structural steel shearhead and c is the square column size. See Fig. 5-24 (Commentary Fig. 11-7) for definitions and location of the critical section,

$$b_o = 4\sqrt{2}\left[\left(\frac{3}{4}\right)\left(l_v - \frac{c}{2}\right) + \frac{c}{2} - \frac{W_s}{2}\right] + 4W_s$$

Solving for l_v and substituting $c = 12''$ and $b_o = 168$ in. and assuming $W_s = 4''$, the required length of each arm of the shearhead is:

$$l_v = \frac{[b_o - 4W_s]}{3\sqrt{2}} - \frac{c}{6} + \frac{2W_s}{3}$$

$$= \frac{[(168) - (4)(4)]}{3\sqrt{2}} - \frac{12}{6} + \frac{(2)(4)}{3}$$

$$= 36.6 \text{ in.}$$

Figure 5-24 shows the location of the critical sections of the structural steel shearhead of our example.

In order to determine the required plastic resisting moment, M_p, of the shearhead (Section 11.11.2.2) it is necessary to assume a trial size structural shape and check if $\alpha_v \geqq 0.15$. α_v is defined as the ratio of the stiffness of one arm of the shearhead to the stiffness of the surrounding composite cracked slab section of width $(c_2 + d)$. To determine the stiffness of the composite cracked slab section, it is necessary to know the amount of reinforcing steel within the slab width $(c_2 + d)$. This amount of reinforcing steel can be determined as follows:

$$M_o = \frac{wl_n^2}{8} = \frac{(0.440)(20)(19)^2}{8} = 397 \text{ ft-kips}$$

$$-M_u \text{ (Int. span col. strip)} = 0.65\, M_o(0.75) = (0.65)(397)(0.75)$$

$$= 193.5 \text{ ft-kips}$$

$$A_s = (\text{for assumed } j = 0.92) = \frac{M_u}{\phi f_y jd} = \frac{(193{,}500)(12)}{(0.90)(60{,}000)(0.92)(6.50)}$$

$$= 7.19 \text{ in.}^2 \quad 23 - \#5 = 7.13 \text{ in.}^2$$

If 50 percent of the column strip top bars are placed within the middle third of the column strip five bars will fall within the width $c + d = 18.5$ in.

The stiffness of the composite cracked slab section with a trial size M4 × 13 structural steel shearhead is shown in Fig. 5-25. The moment of inertia and stiffness of this section can be easily obtained by transforming the concrete into equivalent steel. $2.31\, kd^2/2 - (3.82)(4 - kd) - (1.55)(6.5 - kd) = 0$ and kd = 2.82 in. Knowing kd, the composite section moment of inertia can be calculated as follows:

$$\frac{(2.31)(2.82)^3}{3} = 17.20$$

$$(1.55)(3.68)^2 = 21.00$$

$$\text{M4} \times 13 = 10.40$$

$$(3.82)(1.18)^2 = \underline{5.35}$$

$$I = 53.95 \text{ in.}^4$$

Because the concrete of the composite cracked slab has been transformed

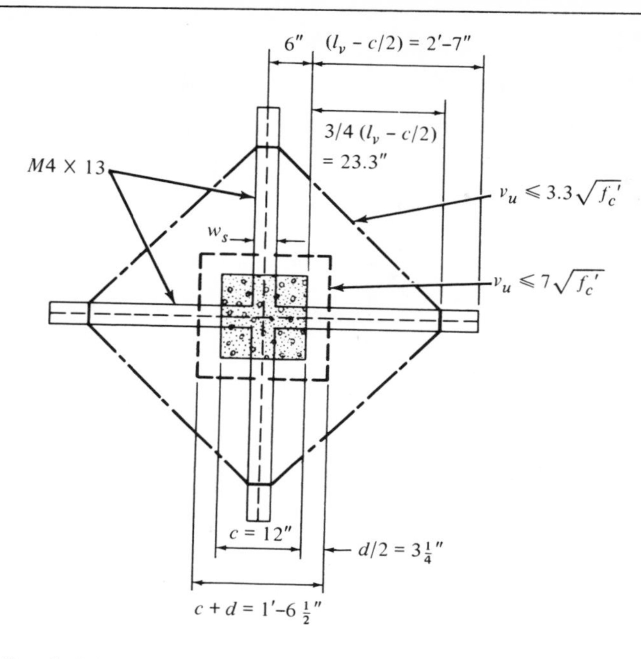

Fig. 5–24 Location of critical sections of steel shearhead.

into equivalent steel, the ratio α_v of the stiffness of one steel shearhead arm to the stiffness of the composite cracked slab section is

$$\alpha_v = \frac{I(\text{M4} \times 13)}{I(\text{composite cracked section})} = \frac{10.5}{53.9}$$

$$= 0.194 \text{ OK} \geq 0.15 \text{ Code (Section 11.11.2.1)}$$

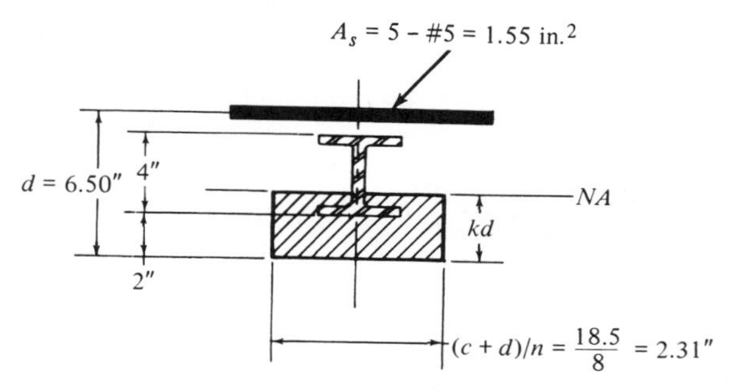

Fig. 5–25 Composite cracked slab.

Moment Resistance for Selecting Shape

The plastic moment of resistance required for each arm can now be calculated (Code Eq. 11-26):

$$M_p = \frac{V_u}{\phi 8}\left[h_v + \alpha_v\left(l_v - \frac{c_1}{2}\right)\right] = \frac{176}{(0.85)(8)}\left[4 + 0.194\left(37 - \frac{12}{2}\right)\right]$$

$$= 235 \text{ in.-kips}$$

From the plastic moment the required plastic section modulus can be determined for structural steel with a yield strength of 36 or 50 ksi.

$$Z = M_p/f_y$$

$$Z(f_y = 36 \text{ ksi}) = 235/36 = 6.52 \text{ in.}^3$$

$$Z(f_y = 50 \text{ ksi}) = 235/50 = 4.70 \text{ in.}^3$$

An M4 $\times$ 13 has a plastic section modulus of 6.1 in.^{3} and if made from a steel with a yield strength of 50 ksi is satisfactory.

Adjustment of slab reinforcement

If sufficient flexural capacity is provided in the shearhead to assure that the shear capacity of the slab is reached before the flexural capacity of the shearhead is exceeded and if the shear capacity of the slab at the critical section of the shearhead is sufficient then the shearhead can be used to reduce the column strip negative reinforcing steel (Section 11.11.2.5).

The shearhead is allowed to contribute a negative resisting moment, M_v, to each column strip:

$$M_v = \frac{\phi \alpha_v V_u}{8}\left(l_v - \frac{c_1}{2}\right) \text{ (Section 11.11.2.5, Eq. 11-27)}$$

$$= \frac{(0.90)(0.194)(176)(37 - 12/2)}{8}$$

$$= 119 \text{ in.-kips}$$

$$= 9.9 \text{ ft-kips}$$

M_v however, cannot exceed 30 percent of the column strip design moment, the change in moment over the length, l_v, of the shearhead, or M_p (Section 11.11.2.5).

30% of column strip moment = (0.30)(258) = 77.4 ft-kips

Change in moment in length, l_v, is $\left[-\frac{V_u}{2}(l_v) + \frac{wl_2l_v^2}{2} \right](0.75)$

$$\Delta M_u = \left[-\frac{176}{2}(3.08) + \frac{(0.440)(20)(3.08)^2}{2} \right](0.75)$$

$$\Delta M_u = - \ (-269 + 42)(0.75)$$

$$\Delta M_u = - - 171 \text{ ft-kips}$$

Plastic moment, $M_p = \frac{235}{12} = 19.6$ ft-kips

The full reduction of 9.9 ft-kips can be used.

BRACKETS OR CAPS. Small brackets or caps at the exterior columns may be used to provide the required shear perimeter, equal to minimum column size, $c_1 = c_2$ in Table 5-2. The code provides limiting ratios of cap geometry.

If the cap is dimensioned as in Fig. 5-26 to maintain the same minimum shear section perimeter $2(c_1 + d/2) + (c_2 + d)$, the thickness of cap, h_c, must be at least equal to the projection of cap, P_c. The Code permits a maximum angle of 45° within which the section is considered effective for shear (Section 13.1.6). Note that the critical section for negative moment in the direction c_1 is now located 1/2 P_c from the face of the reduced column (Section 13.4.2).

For exterior columns only, the critical sections for negative moment do not coincide with ends of the clear span when brackets or capitals are used (Section 13.3.2.2, clear span, and 13.4.2, critical section). The effective clear span, l_n, has been increased by 1/2 P_c which will increase moments. Offsetting this increase, however, is the reduced stiffness of the column itself, K_c, so that the original shear perimeter dimensions for the plate will probably be adequate without calculation. It cannot be assumed, however, that the original panel moments, other than the negative moment at the exterior column, are conservative. Figure 5-26 shows shear caps proportioned for half-size columns. Negative moment at the edge, most sensitive to column stiffness, will be reduced, but both negative moment at first interior column and positive moment will increase. If the ratio of the equivalent column stiffness for the original column to slab stiffness, α_{ec}, was 3.0 or less, it cannot be assumed that reduction in these

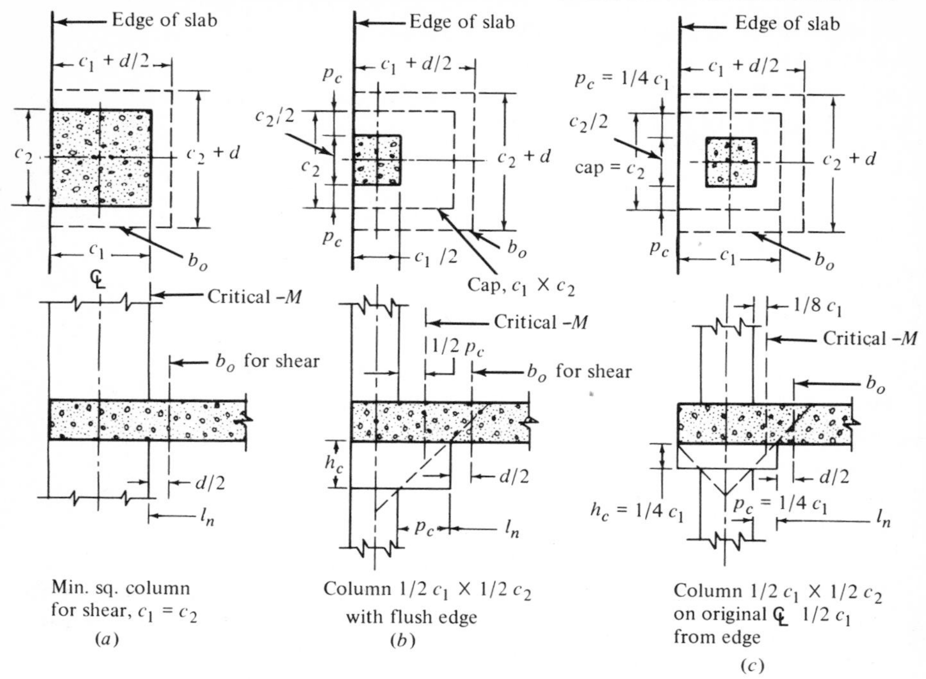

Fig. 5–26 Shear caps for half-size columns.

moments are within the 10 percent adjustment permitted (13.3.4.6) (see Fig. 5-8).

If a cap is employed with a reduced column size because the column required without a shear cap would be too large, it will be necessary to repeat Direct Design Steps 2 through 5 and to adjust the bar selections for the exterior panel accordingly. Computation of K_c will become more complicated by the cap shown and still more complicated if a sloping bracket of variable depth is used.

The Code is ambiguous in prescribing this operation. The moment of inertia "outside of the capital" is based on column cross section (Section 13.4.1.3). The moment of inertia is assumed infinite within the joint depth of the slab-beam (Section 13.4.1.5). It is recommended that the contribution to stiffness, K_c, provided by a capital be taken into account, which is undoubtedly the intent of the Code. A crude approximation which will permit the user to take account of the cap contribution to column stiffness by a simple calculation is to add one-fourth the cap depth to the plate thickness, h, and enter Fig. 5-4 with the large value of h/l_c, computing a new K_c for the reduced column section. The slab stiffness, K_s, need not be changed from that computed for the original large column. The same correction for width of support is provided from the face of column, bracket, or capital (Section 13.4.1.4). The remaining design Steps 3 through 5 can be followed using Figs. 5-6 through 5-8 with the revised value of K_c.

Spandrel Beams

The use of spandrel beams for the reduction of shear stress in heavily loaded flat plates will permit reduction of the exterior column size. The stiffening effect at the edges is also beneficnal in controlling edge deflections and providing moment transfer by torsion from the plate to the edge column. Spandrel beams can also be used with columns as frames to develop resistance to lateral forces and thus may eliminate need of shear walls.

With the many possible permutations of depth, width, L- or T-shapes, location within the depth of the column, proportion of beam width to column width, etc., only a few general suggestions applicable to all can be given here. The 1963 Code required that all stirrups in spandrel or edge beams be closed, and provided with a longitudinal bar in each corner at least equal to the stirrup size but not less than #4 (1963 Section 921). The 1971 Code simply states that *torsion* reinforcement *where required* consist of closed stirrups and longitudinal bars (Section 11.1.6). Torsion effects may be neglected if the nominal torsion stress, V_{tu} is $1.5\sqrt{f_c'}$ or less (Section 11.7.1). Thus, the new code will eliminate many unnecessary closed stirrups but does require additional calculations to determine the torsional stress.

$$v_{tu} = \frac{3T_u}{\phi \Sigma x^2 y} \quad \text{(Section 11.7.2)}$$

Also, the new code specifically permits the use of closed stirrups formed in two pieces, a practice recommended for simplified assembly in the field (Section 12.13.4 and Commentary 11.8.2).

The torsional resistance is calculated as the sum of the x^2y quantities for the rectangular areas contained in the cross section of the spandrel beam above and below the plate and adjoining portions of the plate, x is the smaller of the dimensions (Section 13.0). The portions of plate considered as participating in the resistance to torsion are equal to the largest projection of the spandrel beam, above or below the plate, but not more than $4h_f$ (Section 13.1.5) nor $3h_f$ (Commentary and Section 11.7.2). Conservatively, assume that the *sum* of the plate portions if spandrel is T-shaped, is limited to $4h_f$. If L-shaped, assume the flange on one side is limited to 3 h_f.

For spandrel beams at the edge of a flat plate, "their share of the exterior design moment" must be resisted by torsion (Section 13.3.4.10). Logically, that portion of the unbalanced moment at the exterior column not transferred directly by flexure to the column must be transferred by torsion through the spandrel beams. To apply the expression for the percentage of moment transferred directly (Section 11.13.2), the terms $(c_1 + d)$ and $(c_2 + d)$, plan dimensions of the shear section, are best determined by a sketch of the proposed spandrel and column (see Fig. 5-27). Consider the spandrel depth on either side of the column within a 45° angle as a part of $(c_2 + d)$. Note that a spandrel flush with the inside face of the column will change the shear behavior of the plate from two-way to one-way (beam) shear (Section 11.10.1) (See Figs. 5-28 and 5-29).

The vertical load for the design of the spandrel includes all loads on the tributary areas between the centerline of the span parallel to the spandrel and lines at 45° angles from the corners of the column (see Fig. 5-28). Spandrels for which α_1 $(l_2/l_1) \geqq 1.0$ carry this full tributary load; where $\alpha_1 = 0$, zero load. Interpolation is required for intermediate values of α_1. Also, the spandrel must carry all directly applied loads (Section 13.3.4.7).

EXAMPLE. For the simplest case, consider a spandrel flush with the outer face of the column and equal in width to the column. Let this spandrel be employed to reduce the size required for the exterior column (Table 5-2) in a flat plate floor 8 in. thick, carrying 400 psf (w_l) load on square panels $l_1 = l_2 = 20$ ft. with story height 12 ft. The required column size for a direct connection is tabulated as 44 in. square. Use the spandrel beam to accommodate a maximum column size of 20 in. square (see Fig. 5-27).

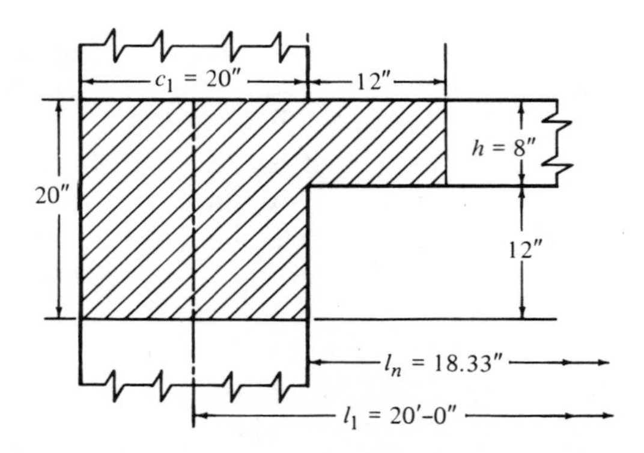

Fig. 5–27 Effective section for torsion—spandrel example.

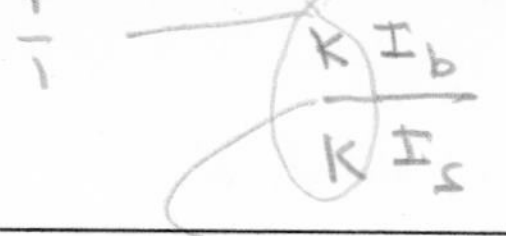

$$\alpha \approx \frac{(20)(20)^3}{(120)(8)^3} \geqq 1.0 \quad \text{(Section 13.0)}$$

and so full tributary area loads are assigned to the spandrel (13.3.4.7). Let the spandrel total load = W (see Fig. 5-28).

$$W = (0.540)(20)(9.17/2) + \left(\frac{20 \times 20}{144}\right)(0.150 \times 1.4) + \left(0.400 \times \frac{20}{12}\right)(18.33)$$

$$= 72.3 \text{ kips in a triangular pattern}$$

$$M_o = \frac{(0.540)(20)(18.33)^2}{8} = 455 \text{ ft-kips}$$

$$C = \left[1 - 0.63(20/20)\right]\frac{(20)(20)^3}{3} + \left[1 - 0.63(8/12)\right]\frac{(12)(8)^3}{3}$$

$$= (0.37)(53{,}330) + (0.58)(2{,}050)$$

$$= 20{,}900$$

$$K_t/E = \frac{(18)(20{,}900)}{(20)(12)\left(1 - \frac{20}{240}\right)^3} = 2{,}070 \quad \text{(Code Eq. 13-6)}$$

See Fig. 5-4:

$$h/l_c = 20/144 = 0.139 \qquad k_c = 1.455$$

$$K_c/E = \frac{(4)(20)(20)^3}{(12)(12 \times 12)}(1.455) = 539 \qquad \Sigma K_c/E = 1{,}078$$

See Fig. 5-5(a)

$$c_1/l_1 = 0.0833 \qquad k_{AB} + 4.13$$

$$K_s/E = (4.13)\frac{(240)(8)^3}{(12)(240)} = 176$$

See Fig. 5-7(c)

$K_{ec}/E = 710$

$\alpha_{ec} = 710/176 = 4$

See Fig. 5-8(*b*)

For $\alpha_{ec} = 4$ Exterior column neg. moment $= -0.52\ M_o$

1st interior column neg. $-M = -0.67\ M_o$

$-M_{\text{ext}} = -236$ ft-kips $\qquad -M_{\text{int}} = -304$ ft-kips

Difference = 68 ft-kips

$$\text{Reduction in shear at exterior column} = \frac{68}{18.33} = 3.7 \text{ kips}$$

The fraction of the exterior moment transferred by flexure directly to the column (Section 11.13.2) can be assumed =

$$\frac{1}{\left(1 + 2/3\sqrt{\dfrac{c_1 + d}{c_2 + c2h}}\right)} = \frac{1}{\left(1 + 2/3\sqrt{\dfrac{26.75}{60}}\right)} = 0.70$$

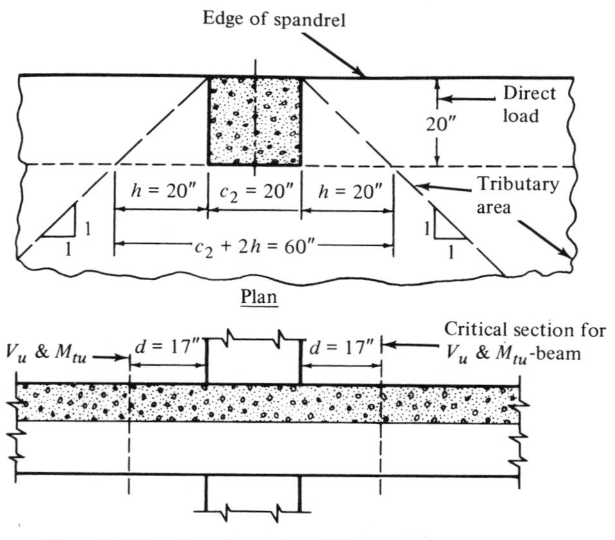

Fig. 5–28 Loads and critical sections—Beam.

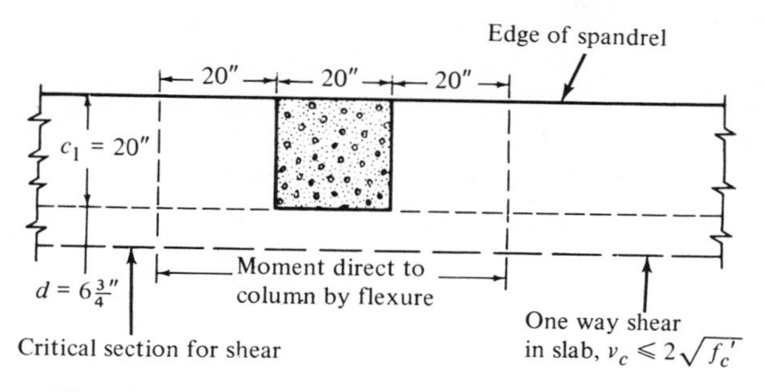

Fig. 5–29 Critical shear in slab with spandrel.

(See Fig. 5-28) moment within the 60 in. plate width centered on the column centerline = 0.70 × 236 = 165 ft-kips.

One-way (beam) shear on a 60 in. width at critical section of plate

$$V_u = \tfrac{1}{2}(0.540)(20)\left(9.17 - \frac{6.75}{12}\right) - 3.7 = 42.7 \text{ kips}$$

Area of shear section, A_c

$$A_c = 60 \times 6.75 = 405 \text{ in.}^2$$

$$v_a = \frac{42{,}700}{(0.85)(405)} = 124 \text{ psi}$$

$$124 < 2\sqrt{f_c'} = 126 \text{ psi allowed}$$

(Sections 11.2.1 and 11.4.1)

Torsional moment per spandrel,

$$T_u = \tfrac{1}{2}\,(0.30)(236) = 35.4 \text{ ft-kips}$$

Torsional stress,

$$v_{tu} = 3\,T_u/\phi\Sigma x^2 y \qquad \text{(Section 11.7.2)}$$

$$v_{tu} = \frac{(3)(35{,}400)(12)}{(0.85)(53{,}330 + 2{,}050)}$$

$$v_{tu} = 27 \text{ psi} < 1.5\,\sqrt{f_c'} = 96 \text{ psi}$$

and so the effect of torsion may be neglected in this spandrel beam (Section 11.7.1).

The spandrel selected for this example illustrates the advantages of selecting a nearly square cross section when code requirements for torsion must be considered. A square shape is, of course, not most efficient for purely vertical loads, but the spandrel is a special case. Where the width is not restricted by architectural considerations, the authors suggest that increased beam width be employed to avoid the design, detailing, inspection, and field placing problems, and ensuing time and expense, involved in the use of torsion reinforcement. In the example used, it will be noted that the vertical shear, $V_u = \frac{1}{2}(72.3) = 36.2$ kips. $v_u = \dfrac{36{,}200}{(0.85)(20)(17)} = 125$ psi, which is more than one-half the allowable shear $2\sqrt{f_c'} = 126$ psi for shear stress greater than one-half the allowable v_c, and minimum stirrups are required (Section 11.1.1 d), except where the beam depth does not exceed $2\frac{1}{2}$ times flange thickness or $\frac{1}{2}$ times the web thickness (Section 11.1.1 c). The example exactly meets both limits on depth and so no stirrups are required. As with previous codes, if stirrups were required even for vertical shear only, it would be good practice to provide closure pieces for the stirrups within a short distance from the column where torsion is largest.

DESIGN NOTES ON CRACK CONTROL IN TWO-WAY SLABS

Top Bars at Intersections of Column Strips and Bottom Bars at Intersections of Middle Strips

	Max. Spacing—Inches—Interior Exposure								
Cover	#3	#4	#5	#6	#7	#8	#9	#10	#11
3/4″	7.3	8.1	8.8	9.4	9.9	10.2	10.6	11.0	11.3
1″	6.4	7.2	7.9	8.4	8.9	9.4	9.7	10.1	10.4

All Other Areas (where principal tension is uni-directional)

Grade 60 Bars; $f_y = 60$ ksi

See Fig. 5-30

Assumed max. $f_s = 0.6\, f_y$ (Section 10.6.2)
$d_c = 1.5''$ for #8 bar with 1″ cover which is conservative for smaller bars or less cover in one-way slabs.

$z = f_s\sqrt[3]{d_c\, A}$ (Section 10.6.2)

$$36\sqrt[3]{(1.5)(3)S_b} = z$$

$$\sqrt[3]{S_b} = \frac{z}{59.44}$$

Limiting z values for crack control (Section 10.6.3)
Beams, interior $z \leq 175$

Beams, exterior $z \leq 145$

z for slabs (Commentary):

Reduced by $\dfrac{1.2}{1.35}$

Slabs, interior, $z \leq 156$
Slabs, exterior, $z \leq 129$

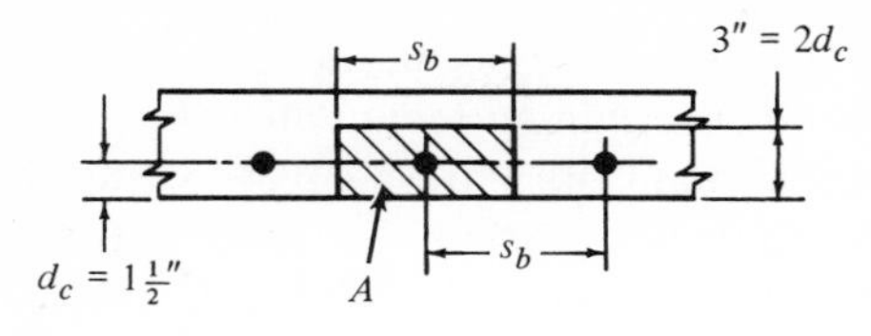

Fig. 5–30 Crack control—flat plates.

Solving for S_b,

Max bar spacings $S_b \leq 18$ in. for interior slabs
$S_b \leq 10$ in. for exterior slabs

Note: Where steel area in excess of that required for moment is furnished, reduce stress, $f_y = 36$, by ratio of A_s required/A_s furnished and calculate larger S_b. The reduction is permitted for one-way stress areas and two-way stress areas.

DESIGN NOTES ON SHEAR IN TWO-WAY SLABS

Shear Due to Vertical Loads and Moment Transfer Combined

Figure 5-31 shows a trace of the computer operations to select a minimum size square exterior column located flush with the edge of the plate. For a

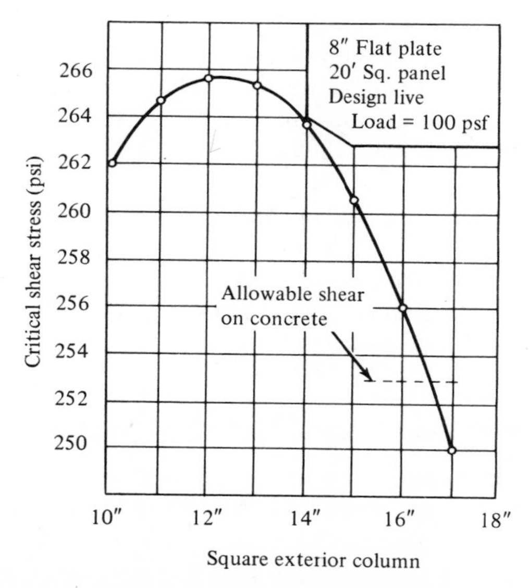

Fig. 5–31 Shear vs. column size.

specified concrete strength = 4,000 psi, allowable shear on the concrete equals 253 psi, which is the critical limit determining minimum size of the column. The shear stresses shown are the sum of the stresses due to gravity loads and transfer of part of the unbalanced slab moment. For each trial size of column, shear is computed for two load conditions: (1) gravity shear maximum and (2) moment transfer shear maximum on interior columns.

This insight into the interior working of a computer operation provides the explanation for the difficulty encountered in a direct manual solution for selection of minimum column size. If the preliminary design selected is too small, successively larger trial sizes do not necessarily reduce shear—the trials do not converge directly to a solution. Note that shear stresses becomes larger for all sizes above 10 in. square to 14.5 in. square columns. The solutions for successively larger sizes do not converge toward the allowable shear until a size above about 12.5 in. is used. This effect results from increasing column stiffness in the ratio $(c + \Delta c)^3/c^3$, with proportionate increase in moment to be transferred by shear to the column while the increase in the section resisting shear is a linear function.

Effect of Column Shape

This insight also provides the engineer a quick guide to reduce further trials required for a solution. The pattern for square columns is shown in Fig. 5-32. If architectural requirements permit increase in the width dimensions, c_2, only, the increase in column stiffness and shear resistance are both linear so that convergent solutions may be expected from the first trial. On the other hand, if architectural requirements proscribe further increases in width, much larger increments in depth, c_1, will be required, since divergence of results for small increases will be more pronounced than for square columns.

See Fig. 5-32, cases (*a*), (*b*), (*c*), and (*g*), for a comparison of shear perimeters for square versus rectangular edge columns and corner columns flush with slab edges. The Code is very specific in defining the critical shear section for these cases (Sections 11.10.1-b, 11.10.2; and 11.12 and corresponding Commentary sections).

Location of Exterior Columns—Cantilevered Edges

For exterior columns at cantilevered edges the Code is somewhat ambiguous in defining the critical shear section perimeter. Openings in the slab wider than the critical section and parallel to it within a distance less than ten times slab thickness (10*h*) (Section 11.12) are to be regarded as a free edge (Commentary 11.12), but the resulting shear perimeter must be

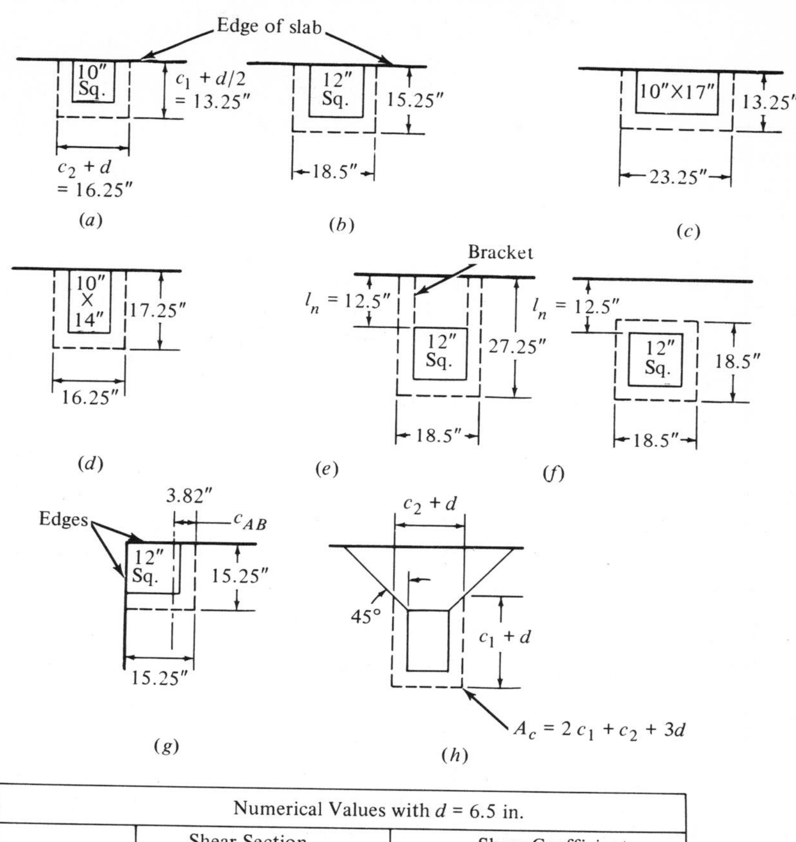

Numerical Values with d = 6.5 in.		
Case	Shear Section Area A_c (in.)^2	Shear Coefficient for Moment Transfer J_c (in.)^4
(a)	43.0 d = 279	6,000
((b)	49.0 d = 319	8,800
(c)	50.0 d = 325	6,650
(d)	51.5 d = 335	11,550
(e)	74.0 d = 481	41,900
(f)	74.0 d = 481	28,300
(g)	30.5 d = 198	5,144

Fig. 5–32 Column shape and location—effect on shear coefficients.

a "minimum" (Section 11.10.2). The reader is also advised that the criterion for using an open 3-sided figure, Fig. 5-32, case (e), or a closed 4-sided figure, case (*f*), is that the perimeter be a minimum in Commentary reference 11.1, the report of ACI Committee 326. This criterion leads to using the closed shape, case (*f*), when the cantilever span $l_n = \frac{1}{2}(c_2 + d)$. (Note that this limit is usually far less than $10h$.) For the numerical example, cases (*e*) and (*f*), note that, although the shear perimeter section areas, A_c, are equal, the moment transfer shear coefficient, J_c, is much larger for the 3-sided figure.

The authors consider both solutions, cases (*e*) and (*f*), unrealistic and likely to overestimate the shear strength since tests have shown that the outer face of a column does not participate equally in resisting shear due to loads on the other side of the column. For a prudent design, it is recommended that the critical section perimeter for computation of both area, A_c, and shear coefficient for moment transfer, J_c, be taken conservatively 3-sided as shown in case (*h*). For small cantilever loads and spans the cantilever dead load moment will, of course, reduce the moment to be transferred to the edge column and resulting shear stress. When the cantilever dead load moment is equal to the dead load moment at an interior column from a span not less than two-thirds the exterior span, it would seem reasonable to use a 4-sided critical shear perimeter for both A_c and J_c, as in case (*f*) The section shown in case (*e*) is, of course, proper and conservative when a bracket as wide as the column is used.

Derivations of Formulas for Shear Constants

The shear coefficient for moment transfer, J_c, is not defined in Code or Commentary. A formula for computing J_c is given (Commentary 11.13, Fig. 11-9) for the case of an interior rectangular column only. J_c is the sum of the moments of inertia about the X-axis and Z-axis for faces of the critical section resisting varying torsional shear and the moments of inertia about the X-axis for the faces resisting uniform vertical shear. See Fig. 5-33 for derivation of formulas for J_c at rectangular interior, edge, and corner columns. (For circular columns, the authors suggest computation of J_c as in Figs. 6-22, 6-23, and 6-24. As a quick approximation, take A_c and J_c for an equivalent square column.)

Case 1: Interior Column

Critical shear section perimeter = $A\text{-}B\text{-}C\text{-}D\text{-}A$

Area of critical shear section = $(AB + BC + CD + DA)$(depth)

$$A_c = (2c_1 + 2c_2 + 4d)(d)$$

By symmetry, c.g. lies at center each way: $C_{AB} = \frac{1}{2}(c_1 + d)$

J = polar moment of inertia of critical shear section

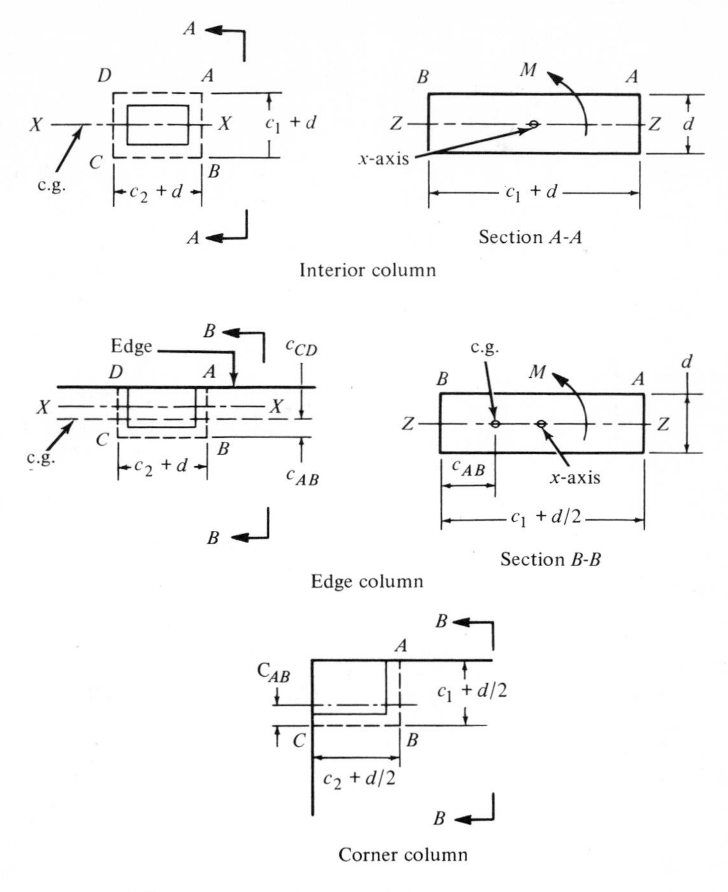

Fig. 5–33 Derivation of shear constants.

For moment about the X-axis,

$J_c = I_{xx} + I_{zz}$ for two faces, $2(c_1 + d) + I_{xx}$ for two faces $2(c_2 + d)$

For faces $(c_1 + d)$:

$$I_{xx} = \frac{2d(c_1 + d)^3}{12} = \frac{d(c_1 + d)^3}{6}$$

$$I_{zz} = \frac{2(c_1 + d)d^3}{12} = \frac{(c_1 + d)d^3}{6}$$

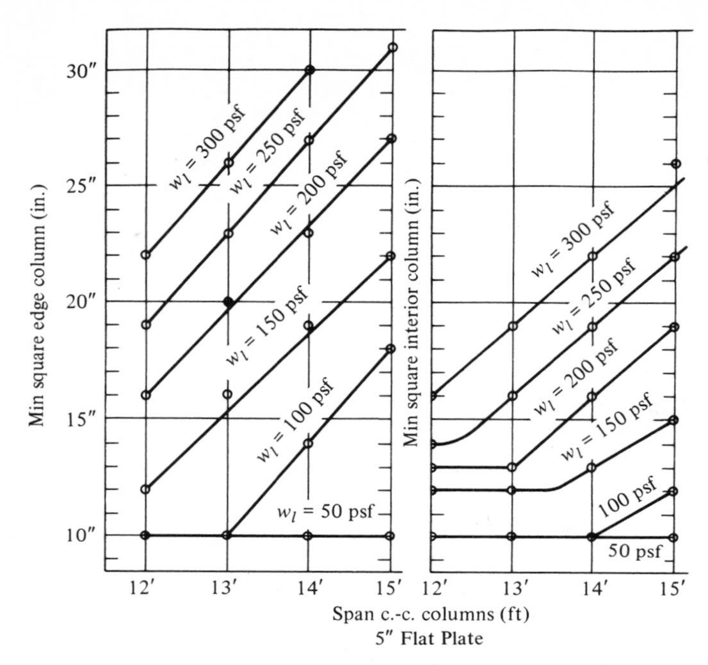

Fig. 5–34 Minimum square columns, 5″ flat plate.

For faces $(c_2 + d)$:

$$I_{xx} = 2(c_2 + d)(d)\left[\frac{c_1 + d}{2}\right]^2$$

$$\boxed{J_c = \frac{d(c_1 + d)^3}{6} + \frac{(c_1 + d)d^3}{6} + \frac{(c_2 + d)(d)(c_1 + d)^2}{2}}$$

(This expression appears as the definition of J_c in the Commentary 11.13.2)

Case 2: **Edge Column**

Critical shear section perimeter (3-sided) $= AB + BC + CD$

$$A_c = (2\,c_1 + c_2 + 2\,d)(d)$$

$$C_{AB} = \frac{(2)(c_1 + \frac{1}{2}\,d)(d)(\frac{1}{2})(c_1 + \frac{1}{2}\,d)}{A_c}$$

$$C_{AB} = \frac{d(c_1 + \frac{1}{2}\,d)^2}{A_c} \qquad C_{CD} = (c_1 + \tfrac{1}{2}\,d) - C_{AB}$$

BC?

$J_c = I_{xx} + I_{zz}$ for two faces, $2(c_1 + \frac{1}{2}\,d) + I_{xx}$ for face $(c_2 + d)$

For (2) faces $(c_1 + \frac{1}{2} d)$:

$$= \frac{(2)(d)(c_1 + \frac{1}{2} d)^3}{12} + (2)(d)(c_1 + \frac{1}{2} d)\left[\frac{(c_1 + \frac{1}{2} d)}{2} - C_{AB}\right]^2$$

$$I_{xx} = \frac{d(c_1 + \frac{1}{2} d)^3}{6} + 2\, d(c_1 + \frac{1}{2} d)\left[\frac{c_1 + \frac{1}{2} d}{2} - C_{AB}\right]^2$$

$$I_{zz} = \frac{2(c_1 + \frac{1}{2} d)d^3}{12} = \frac{(c_1 + \frac{1}{2} d)d^3}{6}$$

For (1) face $(c_2 + d)$:

$$I_{xx} = (c_2 + d)(d)(C_{AB})^2$$

Combining terms,

$$J_c = \frac{(c_1 + \frac{1}{2} d)d^3}{6} + \frac{2\, d}{3}\, [C_{AB})^3 + (C_{CD})^3] + d(c_2 + d)(C_{AB})^2$$

Case 3: **Corner Column**

Critical shear section perimeter (2-sided) $= AB + BC$

$$A_c = (c_1 + c_2 + d)(d)$$

$$C_{AB} = \frac{(c_1 + \frac{1}{2}d)(d)(\frac{1}{2})(c_1 + \frac{1}{2} d)}{A_c} = \frac{d(c_1 + \frac{1}{2} d)^2}{2A_c}$$

$$J_c = I_{xx} \text{ for face } (c_2 + \tfrac{1}{2} d) + I_{xx} + I_{zz} \text{ for face } (c_1 + \tfrac{1}{2} d)$$

For face $(c_2 + \frac{1}{2} d)$:

$$I_{xx} = d(c_2 + \tfrac{1}{2} d)(C_{AB})^2$$

For face $(c_1 + \frac{1}{2} d)$:

$$I_{xx} = \frac{d(c_1 + \frac{1}{2} d)^3}{12} + (c_1 + \tfrac{1}{2} d)(d)\left[\frac{c_1 + \frac{1}{2} d}{2} - C_{AB}\right]^2$$

$$I_{zz} = \frac{(c_1 + \frac{1}{2} d)d^3}{12}$$

$$J_c = \frac{(c_1 + \frac{1}{2} d)d^3}{12} + \frac{d}{3}\, [(C_{AB})^3 + (C_{CD})^3] + d(c_2 + \tfrac{1}{2} d)(C_{AB})^2$$

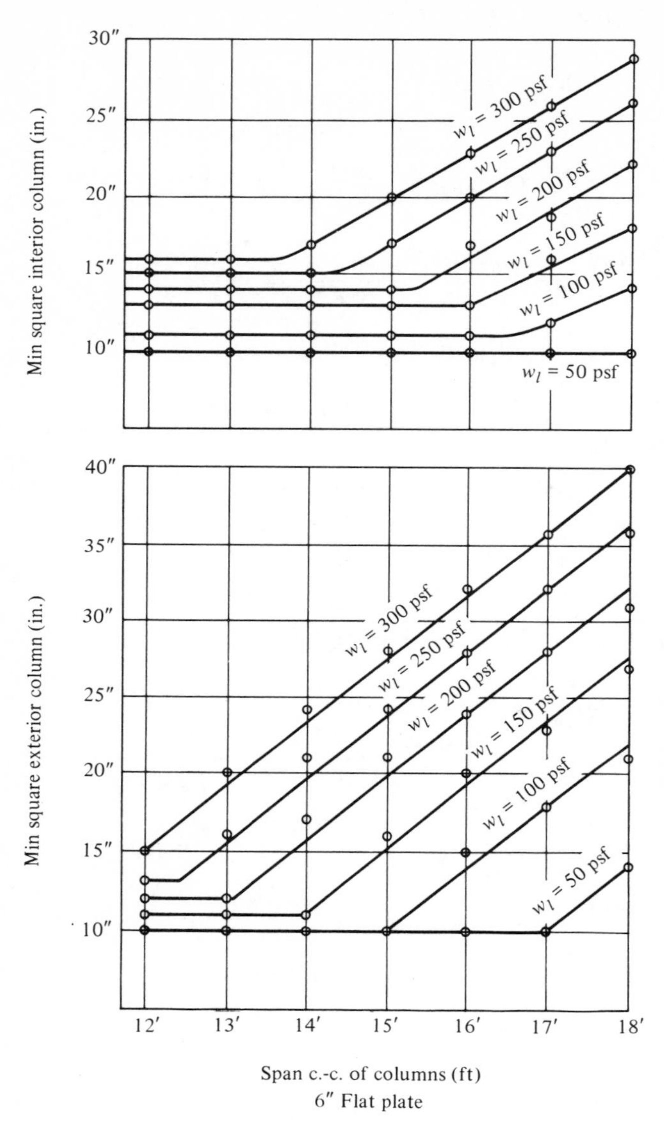

Fig. 5–35 Minimum square columns, 6″ flat plate.

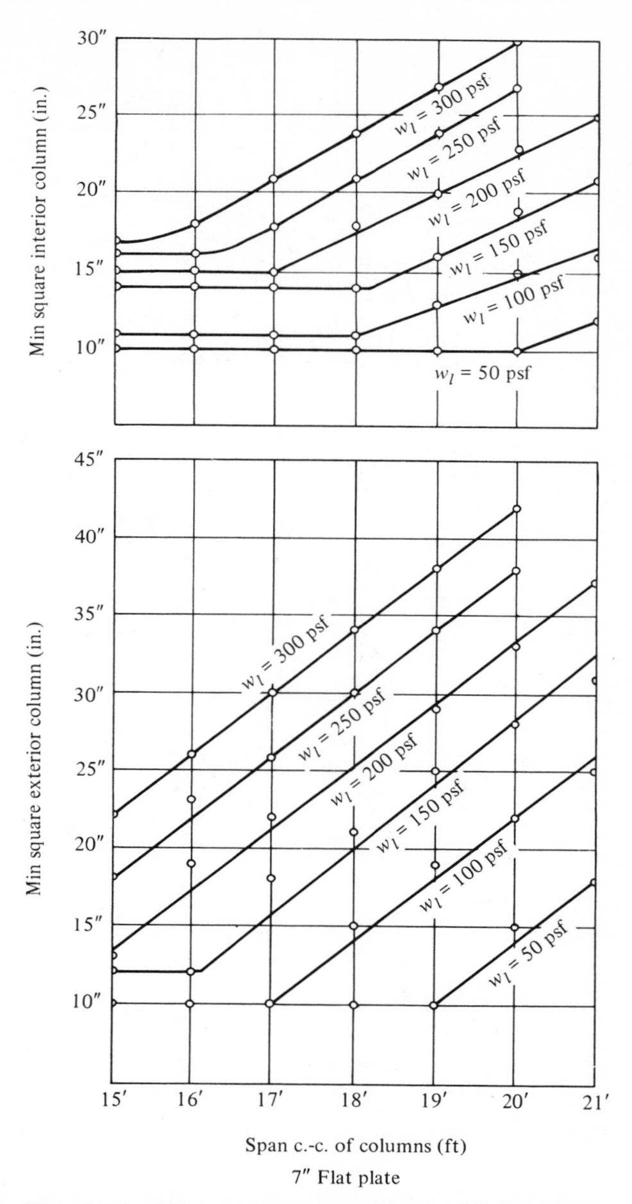

Fig. 5–36 Minimum square columns, 7″ flat plate.

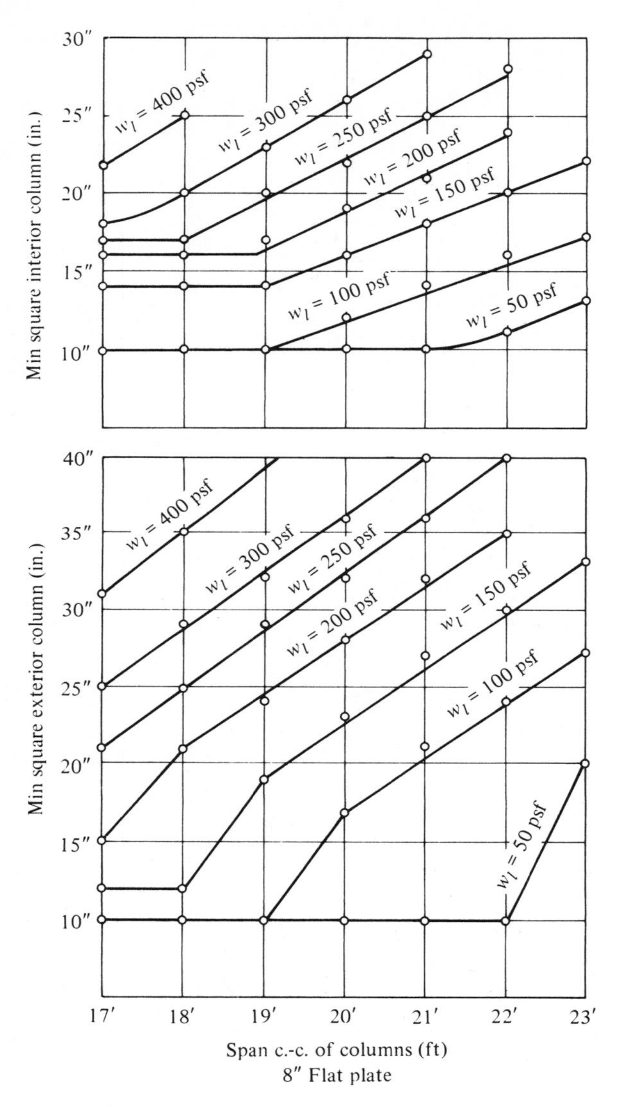

Fig. 5–37 Minimum square columns, 8″ flat plate.

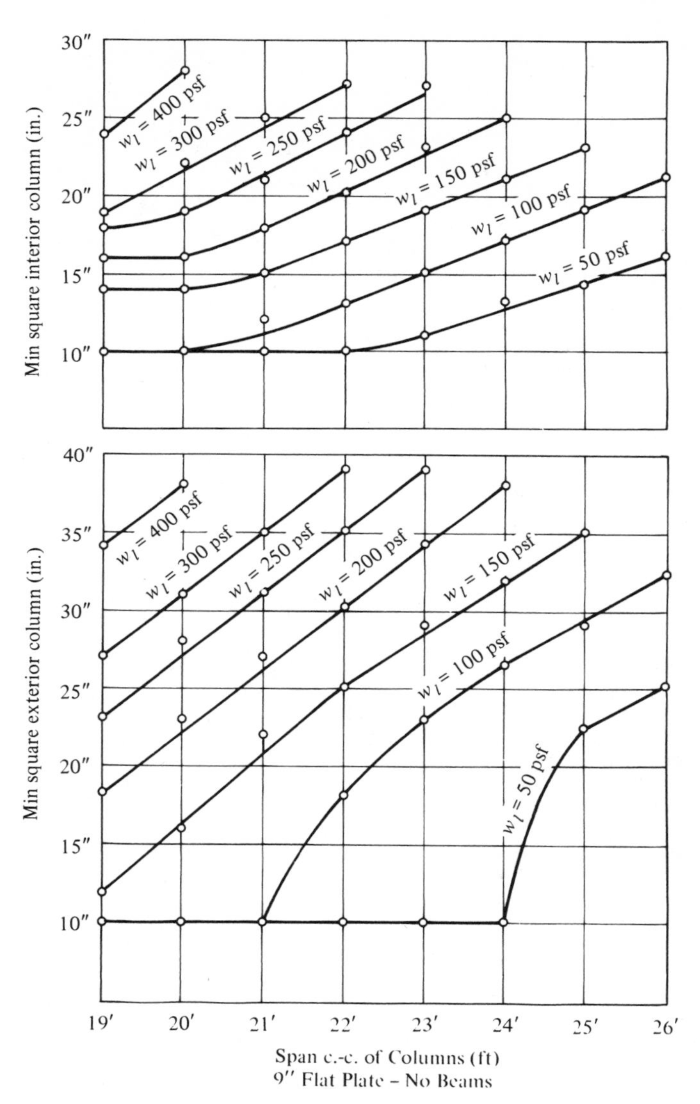

Fig. 5–38 Minimum square columns, 9″ flat plate.

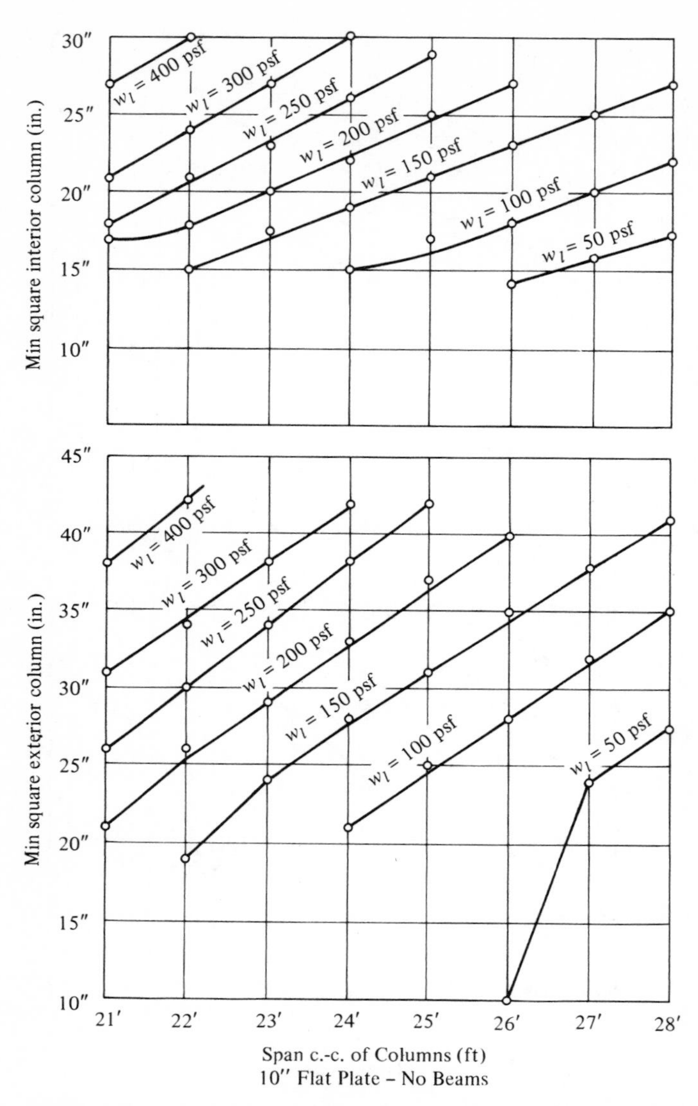

Fig. 5–39 Minimum square columns, 10″ flat plate.

First Trial Column Sizes Versus Load, Span, and Plate Thickness

See Figs. 5-34 through 5-39 for a guide to the selection of first trial square column sizes for columns supporting flat plates without shearheads. See also Fig. 5-32, cases (*a*) through (*h*), for a guide to the effect of varying the column shape and location upon resulting shear coefficients.

6 two-way solid flat slab design

GENERAL

Traditionally, the term "flat slab" has represented a solid uniform depth slab with thicker drop panels supported upon "mushroom head" conical capitals with round columns or various prismatic brackets with square columns. Early codes distinguished between four-way reinforcement and two-way reinforcement systems. The four-way system, intricate by today's standards, was soon abandoned because of the excessive field labor required, particularly with the truss bar systems popular then. Also, effective depth was sacrificed when four layers of the reinforcement intersected at the columns. As forming costs increased, most flat slab design eliminated either the drop panels or the capitals and brackets. Today, the flat plate form is preferred for light loads and short spans; the flat slab with capitals and brackets only, in heavy duty designs where few partitions are used; and the flat slab with drop panels only, for heavy duty design with or without partitions. Edge beams can be used with all three forms of the solid flat slab, but they are usually avoided where not structurally required to avoid the added forming expense. The waffle flat slab (Chapter 7) extends both the span and load limits without excessive dead weight.

LIMITING SIZE OF COLUMNS

Either the drop panel or the capital can be used to reduce the column size required by the limits on allowable two-way shear around the column ("punching shear"). See Table 5-2 for a comparison of column sizes required for shear in flat plate designs. See also Figs. 5-34 through 5-39. If capitals are used, approximately the same total shear periphery as that of the column at the flat plate must be furnished by the capital within a 45° angle to the vertical axis of the column. As the column size below the capital is reduced, the total stiffness of the equivalent column will be reduced and less moment will be required to be transferred by shear at the edge of the column. Thus the size of the capitals required will usually be somewhat smaller than the minimum size exterior columns tabulated.

DROP PANELS

The use of a drop panel will increase the shear capacity of the slab system. It will also increase the relative stiffness of the slab which reduces the unbalanced moments transferred to the columns. The Code requires that the drop panel extend at least one-sixth of the span length in each direction and project below the slab at least one-quarter of the slab thickness in order for the full effective depth of the drop to be used in calculations for negative moment reinforcement (Section 13.5.5). The depth of drop shall not be assumed more than one-fourth of the distance between the edge of the drop and the face of the column or edge of the capital.

In the slab with drop panel examples presented herein, it will be assumed that the designer, faced with a live load and span requiring an excessively large column in a flat plate design, will proceed with a flat slab design as follows: (1) establish a maximum size of column tolerable, say, 0.075 l, (2) assume a drop panel $l/3$ in length, (3) calculate the depth through the drop panel required for vertical shear (Sections 11.10 and 11.13), (4) calculate the minimum slab thickness beyond the drop panel to avoid the deflection calculations otherwise required (Section 9.5.3) and to satisfy the vertical shear at the edge of the drop panel (Section 11.10.1-a).

As a simplification of the forming, if possible, the slab dimension between the drop panels should be an even multiple of 2 ft and preferably of 4 ft. If the span is not an even multiple in feet, it will usually be preferable to add the few odd inches to the drop panel dimension. For edge columns it will usually be necessary to proceed through Steps 3 and 4 twice, since the alternate loading condition requires calculation of the combined vertical load shear and shear developed by transfer of part of the unbalanced moment from the slab to the column.

DIRECT DESIGN METHOD

The best way to explain the application of the new "direct design method" to flat slabs is by numerical examples. It is intended that the user of this guide first proceed through the simpler direct design of flat plates (Chapter 5) before using this section. The following numerical examples are intended to illustrate only the complications added to the direct design method by the drop panels or capitals with the least possible repetition of calculation routines explained in Chapter 5. The chart and short-cut approximate calculations for the slab stiffness of flat plates (Chapter 5) do not apply to flat slabs with drop panels. The tabulations for flat slab stiffness in the

Code Commentary do not apply to drop panels longer than one-third of the span length nor deeper than one and a quarter times the slab depth.

For general use where no available design aids are applicable, as with longer drop panels, diamond shaped drops (oriented at 45° to the column grid), tapered depth drops, waffle slab solid heads, or other variations for architectural effect, a precise moment-area calculation is illustrated in this chapter.* A short cut approximation for trial sections with standard drops is compared to the precise method. All design examples selected to illustrate applications of the 1971 Code can be compared directly to tabulated designs.†

EXAMPLE 1. *Direct design method—flat slab with drop panel—square interior panel*

$l_1 = l_2 = 25'-0''$; $w_l = 400$ psf; maximum column acceptable: square $c_1 = c_2 = 0.075\ l_1$; use $f_c' = 4{,}000$ psi; Grade 60 bars; concrete weight 150 pcf. (See similar design in *CRSI Handbook* under the 1963 Code; minimum column or cap; interior 4.5 ft diam; exterior, 3.5 ft with slab thickness = $9\frac{1}{4}$ in.; drop thickness = $2\frac{1}{2}$ in. Story height, l_c, is 12–'0"; columns above and below slab.

In Fig. 6-1 columns are taken as interior columns past the first line. There are no beams and so $I_b = 0$; $\alpha = 0$ (Section 13.0). For an interior panel, $\beta_s = 1.0$ (Section 9.0); for a square panel, $\beta = 1.0$. For selection of thickness, if the three equations of Section 9.5.3.1, reduced 10 percent as provided in Section 9.5.3.2, are satisfied, no deflection calculations are required by the code.

* *Handook of Frame Constants,* Portland Cement Association, will be a useful reference for stiffness calculations of this sort.

† *CRSI Handbook Ultimate Strength Design 1970* and *CRSI Handbook 1972.*

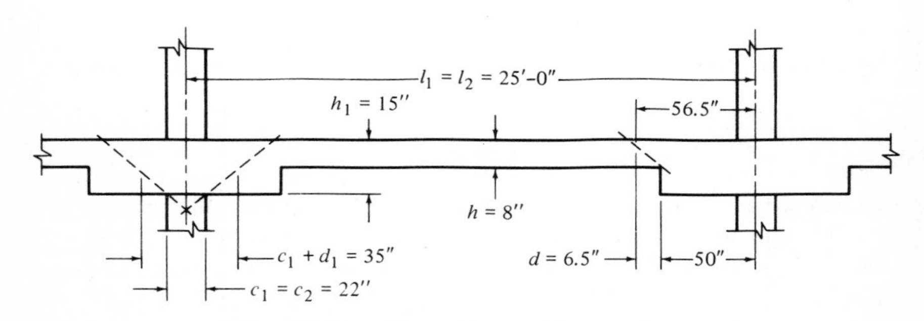

Fig. 6–1 Dimensions—trial section.

Step 1. Thickness to avoid deflection calculations

$$\text{Eq. (9-6): Min } h = \frac{(0.90)(l_n)(800 + 0.005\, f_y)}{36{,}000 + 5000\,(\beta)[\alpha_m - 0.5(1 - \beta_s)(1 + 1/\beta)]}$$

$$h = \frac{(0.90)(1100)l_n}{36{,}000} = l/36.4 \quad \text{(See Table 6-2)}$$

$$\text{Eq. (9-7):} \qquad h \geqq \frac{(0.90)(1100)l_n}{36{,}000 + 5{,}000\,(\beta)(1 + \beta_s)}$$

$$\frac{(0.9)(1100)l}{46{,}000} = l/46.5$$

$$\text{Eq. (9-8): } h \text{ need not be greater than } \frac{(0.90)(1100)l}{36{,}000} = l/36.4$$

Eq. (9-6) controls; clear span = 300 − 22 = 278 in.; the minimum thickness for an interior slab is $h = 278/36.4 = 7.64$ in. For an exterior panel without beams, add 10 percent (Section 9.5.3.3). Since it is not usually justified to vary the thickness by panels or in increments less than $\frac{1}{2}$ in., try $h = 8$ in. Use $d = 6.0$ for shear.

Step 2. Shear for vertical load only

$w_l = 400$ psf ($L = 400/1.7 = 235$);
slab weight = (8/12)(0.150) = 0.100

$$w_d = 0.100 \times 1.4 = 0.140 \text{ ksf};\ w = 0.400 + 0.140 = 0.540 \text{ ksf}$$

Try drop panel, $h_1 = 15$ in. For 1 in. cover and #8 bars, $d_1 = 13$ in.

$l/3 = (1/3)\ (25'\text{–}0'') = 8.33$ ft; $c_1 = c_2 = 22$ in.; $(c + d_1) = 35$ in.

$$\begin{aligned}\text{Total panel shear, } V &= (0.540)[(25)^2 - (35/12)^2] \\ &\quad + (8.33)^2\,(7/12)(0.150)(1.4) \\ &= 333 + 9 \text{ kips} = 342 \text{ kips}\end{aligned}$$

$$\text{Shear at column, } v = \frac{(342 \times 1{,}000)}{(4)(35)(13)(0.85)} = 221 \text{ psi} < 4\sqrt{f_c'} = 253 \text{ psi}$$

Shear at the edge of the drop: $V \doteq 342\ k$ − weight of drop − loads on drop − loads on strip of width d (slab) around drop. Width = 8′–4″ + (2 × 6)″ = 9.33 ft.

$$V = 342 - 9 - (0.540)(9.33)^2 = 287 \text{ kips}$$

$$v = \frac{(287 \times 1{,}000)}{(4)(112)(6)(0.85)} = 125.5 \text{ psi} < 2\sqrt{f_c'} = 126 \text{ psi}$$

(Section 11.4).

Note that the 15 in. total drop panel thickness is about right with some reserve capacity for combined shear. Slab thickness is satisfactory provided that the assumed cover and one and a half bar diameters do not exceed 2 in.* Use: Slab h = 8 in.; drop panel, h_1 = 15 in., 8′–4″ square; columns 22 in.

Step 3. Slab stiffness, K_S

See Fig. 6-2 for dimensions, section properties, etc. for use in calculation of the slab stiffness. At sections between the drop panels the moment of inertia of the slab is

$$I_s = (300)(8)^3/12 = 12{,}800 \text{ in.}^4$$

For the section through the drop, see Fig. 6-3,

$$y_{c.g.} = \frac{(8 \times 300)(11) + (7 \times 100)(3.5)}{(8 \times 300) + (7 \times 100)} = 9.3 \text{ in.}$$

$$I_s = (200)(8)^3/12 + (1600)(1.7)^2 + (100)(15)^3/12 + (1500)(1.8)^2$$

$$= 8540 + 4620 + 28{,}100 + 4860$$

$$= 46{,}100 \text{ in.}^4$$

For the section through the column, between the face and the centerline

* The slab thickness need not be increased even if the one-way shear at the edge of the drop panel were to exceed the allowable shear on concrete (v_c = 126 psi). It will usually be more economical to increase the minimum drop panel width dimension, $l/3$, slightly. In this case, using the minimum dimension, $l/3$ = 8′–4″, the distance between the drop panels is (25′–0″) − (8′–4″) = 16′–8″. As previously noted, it would be desirable to use a drop panel 9′–0″ square so that the distance between drop panels for slab forms becomes a multiple of 4′–0″, the standard width of plywood sheets, especially if many repetitions of this panel will occur.

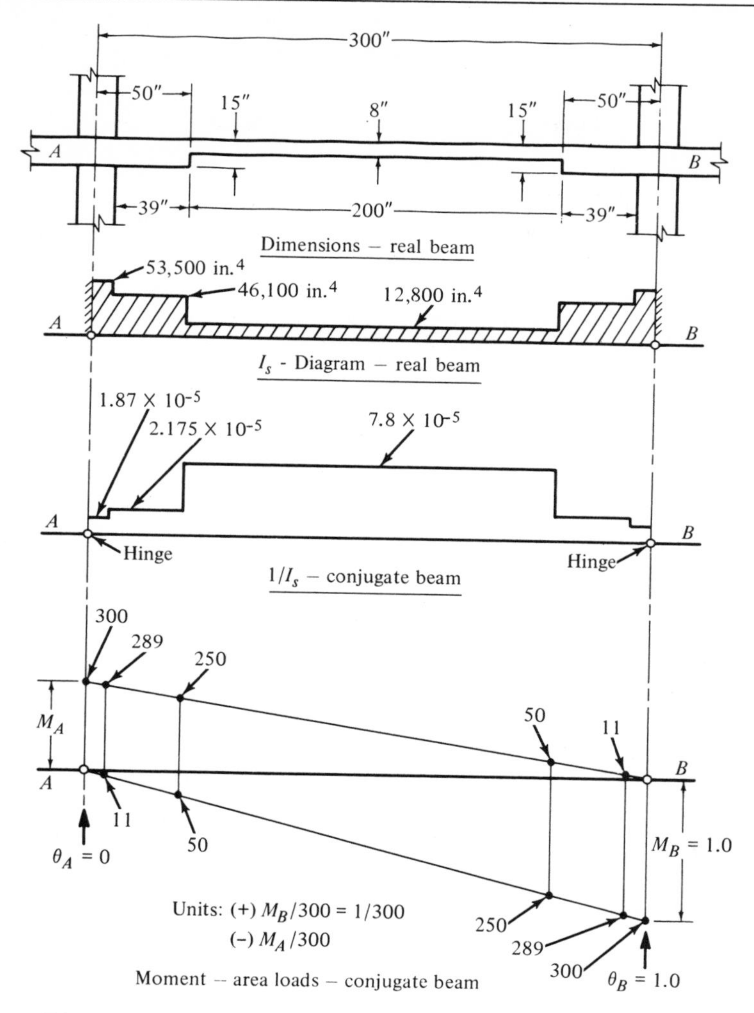

Fig. 6–2 Slab stiffness, K_s/E, by the moment-area method.

$$I_s = \frac{46{,}100}{(1 - 22/300)^2} = 53{,}500 \text{ in.}^4 \quad \text{(Section 13.4.1.4)}$$

The moment-area method gives an "exact" solution for K_S, slab stiffness (see Fig. 6-2). The slab stiffness, $K_s/E = M_B/E$, the moment applied at end B required to cause a unit rotation at B. Since both M_A and M_B are unknowns, two equations are required. First, to solve for M_A in terms of M_B, the sum of the moments of the M/EI areas about $B = 0$, since rotation

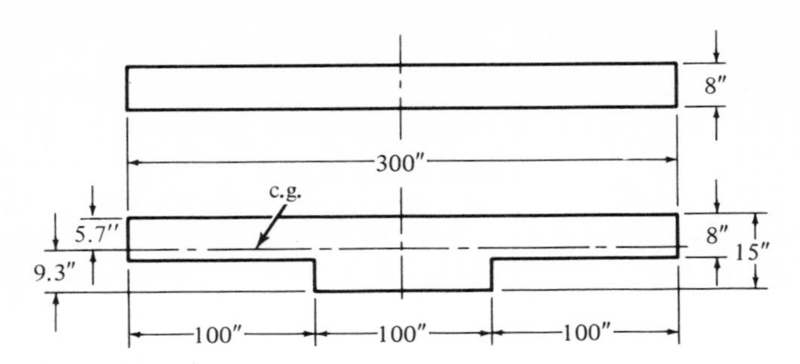

Fig. 6–3 Slab sections—dimensions for Example 1.

θ_A (at the fixed end A) = 0. Second, the sum of the M/EI areas = 1 for unit rotation at end B, θ_B = 1. (The solution for M_A in terms of M_B is the "carry-over factor" from end B to A necessary in design by the equivalent frame analysis method, but not in direct design.)

$$0 = \frac{-M_A}{300E}\left[\frac{(11/2)(11)(297)}{5.35} + \frac{(11)(289)(294.5)}{5.35} + \frac{(39/2)(39)(276)}{4.6} +\right.$$

$$+ \frac{(250)(39)(269.5)}{4.6} + \frac{(200)(200/2)(184)}{1.28} + \frac{(50)(20)(150)}{1.28}$$

$$\left. + \frac{(39/2)(39)(37)}{4.6} + \frac{(11)(39)(30.5)}{4.6} + \frac{(11/2)(11)(8)}{5.35}\right] 10^{-4}$$

$$+ M_B/300E\ [(11.3)(202) + (165)(263) + (93)(269.5) + (7800)(150)$$

$$+ (15{,}600)(116.7) + (2120)(30.5) + (165)(24) + (594)(5.5)$$

$$+ (11.3)(3)]10^{-4}$$

$$0 = -M_A(4{,}745) + M_B(3{,}134) \qquad M_A = 0.66\ M_B$$

Using the (M/EI) area terms from the expressions above and noting that symmetry permits subtraction of M_A terms as a group,

$$1 = \frac{(1.00 - 0.66)}{300E} M_B[11.3 + 594 + 165 + 2120 + 15{,}600$$

$$+ 7800 + 165 + 93 + 11.3]10^{-4}$$

$$M_B/E = K_s/E = \frac{300}{(0.34)(2.65)} = 333 \text{ in.}^3$$

(An "exact" method solution preferred by many engineers is the column analogy; for an example of this method see Fig. 6-16.)

Note: As a fast approximation satisfactory for most purposes, a simple average I_s may be calculated for the slab with drop panel for the fractions of the span length at each depth as follows:

$$\begin{aligned} \text{drop panel } \tfrac{1}{2}(53{,}500 + 46{,}100) \times (100/300) &= 16{,}600 \text{ in.}^4 \\ \text{slab } (12{,}800) \times (200/300) &= 8{,}500 \text{ in.}^4 \\ \hline \text{Average } I &= 25{,}100 \text{ in.}^4 \end{aligned}$$

$$\text{Approximate } K_s/E = \frac{(4)I\text{av.}}{l} = \frac{(4)(25{,}100)}{300} = 334 \text{ in.}^3$$

Step 4. Column stiffness, K_c

Column stiffness, K_c/E, is determined in exactly the same manner as with the flat plate, using the drop panel thickness, h_1, instead of the plate thickness, h. See chart Fig. 5-4 for $h/l_c = 15/144 = 0.104$. Read $k_c = 1.31$. $I_c = (22)(22)^3/12 = 19{,}500$ in.^{4}

$$K_c/E = (k_c)\,\frac{(4)(I_c)}{l_c} = (1.31)(4)(19{,}500)/144 = 710 \text{ in.}^3$$

If the column analogy formula is used for an exact solution,

$$K_c/E = \left(\frac{I_c}{l_c - h}\right)\left[1 + \frac{3h^2}{(l_c - h)^2}\right] = \frac{19{,}500}{129}\left[1 - \frac{(3)(144)^2}{(129)^2}\right] = 719 \text{ in.}^3$$

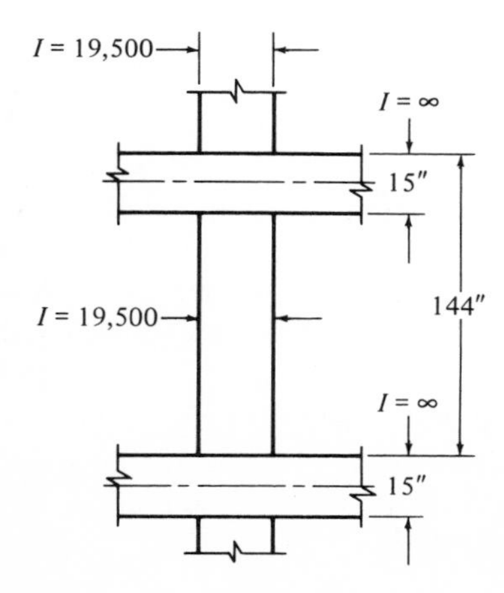

Fig. 6–4 Column dimensions and moments of inertia.

Interior Column $\alpha_c = \Sigma K_c / \Sigma K_s$ (Section 13.0)

$$\alpha_c = (2)(710)/(2)(333) = 2.13$$

Unless $\alpha_c \geqq \alpha_{min}$ calculated positive moments must be increased (Section 13.3.6.1). The required α_{min} for the live load to dead load ratio, $\frac{(400/1.7)}{100} = 2.35$, and $l_2/l_1 = 1.0$ may be read off Fig. 5-3 or calculated as follows (Section 13.3.6.1-a):

$$\alpha_{min} = 1.6 + (2.3 - 1.6)\left[\frac{2.35 - 2.00}{3.00 - 2.00}\right] = 1.845 < 2.35$$

The requirement that $\alpha_c \geqq \alpha_{min}$ is satisfied so that no increase in the positive moment reinforcement is required (Section 13.3.6.1-b).

Note that the Code would permit use of smaller columns than those selected for this example. The effect of the reduction in the interior column stiffness would be to increase the positive moments in all panels.

For the example here, where the same size of columns were used above and below the slab, sufficient stiffness is provided by 22 in. square columns. If there were no columns above the slab and the top slab were designed for the same live load, the 22 in. square columns below it would not satisfy the Code (Section 13.3.6.1). More positive moment reinforcement would be required in the top slab. α_c would become less than $\alpha_{min} \cdot \alpha_c = \frac{2.13}{2} = 1.07 < 1.845$, and the positive moments and bottom steel would have to be increased by the factor, F.

$$F = 1 + \frac{2 - D/L}{4 + D/L}(1 - \alpha_c/\alpha_{min}) \quad \text{Code Eq. (13-4)}$$

$$= 1 + \frac{2 - 100/235}{4 + 100/235}(1 - 1.07/1.845) = 1.15$$

In Fig. 6-5, note the increase required in bottom reinforcement as live load is increased or as column stiffness is reduced. It should be noted that there is some reduction in the negative moment reinforcement when this increase is required in the positive moment reinforcement (Section 13.3.4.6).

Note also for this example, if the top slab were designed only for normal roof loads, say, 40 psf total, the ratio of the dead to live loads becomes $100/40 = 2.5$. The requirement for a minimum α_c does not apply for ratios of dead to live load equal or larger than 2.0.

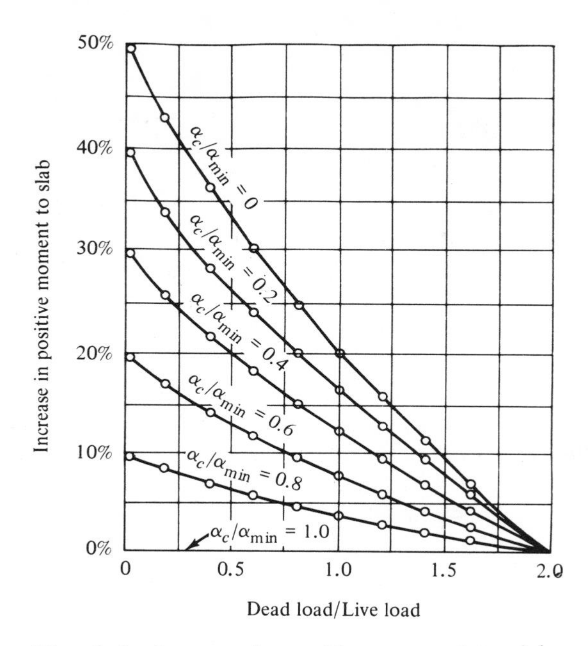

Fig. 6–5 Increase in positive moment to slab.

Step 5. Combined shear due to load and unbalanced moment

Up to this point in the example, all calculations have been based on the tentative trial section dimensions. These dimensions for the slab depth and the column size have been checked and satisfy all requirements except that of two-way shear at a distance $d/2$ from the columns due to combined vertical loads and transfer of unbalanced moments by shear. Note that the drop panel depth selected was ample to allow about 10 percent excess allowable shear capacity for vertical loads only at the section $d/2$ from the column. Since unbalanced moment to interior columns with equal spans on each side is not large, this capacity will probably suffice for the combined condition, at interior columns beyond the first line, and perhaps for the first interior columns also.

In order to investigate the condition of combined shear, it is necessary to determine the unbalanced moment which must be transferred to the column (Section 13.3.5.2, Eq. 13-3) and the proportion of that moment which is assumed to be transferred by shear (Section 11.13.2). To solve Eq. 13-3, the term α_{ec} must be determined, which in turn, requires K_c, K_s, K_t, and K_{ec}. The terms K_c and K_s have already been calculated for the determination of α_{ec}.

DETERMINATION OF TORSIONAL STIFFNESS, K_t, FOR SLABS WITH DROP PANELS. The calculation of the term, K_t, is complicated by the drop panel. (Charts in Chapter 5 do not apply.) A constant cross section *may* be assumed equal to the width of the column and depth of the slab at the face of the column (drop thickness) (Section 13.4.1.5-a). Even for early trials, the authors prefer to consider the actual drop panel-slab dimensions involved since increases in the quantity, K_t, increase the assumed equivalent column stiffness, K_{ec}, the ratio, α_{ec}, and the unbalanced moment to the interior columns. It is felt that the assumption of a uniform depth as in the drop across the full span would be unrealistic for the determination of moments in the slab.

Approximate short-cut solution for K_t

The following example presents a reasonably accurate method of evaluating the torsional stiffness by averaging the depths of the drop panel and the slab. See Fig. 6-6 for a method of weighting the stiffness for both length of span and moment affected.

For an assumed $T_u = 1.00$, area (1) $= 1/2\ (0.667)(100) = 33$; area (2) $= 31$.

Total area under one half the T_u curve $33 + 31 = 64$

$$C = \Sigma[1.00 - 0.63(x/y)](x)^3y/3 \quad \text{(Section 13.4.1.5-c, Eq. 13-7),}$$

where x is the smaller dimension of the torsional cross section, xy,

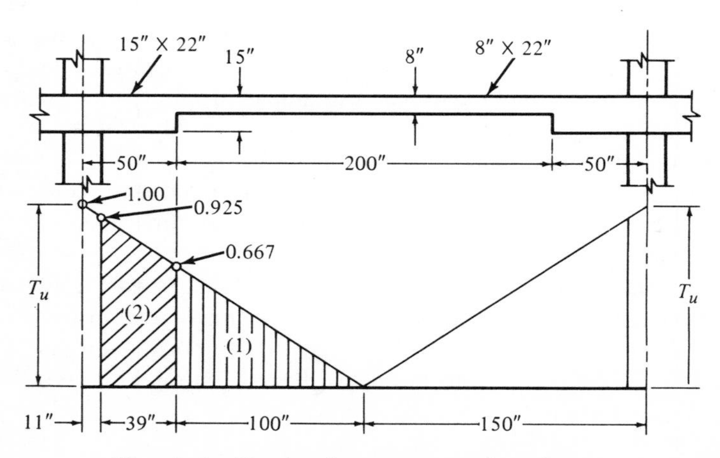

Fig. 6–6 Torsional moments and sections.

$$= [1 - 0.63(15/22)](15)^3(22/3) = 14{,}120 \text{ at the drop panel}$$

$$= [1 - 0.63(8/22)](8)^3(22/3) \quad = \quad 2{,}890 \text{ at the slab}$$

Weighted average $C = (14{,}120)(31/64) + (2{,}890)(33/64) = 8{,}330$ in.^{4}

$$K_t = \frac{\Sigma 9\ EC}{l_2(1 - c_2/l_2)^3} \quad \text{(Section 13.4.1.5 Eq. 13-6)}$$

$$K_t/E = \frac{(2)(9)(8{,}330)}{(300)(1 - 22/300)^3} = 628 \text{ in.}^3$$

(The exact solution by computer for K_t/E results in $K_t/E = 787$ in.^{3}, and so accuracy of the approximate method illustrated here seems satisfactory.)

$$1/K_{ec} = 1/\Sigma K_c + 1/K_t \quad \text{(Eq. 13-5)}$$

$$= 1/(2)(710) + 1/628$$

$$K_{ec} = 435 \text{ in.}^3 \quad \alpha_{ec} = K_{ec}/K_s \quad \text{Section 13.0}$$

$\alpha_{ec} = 435/(2 \times 334) = 0.651$ for interior columns with slabs on both sides and floors with columns above and below. For square panels, the prescribed formula for unbalanced moment to the columns reduces to

$$M_{un} = \frac{(0.08)0.5(w_l)l_2 l_n^2}{(1 + 1/\alpha_{ec})} \quad \text{(Section 13.3.5.2, Eq. 13-3)}$$

$$M_{un} = \frac{(0.08)(0.5)(0.400)(25)(25 - 22/12)^2}{(1 + 1.536)} = 85 \text{ ft-kips}$$

to the interior columns beyond the first line.

A comparison of the results for equivalent column stiffness and the

TABLE 6-1 Effect of Various Code Interpretations for Calculation of Torsional Stiffness for Flat Slabs with Drop Panel

Method	K_t	K_{ec}	α_{ec}	M_{un}
(1) Drop panel, $t_1 = 15$ in.	1065	609	0.912	102′k
(2) "Exact" solution	787	507	0.758	93
(3) Approximate average	628	435	0.651	85
(4) Slab depth, $t = 8$ in.*	218	189	0.283	47

* Neglect drop panel in K_t only.

effect upon unbalanced moments assumed to be transferred to the interior columns (Section 13.3.5.2) using different assumptions for calculating the torsional stiffness is given in Table 6-1. Four interpretations are compared: (1) using the drop panel depth for the full panel width, (2) the exact solution, (3) an approximate average for the effect of the two depths, and (4) using the slab thickness neglecting the drop panel.

Combined shear

The fraction of the unbalanced moment to be transferred by shear is prescribed by the following formula (Section 11.13.2):

$$K_v = \left(1 - \frac{1}{1 + \frac{2}{3}\sqrt{\frac{c_1 + d}{c_2 + d}}}\right) = \left(1 - \frac{1}{1.67}\right) = 0.40$$

The moment transferred by shear, $M_s = (0.4)(85) = 34$ ft-kips. The moment of inertia of the shear section (Fig. 6-7), J_c is

$$J_c = \frac{(2)(35)(13)^3}{12} + \frac{(2)(13)(35)^3}{12} + (2)(35)(13)\left(\frac{35}{2}\right)^2$$

$$= 384{,}000 \text{ in.}^4$$

Unit shear caused by transfer of moment, v_{cs}, is

$$v_{cs} = \frac{Mc}{I_{cs}} = \pm\frac{(34{,}000)(12)(17.5)}{384{,}000} = 19 \text{ psi}$$

Since the design live load must be on alternate spans only, for the un-

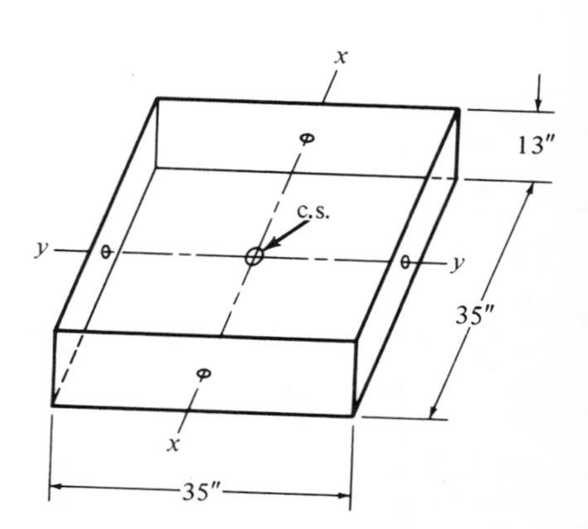

Fig. 6–7 Shear section dimensions.

balanced moment condition to occur, the full shear for all panels loaded as previously calculated will be reduced for combination with the shear due to moment transfer. In this case no further calculation is required since the shear under full load $v_u = 221$ plus the moment shear ($v_{cs} = 19$) = 240 psi which is less than the allowable shear $v_c = 4\sqrt{f_c'} = 253$ psi. *(Note:* see Chapter 5; this calculation will be slightly more critical at the first interior columns and much more critical at all exterior columns.)

The completion of the design for this flat slab example is very similar to that for flat plate designs (see examples in Chapter 5), and will not be repeated here. Bar length detail requirements for the flat slab with drop panels are similar to those for flat plates also, and will not be repeated.

EXAMPLE 2. *Thickness of edge and corner panels with flush edge*

For a concise summary of the three minimum thickness equations (Section 9.5.3) and the reductions in minimum thickness required for slabs with drop panels or increases required for panels with discontinuous edges, see Table 6-2. If these minimum thickness requirements are satisfied, no computations for deflection are required to show that computed deflection limits are not exceeded (Sections 9.5.3.4 and Table 9.5-b). Note that edge and corner panels require edge beams or a 10 percent greater thickness than interior panels (Section 9.5.3.3). For a square edge or corner panel, center-to-center spans, $l_1 = l_2 = 25' -0''$, and square columns $c_1 = c_2 = 22$ in. as in the preceding example for an interior column, and the minimum thickness, $h = \dfrac{25 - (22/12)(12)}{32.73} = 8.4$ in. The design of edge and corner panels for flat slabs with drop panels, flush edges, and no beams is similar to that for flat plates and will not be repeated here.

If the outer face of the exterior columns is to be flush with the outer face of the slab and it is desired to maintain the same slab thickness, $h = 8$ in., as selected for the interior panel example, an edge beam could be used (see Table 6-2). With edge beams such that $\alpha \geq 0.8$, the minimum slab thickness will be $h = \dfrac{1}{37.88} \times$ clear span, and min $h = 7.35$ in. so that the 8 in. thickness may be used.

EXAMPLE 3. *Minimum edge beams for uniform slab thickness in all panels*

The ratio of the flexural stiffness of the beam section to the flexural stiffness of the slab, α, is defined as $\alpha = I_b/I_s$ (Sections 9.5.3.3, 9.0, and 13.0). In this case the moment of inertia of the gross section of a slab one-half panel in width with a thickness equal to the *computed* minimum must be used to establish the required minimum value for the ratio, $\alpha \geq$

TABLE 6-2 Minimum Thickness of Two-Way Slabs*

(expressed as fractions of longer clear span)

Two-Way Construction	Minimum $h =$ Eq. 9–6	Minimum $h =$ Eq. 9–7	—But h Need Not exceed Eq. 9–8
Solid Flat Plate			
Square interior panel (Min $h = 5''$)	$\ell/32.73$[a]	$\ell/41.8$	$\ell/32.73$[a]
Square edge panel (Min $h = 5''$)	$\ell/28.7$	$\ell/36.98$	$\ell/32.73$[a]
Square corner panel (Min $h = 5''$)	$\ell/27.7$	$\ell/35.95$	$\ell/32.73$[a]
Solid Flat Plate with Edge Beams (Stiffness such that $\alpha = 0.8$)			
Square edge panel	$\ell/35.23$[a]	$\ell/40.68$	$\ell/32.73$
Square corner panel	$\ell/35.00$[a]	$\ell/39.55$	$\ell/32.73$
Solid Flat Slab with Drop Panels (Length $> \ell/3$; Depth $> 1.25\ h$)			
Square interior panel (Min $h = 4''$)	$\ell/36.37$[a]	$\ell/46.56$	$\ell/32.73$
Square edge panel (Min $h = 4''$)	$\ell/31.59$	$\ell/40.68$	$\ell/32.73$[a]
Square corner panel (Min $h = 4''$)	$\ell/30.45$	$\ell/39.55$	$\ell/32.73$[a]
Solid Flat Slab with Drop Panels and Edge Beam $\alpha = 0.8$			
Square edge panel	$\ell/39.14$[a]	$\ell/45.20$	$\ell/32.73$
Square corner panel	$\ell/37.88$[a]	$\ell/43.94$	$\ell/32.73$
Solid Flat Plate $\beta = \ell_{n1}/\ell_{n2} = 2.0$			
Rectangular interior panel (Min $h = 5''$)	$\ell/32.73$[a]	$\ell/50.91$	$\ell/32.73$[a]
Rectangular edge panel (Min $h = 5''$)	$\ell/28.20$	$\ell/44.21$	$\ell/32.73$[a]
Rectangular corner panel (Min $h = 5''$)	$\ell/26.65$	$\ell/42.15$	$\ell/32.73$[a]
Solid Flat Plate with Edge Beams (Stiffness such that $\alpha = 0.8$) $\beta = \ell_{n1}/\ell_{n2} = 2.0$			
Rectangular edge panel	$\ell/38.30$[a]	$\ell/48.64$	$\ell/32.73$
Rectangular corner panel	$\ell/36.59$[a]	$\ell/46.36$	$\ell/32.73$
Solid Flat Slab with Drop Panels (Length $\ell/3$; Depth $1.25\ h$) $\beta = \ell_{n1}/\ell_{n2} = 2.0$			
Rectanguler interior panel (Min $h = 4''$)	$\ell/36.37$[a]	$\ell/56.57$	$\ell/32.73$
Rectangular edge panel (Min $h = 4''$)	$\ell/31.02$	$\ell/48.64$	$\ell/32.73$[a]
Rectangular corner panel (Min $h = 4''$)	$\ell/29.32$	$\ell/46.36$	$\ell/32.73$[a]
Solid Flat Slab with Drop Panels and Edge Beam $\alpha = 0.8$ $\beta = \ell_{n1}/\ell_{n2} = 2.0$			
Rectangular edge panel	$\ell/42.55$[a]	$\ell/54.04$	$\ell/32.73$
Rectangular corner panel	$\ell/40.66$[a]	$\ell/51.52$	$\ell/32.73$

[a] Controlling limit.

* Minimum thickness unless deflection is computed, and is within limits of maximum allowable computed deflection (Table 9.5-b).

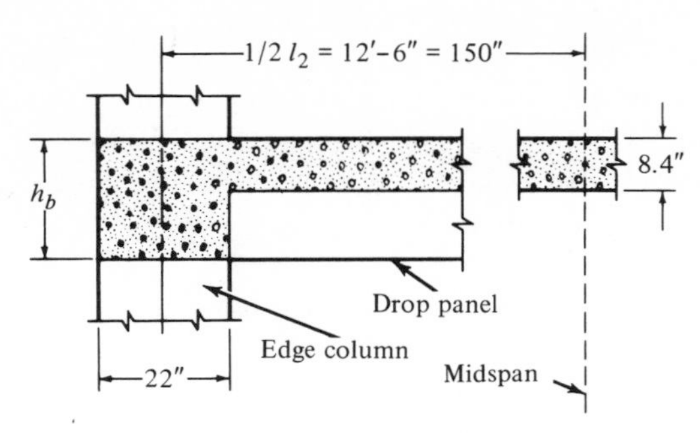

Fig. 6–8 Minimum edge beam dimensions for uniform slab thickness in all panels.

0.8. From Example 2, minimum computed h_s = 8.4 in. See Fig. 6-8 for section dimensions used.

Determine the minimum depth of the edge beam, h_b, for a width of 22 in. centered on the edge column centerline and flush with the face of the edge column. $I_b = 0.8\ I_s$.

$$(22)(h_b)^3 = (0.8)(150)(8.4)^3$$

$$\text{Min } h_b = 14.8 \text{ in.}$$

For simplicity of the formwork, the edge beams will, of course, be made 15 in. deep—the same as the depth of the drop panels. For design examples of slabs with edge beams, see Chapter 8. For design of beams see Chapter 9.

Cantilevers

No specific provisions are made for the use of cantilever slabs under direct design. If a cantilever of sufficient span is used, however, it can replace the edge beam required, with the outer edges flush. The term β_s, in the thickness equations, is defined (Section 9.0) as the ratio of the length of continuous edges to the total perimeter of the panel. For a square panel with slab edges flush with column, $\beta_s = 0.5$. If the cantilever span provides a negative moment equal to that of the minimum adjacent span, l_1, permitted under direct design, (Section 13.3.1.3) $l_n' = 2/3 l_n$, the edges may be considered continuous and $\beta_s = 1.0$ as for interior spans. The 10 percent increase in minimum thickness required in panels with discontinuous edges and no edge beam is then no longer applicable (Section 9.5.3.3).

EXAMPLE 4. *Cantitlever for continuous edge*

The minimum cantilever span necessary to provide such continuity is easily computed. The maximum interior panel negative moment, $-M_u = 0.65\ M_o$ (Section 13.3.3.2). Equate the negative moment of a minimum interior panel with a clear span $= \frac{2}{3}\ l_n$, to that for cantilever uniformly loaded with a clear span equal to l_n'.

l_n = clear span for the first interior panel.

$$\frac{wl_2}{8}\left(\frac{2}{3}\ l_n\right)^2 (0.65) = \frac{wl_2(l_n')^2}{2}$$

Minimum cantilever span, $l_n' \geq 0.27\ l_n$ for "continuous edges." The minimum square corner and edge panel thickness with $\beta_s = 1.0$ becomes $h = l_n/36.37$ (Section 9.5.3.1, Eq. 9-6).

Matching this minimum cantilever span with the interior panel slab design in the previous example, the edge panel clear span $l_n = 23 \times 12 = 276$ in. The minimum cantilever $l_n = 0.27 \times 276$ in. $= 75$ in. (see Fig. 6-9).

Recommendations for Selection of Flat Slab Thickness

To avoid complex calculations, the authors suggest that selection of slab thickness conform to the minimums, except for elements such as cantilevers (see Chapter 2). The application of concepts such as M_{cr}/M_a (Section 9.0) ratio, cracked moment of inertia, etc. (Section 9.5.2.2) to deflection calculations for two-way slabs complicated by drop panels, plus lack of specific code requirements for treatment of same creates special problems outside the scope of this guide. See Table 6-2, for solutions of the Code formulas for minimum thicknesses without deflection computations.

If a thickness less than the minimum is desired, however, deflections,

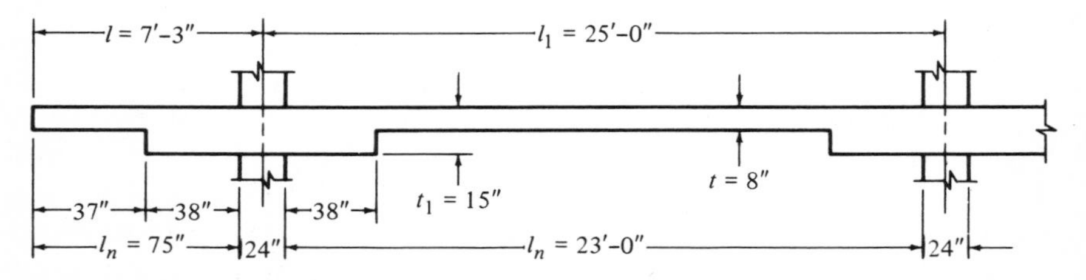

Fig. 6–9 Dimensions of edge panel with minimum cantilever.

both immediate and long-time, must be calculated (Section 9.5.3.4). Such computed immediate deflections for live load must not exceed $l_n/180$ for roofs nor $l_n/360$ for floors; long-time deflections for roofs or floors supporting or attached to nonstructural elements likely to be damaged, $l_n/480$, and if not attached to such elements, $l_n/240$ (Table 9.5.-b). Where approximately equal spans are employed without edge beams, the minimum thickness for corner panels will control. For large areas with a high number of interior panels, the panel minimum thicknesses may be employed throughout if edge beams are sufficiently stiff, $\alpha \geq 0.80$ (Section 9.5.3.3).

For a three-bay wide structure with unequal spans, and the exterior spans larger, the minimum thickness will usually be established by the corner panel.

Special Conditions

OPENINGS. Openings of any size in a two-way slab are permitted (Section 13.6) if all applicable general requirements for strength and serviceability are satisfied (Section 13.6.1). The code specifically permits openings within the following limits without a special analysis (Section 13.6.2):

(a) Any size within the area common to both middle strips
(b) One-eighth of the width of each column strip in the area common to both column strips
(c) One-quarter of the width of either strip in the area common to one column and one middle strip

Significant variations in the wording of the provisions for cases (a), (b), (c) will be noted. For case (a) the "total amount of reinforcement required for the panel *without* the opening" must be maintained. Thus, if the entire center quarter panel were omitted, the center half span of both middle strips, reinforcement for the loads removed as well as the loads remaining must be supplied in the column strips. In case (b), "the equivalent of the reinforcement interrupted shall be added on all sides of the openings." In case (c), the actual wording is that "not more than one-quarter of the reinforcement in either strip shall be interrupted by the opening. The equivalent of reinforcement interrupted shall be added on all sides of the openings." The single specific requirement for analysis of the effect of openings meeting limitations of cases (a), (b), and (c) is that shear requirements must be met. The special analysis for openings not within the three limitations, of course, includes shear (Section 11.12) and deflection limits (Section 9.5.3.4).

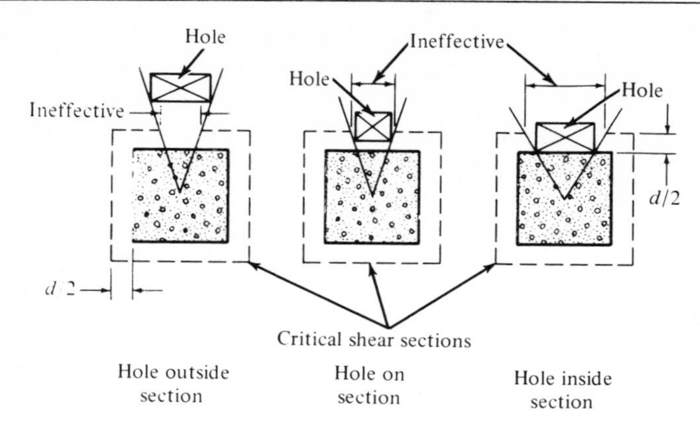

Fig. 6–10 Critical shear sections reduced by openings.

Effect of openings on shear capacity

The shear requirements related to openings in flat slabs are quite specific (Section 11.12). When the openings are located within a column strip, cases (b) and (c), "that part of the periphery of the critical shear section (around the column head) which is enclosed by radial projections of the opening to the centroid of the loaded area (column) shall be considered ineffective" (Section 11.12-a). "For slabs with shearheads, one-half of the periphery specified in (a) shall be considered ineffective." Note that *all* shear capacity constants (A_c, J_c) are affected (see Fig. 6-10; see also Fig. 7-4 for the effect of waffle slab voids near the column).

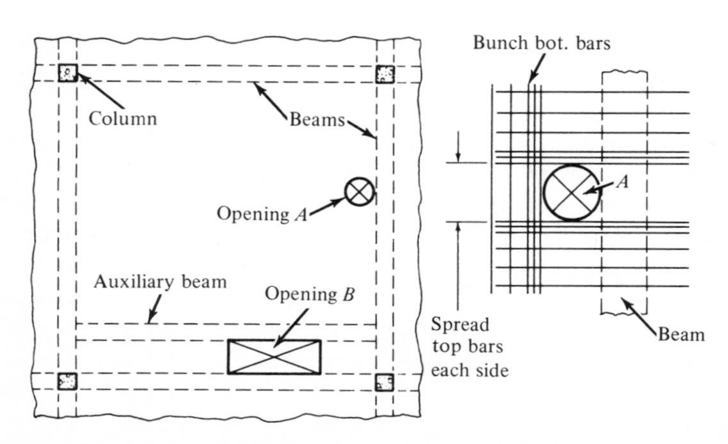

Fig. 6–11 Openings in two-way slabs with beams.

Openings in waffle flat slabs

For the special case of waffle flat slabs, any opening entirely in the top slab, that is between joists, can be regarded as satisfying all requirements without special analysis. Any top bars which would otherwise be interrupted can simply be spread on both sides of the hole. Where larger openings are required in waffle slabs, cases (b) and (c), and the joist ribs must be interrupted, the moment requirements can be satisfied by adding bottom steel equivalent to that interrupted in the adjacent ribs.

Openings in two-way slabs on beams

For two-way slabs on beams, openings in the slabs larger than those around which the reinforcement can be spread without interruption, special analysis may be avoided by framing the opening with small auxiliary beams to carry tributary loads to the main two-way beams (see Fig. 6-11).

TRANSFER OF MOMENT AND SHEAR TO CORNER COLUMNS

A special problem exists in the design for transfer of unbalanced moment and shear to corner columns where no edge beam or cantilever is present. The prescribed critical shear section for two-way shear is two-sided, so that the centroid is off the column center in both directions, and the column center itself may be off center in the half-column strip. Both conditions are inefficient. The code provides the same critical sections as for all columns: for moment, at the face of the column or capital; for shear, a distance $d/2$ from the column face for two-way shear, and a section across the half-column strip and half the middle strip for one-way (beam) shear. For two-way slabs with beams, $\alpha > 1.0$, in each direction additional steel is required top and bottom (Section 13.5.4).

For the special case of a flat slab exterior corner with or without drop panel, and no edge beams, the authors suggest that the user investigate a critical section for moment on a diagonal in addition to the orthogonal critical sections specified in the code (see Fig. 6-12).

The additional analysis is intended to insure that sufficient reinforcement is provided and effectively developed to avoid negative moment (top) diagonal cracks at the diagonal section. The sections prescribed for one-way shear parallel to the edges extending across the entire column half-strip and half the middle strip seem unrealistic.

The exterior moments assigned to the two column half-strips can be combined and oriented at right angles to a critical section (at 45° for a square panel). Similarly, the components of force carried in the two sets of orthogonal top bars can be combined. Only those bars fully developed beyond the critical section can be considered fully effective. Usually, some additional top steel will be found desirable. To avoid building up additional

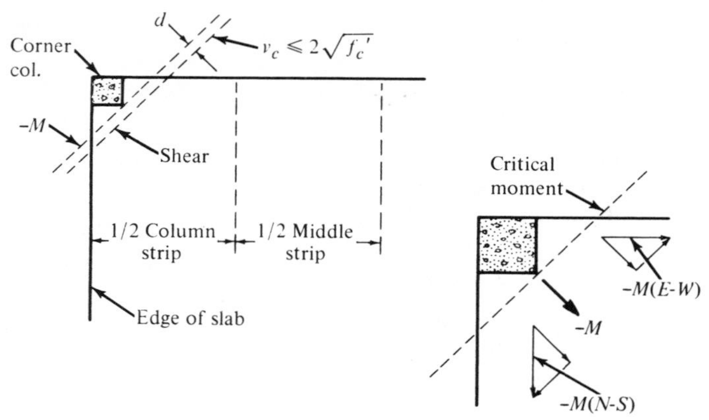

Fig. 6–12 Additional critical sections at exterior corners.

layers of top steel and the consequent decrease in the effective depth, such additional bars can be placed parallel to the edges within a width equal to the column side, c, plus $1/2\ h$ on one side of the column (Section 13.2.4).

In addition to the beam (one-way) shear investigation on the suggested diagonal section, some adjustment in the fraction of the unbalanced moment transferred by shear may be required (Section 11.13.2).

ROUND COLUMNS AND CONICAL CAPITALS. Round columns with conical capitals using standard reusable steel forms were very commonly used with the original flat slab and drop panel system. Single use standard round column forms are now available, and the round columns are frequently used with capitals *or* drop panels.

Two special problems with the use of round columns and round capitals arise under the 1971 Code: (1) determination of the torsional shear constant, J_c, of a circular section for moment transfer by shear (Section 11.13.2), and (2) determination of the column stiffnesses, K_c, when a conical capital is furnished at the upper end (Section 13.4.1.3).

Shear transfer constant, J_c, for round interior columns

For round interior columns, $J_c = 4\ dr^3 + \dfrac{rd^3}{3}$, where d = the effective depth of the slab or drop panel above, r = the radius of the round shear section $= 1/2\ (t + d)$, and t = diameter of the round column (or capital) at the bottom of the slab or drop panel above (for derivation, see Figs. 6-21, -22, -23, and -24 at the end of this chapter).

Round and square columns, without capitals or brackets, may be considered equivalent when the diameter of the round column equals 1.14

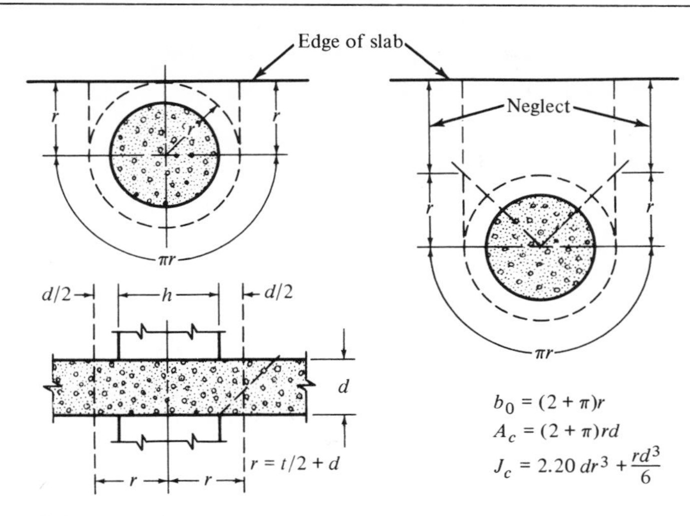

Fig. 6–13 Torsional constant, J_c, for round edge exterior columns.

times the side dimension of the square column. Cross-sectional area, flexural stiffness, and the shear constant, J_c, will be closely equivalent. See Figs. 6-25 through 6-29 for minimum size square columns with flat slabs.

SHEAR TRANSFER CONSTANT, J_c, FOR ROUND EXTERIOR COLUMNS. For round columns (or capitals) at or near the edge of the slab, the exact solution for J_c involves a semicircular shear section plus two straight tangents perpendicular to and extending to the edge (see Fig. 6-13). A conservative value for this open section should be used. The authors suggest $J_c = 7/6\ dr^3 + rd^3/3$ for all round edge columns whether tangent to the edge or set back slightly, until the cantilever setback develops sufficient moment so that the column can be considered as an interior column. See Fig. 5-32(*h*) for the similar case of square edge columns.

Effect of Capitals or Brackets upon Column Stiffness, K_c

An exact solution can be obtained using either the moment-area or column-analogy method. In the exact solution, an infinite moment of inertia within the depth of the slab or drop panel, a variable moment of inertia for the depth of the capital, and a constant moment of inertia for the gross cross section of the column for the height from the floor line to the bottom of the capital must be considered (Section 13.4.1.3). The exact solution will be complicated by the variable moment of inertia portion and its

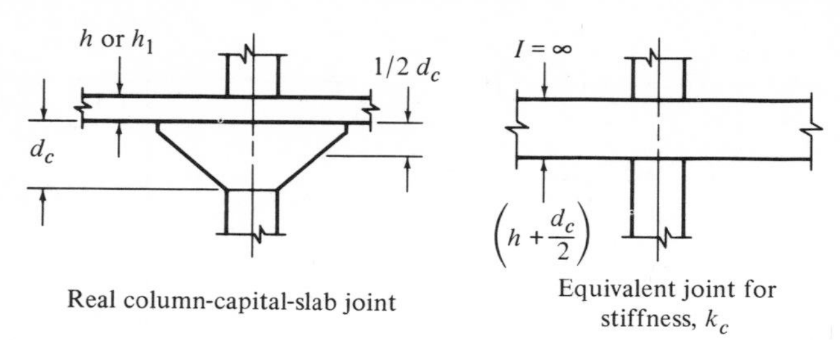

Fig. 6–14 Column stiffness, K_c, for round columns with capitals.

eccentric location from the midpoint of the column length. As a result of this eccentricity, separate carry-over factors will result at the top and bottom of the column for moment distribution in the equivalent frame method. Separate stiffness factors, K_c, will result for use with both the direct design and equivalent frame methods.

An approximate average value for K_c, satisfactory for most practical design applications, can be quickly determined (see Fig. 6-14). Replace the actual column capital with an "equivalent drop panel" equal to the depth of slab (or actual drop panel if used) plus *one-half* the depth of the capital. See Example 1, Step 4, for the column-analogy formula. Use the "equivalent drop panel" depth for t and solve for an average K_c which may be used at the joints above and below the column. It will be necessary to use the column-analogy formula since the ratio of "equivalent drop panel depth" to story height will often exceed the range of the chart (Fig. 5-4).

This approximation assumes an infinite stiffness within the joint to the midpoint of the capital depth and neglects the smaller bottom half of the capital. The approximation would seem justifiable since the Code does not specifically prescribe the exact solution: "The moment of inertia . . . of the column . . . *outside* of the joint or capital *may* be based on the cross section area. . . ." "Variation in the moments of inertia of the . . . columns along their axes shall be taken into account" "Section 13.4.1.3). This approximation does "take into account" the addition to column stiffness contributed by a capital as well as the proportions of the capital itself.

EQUIVALENT FRAME METHOD

The equivalent frame method of analysis, for flat slabs with drop panels is complicated by the presence of the drop panel, which causes a variation in the flexural and torsional stiffness of the slab. An example of three spans with cantilevers and the same concrete dimensions and loads as the interior

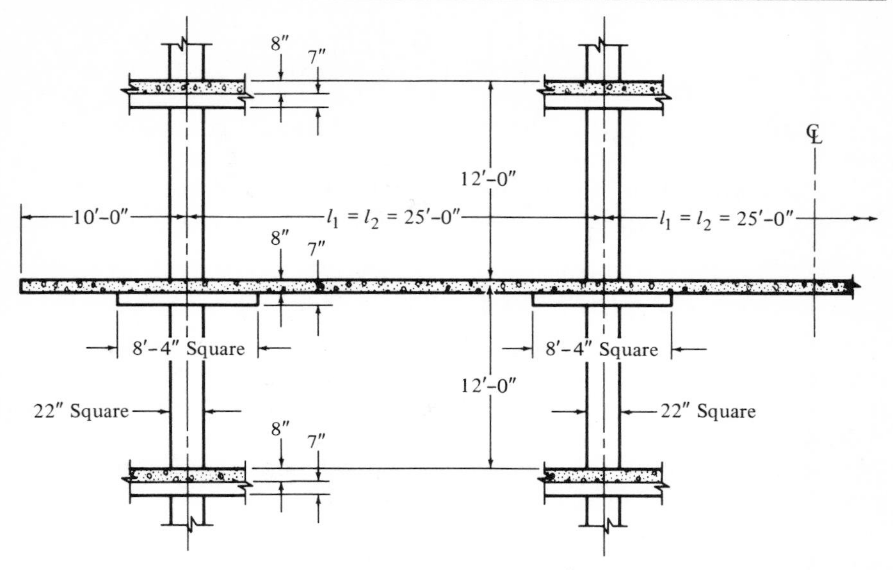

Fig. 6–15 Flat slab equivalent frame.

span designed by the direct design approach will be shown to demonstrate this method.

EXAMPLE. Figure 6-15 shows the dimensions of the structure with square panels.

Step 1. Determine the slab thickness, h

The Code makes no recommendations for minimum slab thickness for slabs with cantilevers when deflections are not calculated. If the cantilever moment approaches 0.65 M_o, the minimum slab thickness can be based on that for an interior span (see Table 6-2).

$$h \text{ (minimum)} = l_n/36.67 = \frac{300 - 22}{36.37} = 7.65'' \leqq 8''$$

Step 2. Trial columns dimensions

Use square columns with $c_1 = c_2 = 22$ in. which is the same as the direct design interior panel example.

Step 3. Determine column stiffness, K_c

Column stiffness can be determined as in Step 4 of the equivalent frame method of analysis for the flat plate or from Fig. 5-4 used with the direct design method.

$$K_c/E_c = \frac{I_c}{(l_c - h)}\left[1 + \frac{3l_c^2}{(l_c - h)^2}\right] = \left[\frac{(0.0833)(22)^4}{(144 - 15)}\right]\left[1 + \frac{(3)(144)^2}{(144 - 15)^2}\right]$$

$$= 717 \text{ in.}^3 \quad (710 \text{ in.}^3 \text{ by Fig. 5-4 and direct design method})$$

This method of determining the column stiffness assumes that the column has an infinite moment of inertia for the depth of the drop panel (Section 13.4.1.5).

Step 4. Determine flexural slab stiffness, K_s

The slab stiffness will be determined by the column analogy method as in Step 4 for the flat plate. The results will be compared with those obtained by the moment area method in the direct design example. Refer to the flat slab with drop panel example by direct design method for the moment of inertia of the slab at the column, drop panel, and between the drop panels. See Fig. 6-16 for slab analogous column.

$$\begin{aligned} A_{ac} &= (0.0000782)(16.67)(12) &= 0.0156 \\ &+ (0.0000218)(3.25)(12)(2) &= 0.0017 \\ &+ (0.0000186)(0.917)(12)(2) &= \underline{0.0004} \\ && 0.0177 \end{aligned}$$

$$I_{ac} = \frac{(0.0000186)(300)^3}{12} + \frac{(0.0000032)(278)^3}{12} + \frac{(0.0000564)(200)^3}{12}$$

$$= 41.8 + 5.7 + 37.6 = 85.1$$

$$K_s/E_c = \frac{1}{A_{ac}} + \frac{(1)(l_1/2)(l_1/2)}{I_{ac}}$$

$$= \frac{1}{0.0177} + \frac{(1)(150)(150)}{85.1} = 56.5 + 264.4$$

$= 321$ in.^{3} (334 in.^{3} by moment area method used with direct design method.)

Step 5. Torsional stiffness of the slab, K_t

The member considered for torsional stiffness is one that consists of a part of the slab perpendicular to the direction in which moments are being considered. This portion of slab is considered to have a width c_1 (width of column in direction moments are being considered) and a constant depth equal to that of the slab (Section 13.4.1.5). K_t can vary considerably depending on the depth used in calculating the torsional constant, C. Calculated values for K_t/E_c, based on a C-value, using the depth of slab equal to that between drop panels, a varying depth of slab, an average depth of slab (direct design example), or the depth of the slab at the drop panel, are compared in the direct design example. The "exact" solution, K_t, of a varying depth torsional member was based on the average effective angle of rotation and was calculated as shown by Eberhardt and Hoffman.*

$$\frac{1}{K_t} = \theta_t = \frac{l_2(1 - c_2/l_2)^5}{40GC_D} + l_2(1 - 2x_2/l_2)^5 \left[\frac{1}{C_s} - \frac{1}{C_D}\right] \quad \text{Eq. (6-1)}$$

where

x_2 = distance from centerline of span to edge of drop panel

C_s = torsional constant for slab cross section of width c_1 and depth equal to slab thickness between drop panels

C_D = torsional constant for slab cross section of width c_1 and depth equal to total slab thickness at drop panel

If E_c is assumed equal to 2 G, then Eq. (6-1) becomes

$$\frac{E_c}{K_t} - \frac{l_2(1 - c_2/l_2)^5}{20C_D} + \frac{l_2(1 - 2x_2/l_2)^5}{20}\left[\frac{1}{C_s} - \frac{1}{C_D}\right] \quad \text{Eq. (6-2)}$$

$$C_s = \left[(1 - 0.63)\frac{(x)}{y}\right]\frac{x^3(y)}{3} = \left[(1 - 0.63)\frac{(8)}{22}\right]\frac{(8)^3(22)}{3} = 2890 \text{ in.}^4$$

$$C_D = \left[(1 - 0.63)\frac{(15)}{22}\right]\frac{(15)^3(22)}{3} = 14{,}120 \text{ in.}^4$$

$$\frac{E_c}{K_t} = \frac{300\,(1 - 22/300)^5}{(20)(14120)} + \frac{300(1 - 100/300)^5}{20}\left[\frac{1}{2890} - \frac{1}{14120}\right]$$

$$= 0.00127$$

$K_t/E_c = 787 \text{ in.}^3$

---

* Arthur C. Eberhardt and Edward S. Hoffman, "Equivalent Frame Analysis for Slab Design," *ACI Journal*, May 1971.

Step 6. Stiffness of equivalent columns

$$K_{ec}/E_c = \frac{\Sigma K_c}{1 + \Sigma K_c/K_t} = \frac{(2)(717)}{1 + \dfrac{(2)(717)}{787}}$$

$$= 508 \text{ in.}^3 \text{ (435 in.}^3 \text{ in direct design example).}$$

Step 7. Design loads

The service live load is 400/1.7 or 235 psf. The dead load of the slab beyond the drop panel is 8/12 × 150 or 100 psf. The live load exceeds three-quarters of the dead load [$235 < 3/4\ (100)$]. When live load is greater than three-quarters of the dead load, the equivalent frame method requires that moments be determined for two patterns of loading, using three-quarters of the design live load. Alternate panels are loaded for maximum and minimum positive moments. Two adjacent panels only are loaded for maximum negative moments. In no case shall design moments be less than would occur with full design live load, w_l, on all spans (Section 13.4.1.8).

$$w_d \text{ (slab beyond drop panel)} = \frac{8}{12}\,\frac{(150)(1.4)(25)}{(1000)} = 3.50 \text{ klf}$$

$$w_d \text{ (drop panel)} = \frac{7}{12}\,\frac{(150)(1.4)(8.33)}{(1000)} = 1.02 \text{ klf}$$

$$100\%\ w_l = \frac{(400)(25)}{1000} = 10.00 \text{ klf}$$

$$0.75\ w_l = \frac{(0.75)(400)(25)}{1000} = 7.50 \text{ klf}$$

Step 8. Carry-over factors for moment distribution (see Fig. 6-16)

$$COF = \frac{1/A_{ac} - M_C/I_{ac}}{1/A_{ac} + M_C/I_{ac}} = \frac{\dfrac{1}{0.0177} - \dfrac{(150)(150)}{85.1}}{\dfrac{1}{0.0177} + \dfrac{(150)(150)}{85.1}} = \frac{56.5 - 264.4}{56.5 + 264.4} = \frac{-207.9}{320.9}$$

$$= -0.648$$

Step 9. Fixed end moments

The fixed end moments will be determined by the column-analogy method for the equivalent frame (see Fig. 6-16).

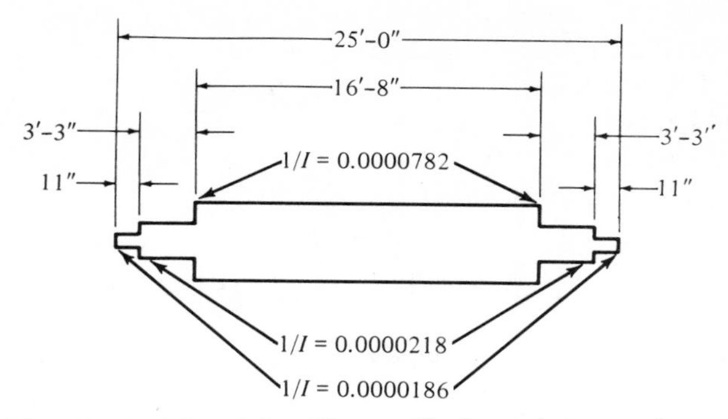

Fig. 6–16 Flat slab stiffness, K_s, by column analogy.

$$
\begin{aligned}
A_{ac} = {} & (1)(16.67) & = 16.67 \\
& (2)(0.279)(3.25) & 1.81 \\
& (2)(0.238)(0.92) & \underline{0.44} \\
& & 18.92
\end{aligned}
$$

Area of M/I diagram for uniform load of 1.0 klf (see Fig. 6-17).

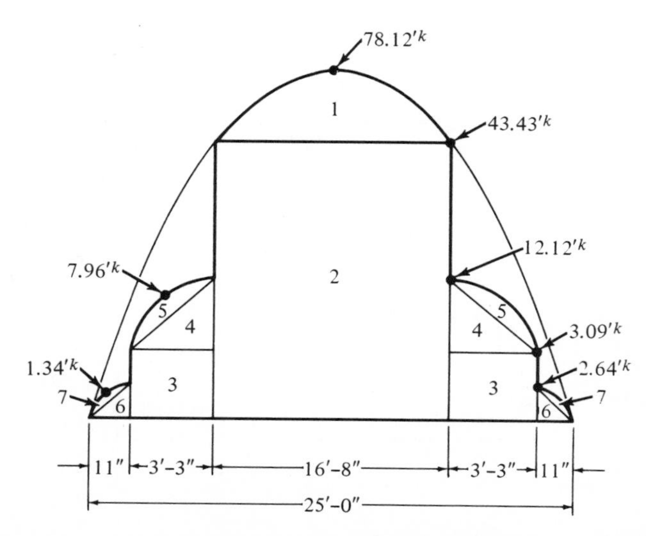

Fig. 6–17 M/I diagram for uniform load of 1.0 klf on flat slab.

(1) (2/3)(34.7)(16.67) = 386
(2) (43.4)(16.67) = 724
(3) (3.09)(3.25)(2) = 20
(4) (1/2)(9.13)(3.25)(2) = 30
(5) (2/3)(0.36)(3.25)(2) = 2
(6) (1/2)(2.64)(0.92)(2) = 2
(7) = –

1174

Area of M/I diagram for drop panel design dead load of 1.02 klf (see Fig. 6-18).

(1) (8.86)(16.67) = 148
(2) (0.97)(3.25)(2) = 6
(3) (1/2)(1.50)(3.25)(2) = 5
(4) (2/3)(0.38)(3.25)(2) = 2
(5) (1/2)(0.83)(0.92)(2) = 1
(6) (2/3)(0.02)(0.92)(2) = –

162

The fixed end moment for design dead load is

$$FEM\ (w_d) = \frac{(1174)(3.5) + 162}{18.92} = 226 \text{ ft-kips}$$

The fixed end moment for design dead load plus full design live load is

$$FEM\ (w_d + 0.75\ w_l) = 226 + \frac{(1174)(7.5)}{18.92} = 226 + 465 = 691 \text{ ft-kips}$$

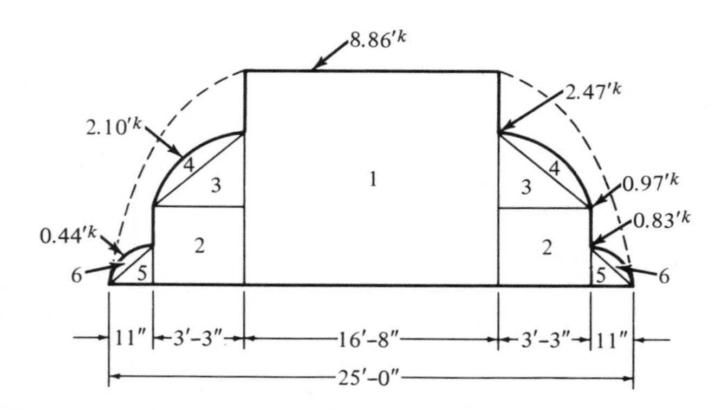

Fig. 6–18 M/I Diagram for additional dead load of drop panel.

Step 10. Moment distribution factors

For distributing slab moments at exterior and interior columns.

$$DF\ (\text{Ext. Col.}) = \frac{K_s}{K_s + K_{ec}} = \frac{321}{321 + 508} = 0.387$$

$$DF\ (\text{In. Col.}) = \frac{K_s}{\Sigma K_s + K_{ec}} = \frac{321}{(2)(321) + 508} = 0.279$$

Elastic analyses of the equivalent frame for the structure shown in Fig. 6-19

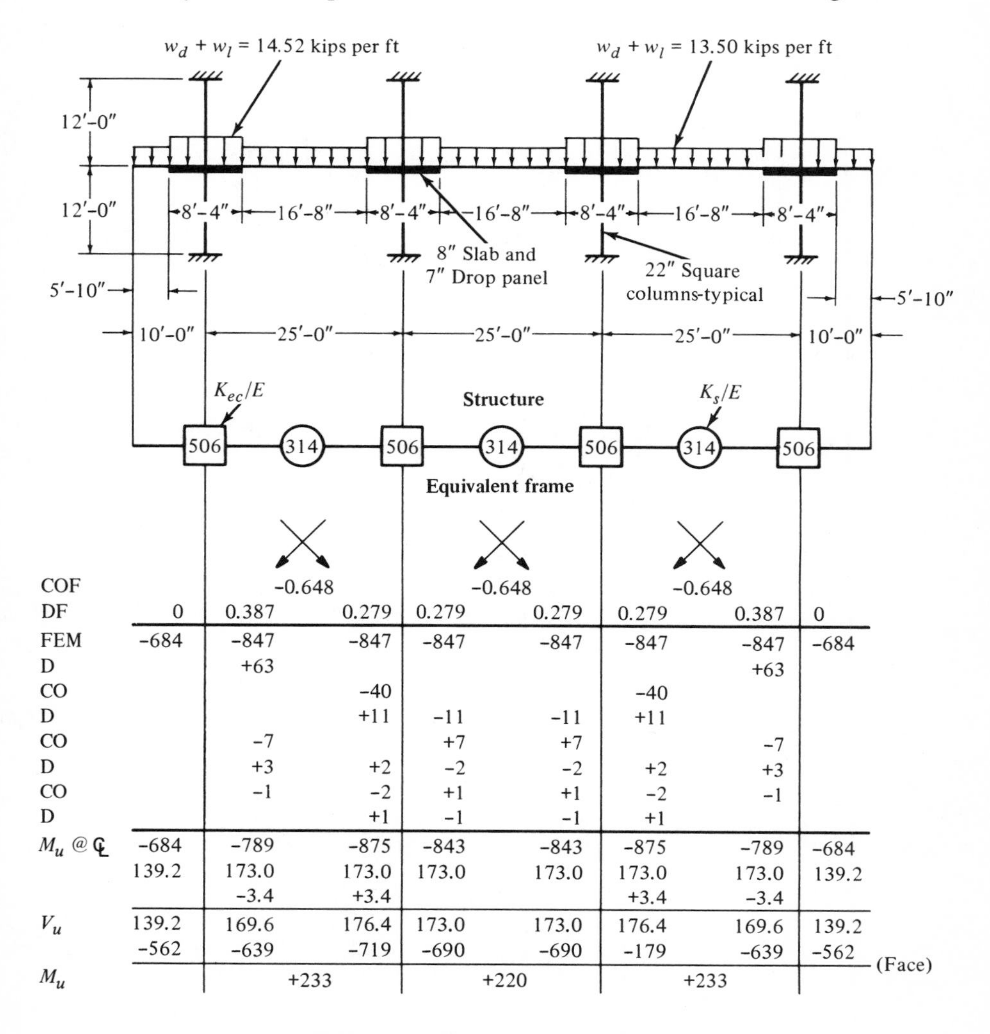

COF			-0.648		-0.648		-0.648	
DF	0	0.387	0.279	0.279	0.279	0.279	0.387	0
FEM	-684	-847	-847	-847	-847	-847	-847	-684
D		+63					+63	
CO			-40			-40		
D			+11	-11	-11	+11		
CO		-7		+7	+7		-7	
D		+3	+2	-2	-2	+2	+3	
CO		-1	-2	+1	+1	-2	-1	
D			+1	-1	-1	+1		
M_u @ ℄	-684	-789	-875	-843	-843	-875	-789	-684
	139.2	173.0	173.0	173.0	173.0	173.0	173.0	139.2
		-3.4	+3.4			+3.4	-3.4	
V_u	139.2	169.6	176.4	173.0	173.0	176.4	169.6	139.2
	-562	-639	-719	-690	-690	-179	-639	-562 (Face)
M_u		+233		+220		+233		

Fig. 6–19 Elastic analysis—full design load all panels.

TABLE 6-3 Comparison of Moments by the Equivalent Frame and Direct Design Methods

Design Method and Loading	TOTAL PANEL MOMENTS					
	MAXIMUM NEGATIVE MOMENT AT FACE OF COLUMN (ft-kips)		MAXIMUM POSITIVE MOMENT (ft-kips)		MINIMUM POSITIVE MOMENT (ft-kips)	
	Exterior column	Interior column	Exterior span	Interior span	Exterior span	Interior span
Equivalent frame $w_d + w_\ell$ on all spans	−639	−719	+233	+220	+233	+220
Equivalent frame $w_d + 0.75w_\ell$ pattern loading (Section 13.4.1.8)	−570	−670	+258	+249	−10	−14
Direct design interior Span with $-M_\mu = -0.65\ M_o$ and $+M_\mu = +0.35\ M_o$	—	−592	—	+318	0	0

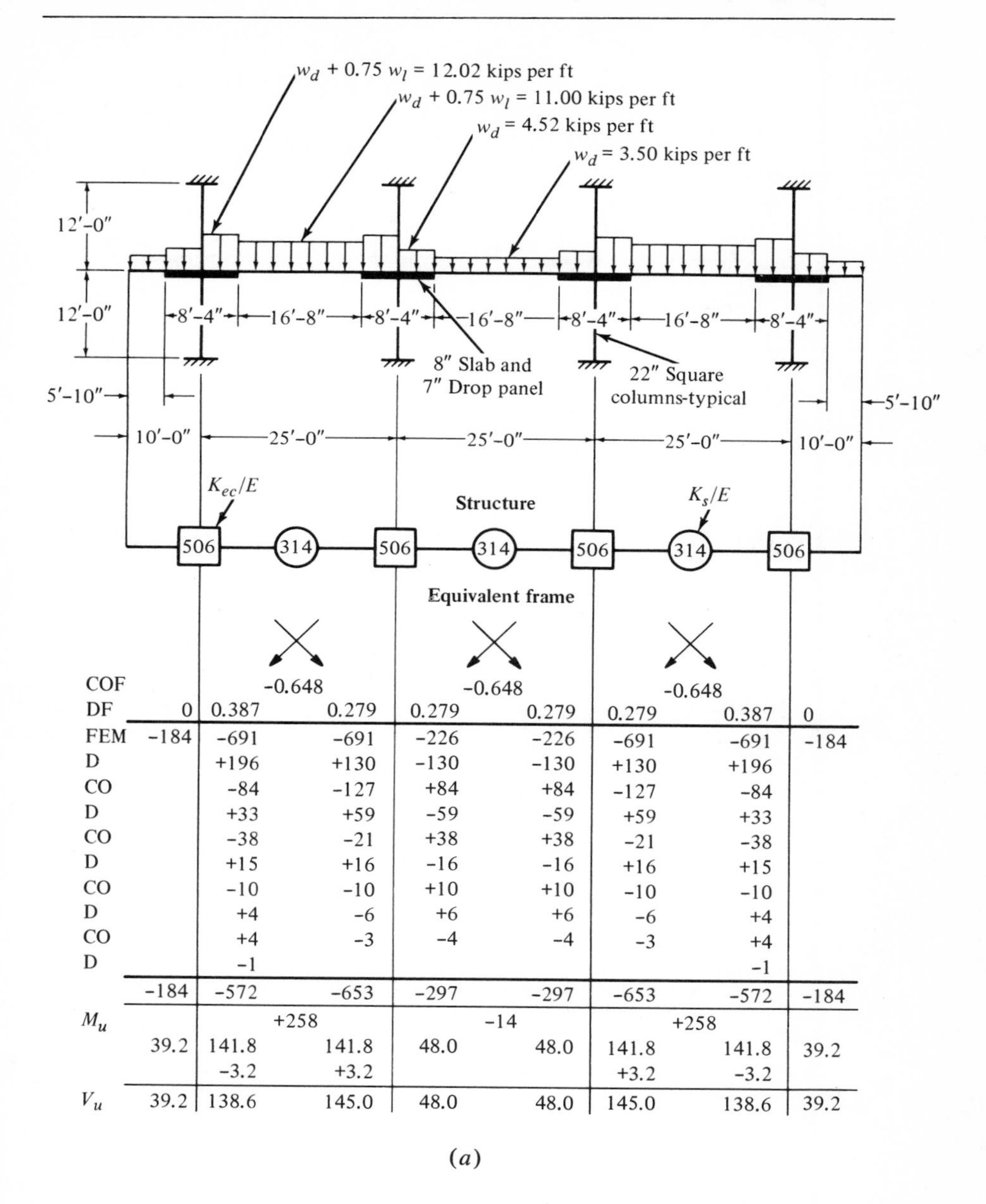

COF		-0.648		-0.648		-0.648		
DF	0	0.387	0.279	0.279	0.279	0.279	0.387	0
FEM	-184	-691	-691	-226	-226	-691	-691	-184
D		+196	+130	-130	-130	+130	+196	
CO		-84	-127	+84	+84	-127	-84	
D		+33	+59	-59	-59	+59	+33	
CO		-38	-21	+38	+38	-21	-38	
D		+15	+16	-16	-16	+16	+15	
CO		-10	-10	+10	+10	-10	-10	
D		+4	-6	+6	+6	-6	+4	
CO		+4	-3	-4	-4	-3	+4	
D		-1					-1	
	-184	-572	-653	-297	-297	-653	-572	-184
M_u		+258		-14		+258		
	39.2	141.8	141.8	48.0	48.0	141.8	141.8	39.2
		-3.2	+3.2			+3.2	-3.2	
V_u	39.2	138.6	145.0	48.0	48.0	145.0	138.6	39.2

(*a*)

Fig. 6–20 (*a*) Elastic analysis load pattern for maximum positive moment; (*b*) Elastic analysis load pattern for maximum positive moment; (*c*) Elastic analysis load pattern for maximum negative moment; (*d*) elastic analysis load pattern for maximum negative moment.

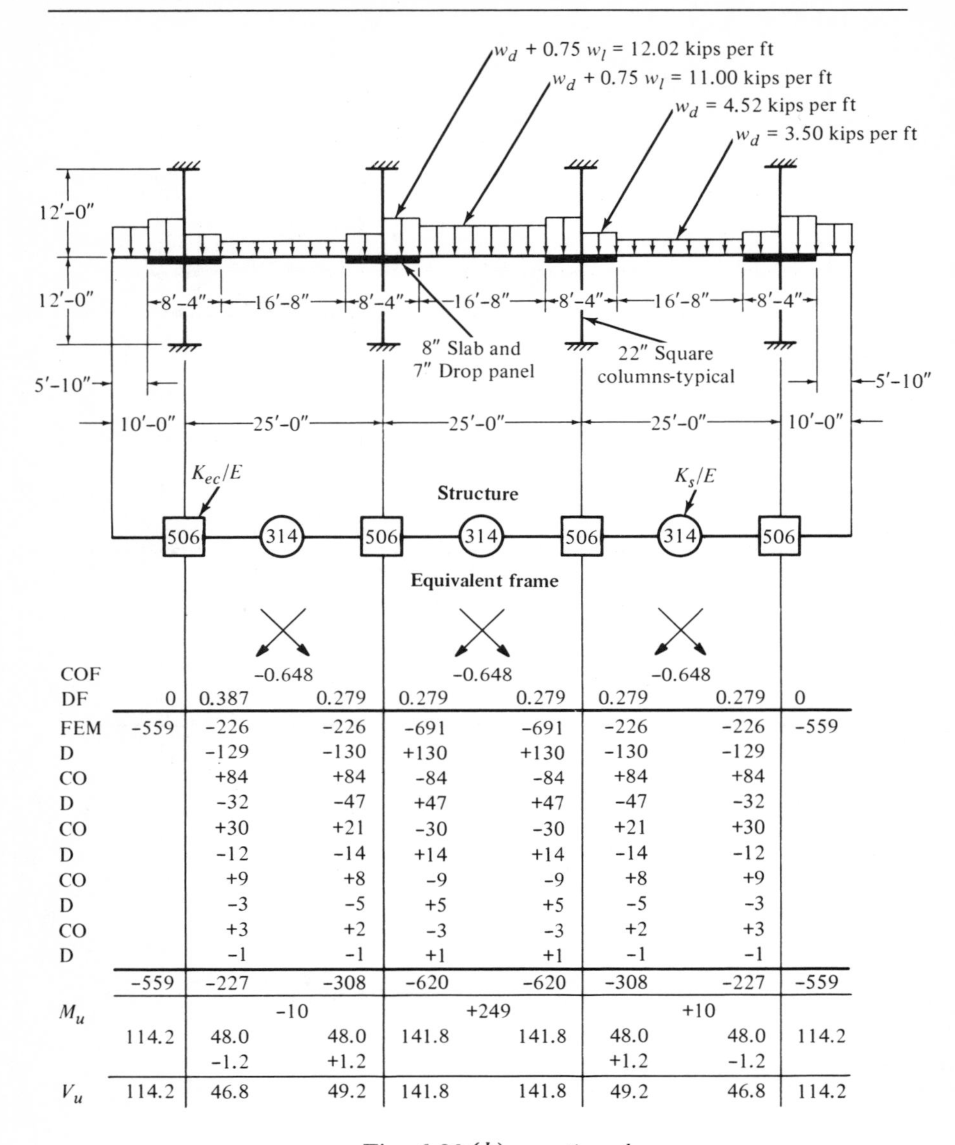

COF			−0.648		−0.648		−0.648	
DF	0	0.387	0.279	0.279	0.279	0.279	0.279	0
FEM	−559	−226	−226	−691	−691	−226	−226	−559
D		−129	−130	+130	+130	−130	−129	
CO		+84	+84	−84	−84	+84	+84	
D		−32	−47	+47	+47	−47	−32	
CO		+30	+21	−30	−30	+21	+30	
D		−12	−14	+14	+14	−14	−12	
CO		+9	+8	−9	−9	+8	+9	
D		−3	−5	+5	+5	−5	−3	
CO		+3	+2	−3	−3	+2	+3	
D		−1	−1	+1	+1	−1	−1	
	−559	−227	−308	−620	−620	−308	−227	−559
M_u		−10		+249		+10		
	114.2	48.0	48.0	141.8	141.8	48.0	48.0	114.2
		−1.2	+1.2			+1.2	−1.2	
V_u	114.2	46.8	49.2	141.8	141.8	49.2	46.8	114.2

Fig. 6-20 (*b*) *continued*

can be made by moment distribution using the carry-over factor of Step 8, the fixed end moments of Step 9, and the distribution factors of Step 10. To determine the design moments in accordance with the Code, it is necessary to analyze the structure with $w_d + w_l$ on all spans, with w_d +

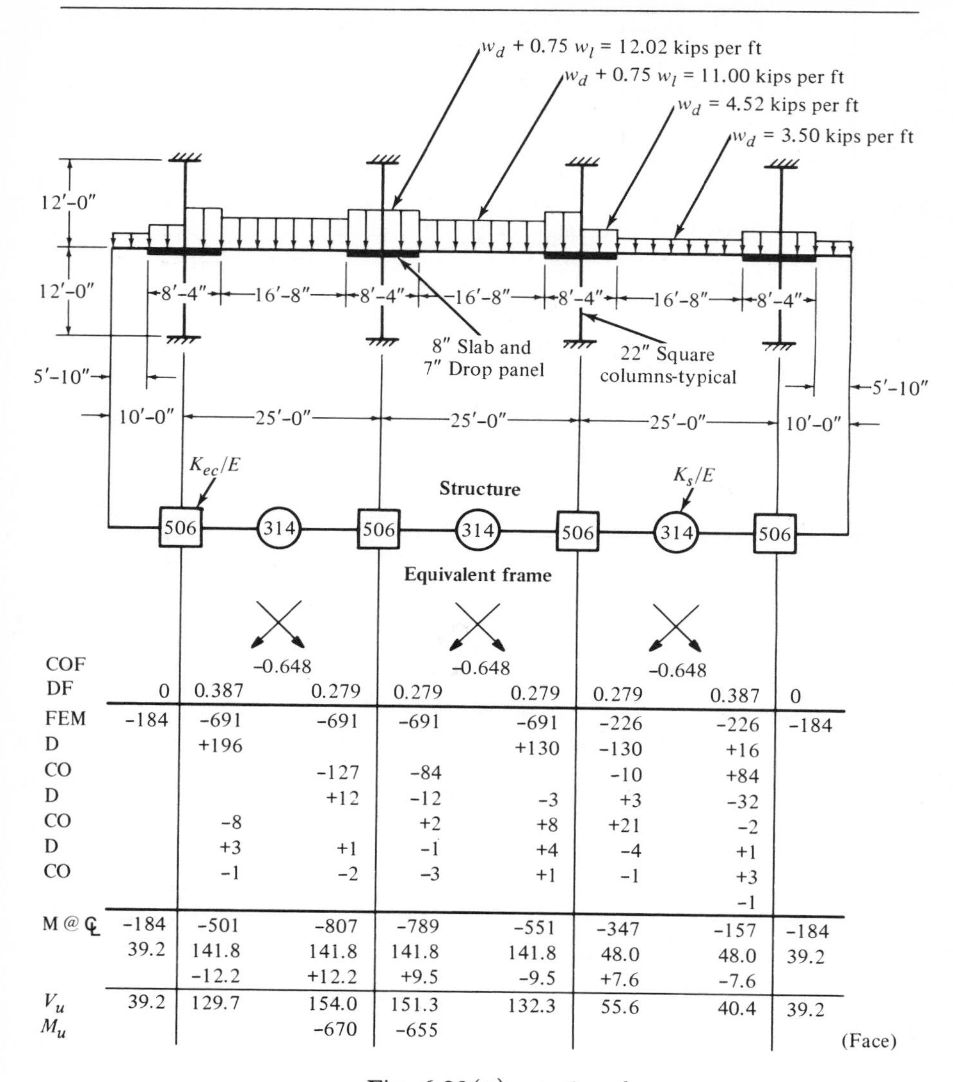

COF			−0.648		−0.648		−0.648	
DF	0	0.387	0.279	0.279	0.279	0.279	0.387	0
FEM	-184	-691	-691	-691	-691	-226	-226	-184
D		+196			+130	-130	+16	
CO			-127	-84		-10	+84	
D			+12	-12	-3	+3	-32	
CO		-8		+2	+8	+21	-2	
D		+3	+1	-1	+4	-4	+1	
CO		-1	-2	-3	+1	-1	+3	
							-1	
M @ ℄	-184	-501	-807	-789	-551	-347	-157	-184
	39.2	141.8	141.8	141.8	141.8	48.0	48.0	39.2
		-12.2	+12.2	+9.5	-9.5	+7.6	-7.6	
V_u	39.2	129.7	154.0	151.3	132.3	55.6	40.4	39.2
M_u			-670	-655				(Face)

Fig. 6-20(*c*) *continued*

0.75 w_l on alternate spans for maximum (and minimum) positive moments, and with $w_d + 0.75\ w_l$ on adjacent spans for maximum negative moment (Section 13.4.1.9).

The results of these five analyses are shown in Table 6-3, along with moments for an interior span obtained by the direct design method. Note that $w_d + w_l$ on all spans produces the maximum design negative

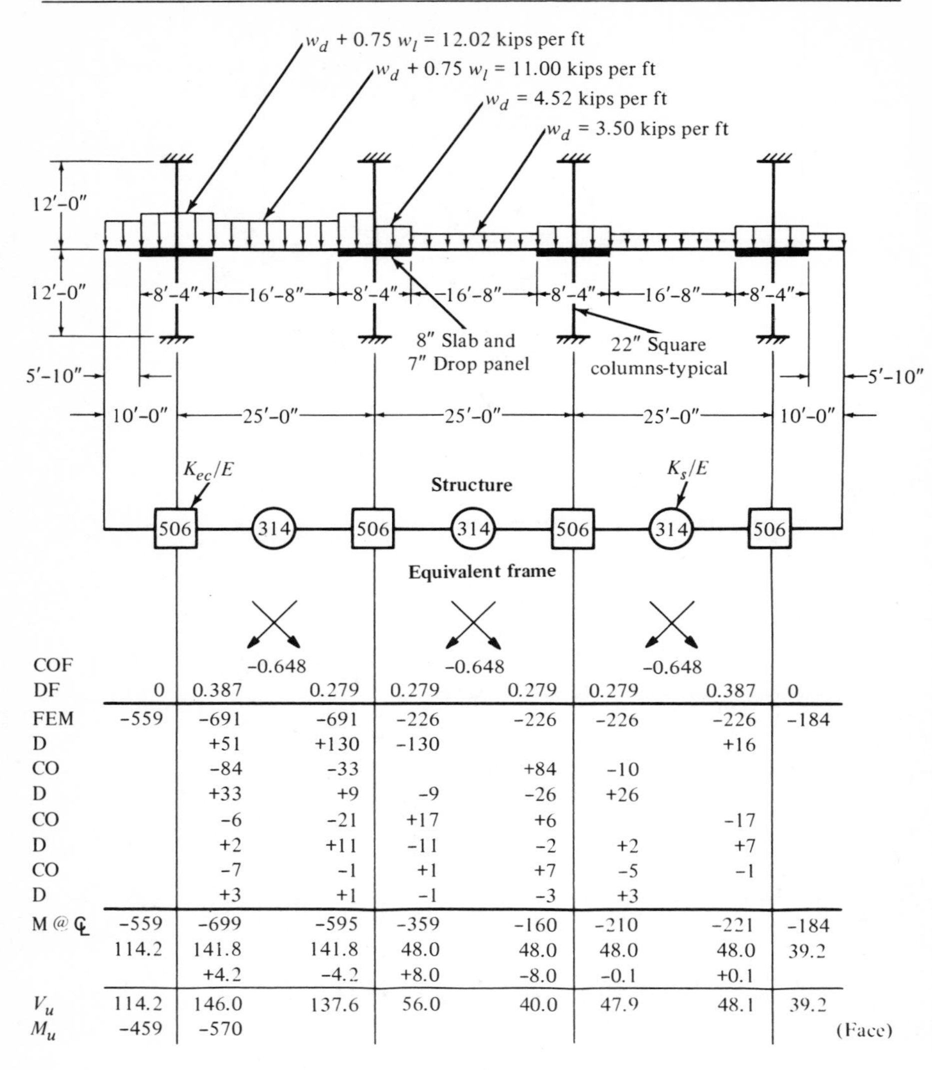

COF		-0.648			-0.648		-0.648	
DF	0	0.387	0.279	0.279	0.279	0.279	0.387	0
FEM	-559	-691	-691	-226	-226	-226	-226	-184
D		+51	+130	-130			+16	
CO		-84	-33		+84	-10		
D		+33	+9	-9	-26	+26		
CO		-6	-21	+17	+6		-17	
D		+2	+11	-11	-2	+2	+7	
CO		-7	-1	+1	+7	-5	-1	
D		+3	+1	-1	-3	+3		
M @ ℄	-559	-699	-595	-359	-160	-210	-221	-184
	114.2	141.8	141.8	48.0	48.0	48.0	48.0	39.2
		+4.2	-4.2	+8.0	-8.0	-0.1	+0.1	
V_u	114.2	146.0	137.6	56.0	40.0	47.9	48.1	39.2
M_u	-459	-570						

(Face)

Fig. 6-20 (*d*) *continued*

moment at the face of the column and that $w_d + 0.75\ w_l$ on alternate spans produces the maximum design positive moment. Note also that the direct design method gives a 45 percent larger positive moment and a 16 percent smaller negative moment than the equivalent frame method for an interior span. This comparison also shows that negative moments develop at the midspan of unloaded panels. See "Approxi-

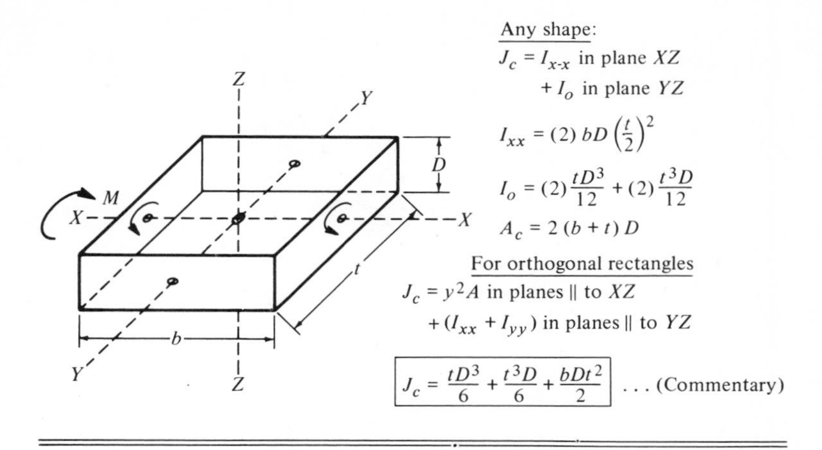

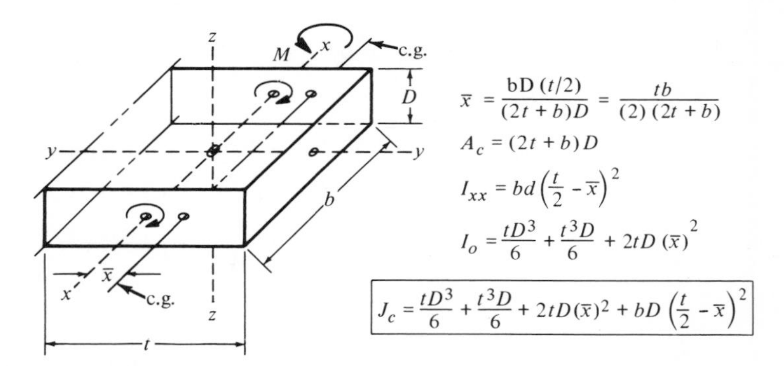

Fig. 6–21 Shear constants—derivation for rectangular columns.

mate Method versus Frame Analysis," Chapter 2, for a similar comparison applied to one-way construction with beam-column or slab and beam frames. The authors' recommendations in Chapter 2 are equally applicable to the use of the approximate direct design method versus the "exact" equivalent frame method and will not be repeated here.

SUPPLEMENTAL FLAT SLAB DESIGN DATA

Shear Constants

Brief derivations of shear constants for the area of the critical shear section, A_c, and the flexure-torsion moment of inertia, J_c, are given in Figs.

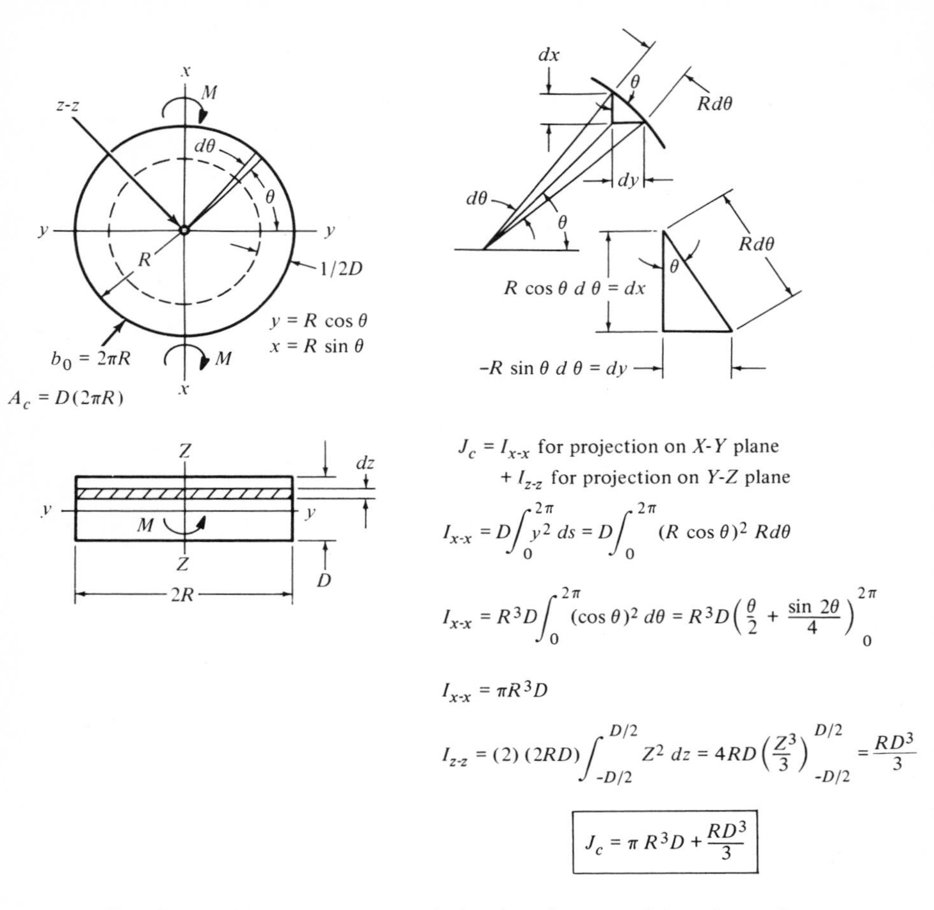

$J_c = I_{x\text{-}x}$ for projection on X-Y plane
$+ I_{z\text{-}z}$ for projection on Y-Z plane

$$I_{x\text{-}x} = D\int_0^{2\pi} y^2\, ds = D\int_0^{2\pi} (R\cos\theta)^2\, Rd\theta$$

$$I_{x\text{-}x} = R^3 D\int_0^{2\pi} (\cos\theta)^2\, d\theta = R^3 D\left(\frac{\theta}{2} + \frac{\sin 2\theta}{4}\right)_0^{2\pi}$$

$$I_{x\text{-}x} = \pi R^3 D$$

$$I_{z\text{-}z} = (2)(2RD)\int_{-D/2}^{D/2} Z^2\, dz = 4RD\left(\frac{Z^3}{3}\right)_{-D/2}^{D/2} = \frac{RD^3}{3}$$

$$\boxed{J_c = \pi R^3 D + \frac{RD^3}{3}}$$

Fig. 6–22 Shear constants—derivation for round interior columns.

6-21, -22, -23, and -24, for rectangular and round columns. Since the constants are reduced near openings or edges, values for various locations of the columns near edges and corners are given. The authors' recommendation (page 123) for a conservative upper limit on the shear constants for an open critical shear section with square columns should be extended to round columns near edges. Unless the cantilever edge moment is sufficient to create an interior column condition (closed shear section), the portions of the critical shear section on the cantilever side of the

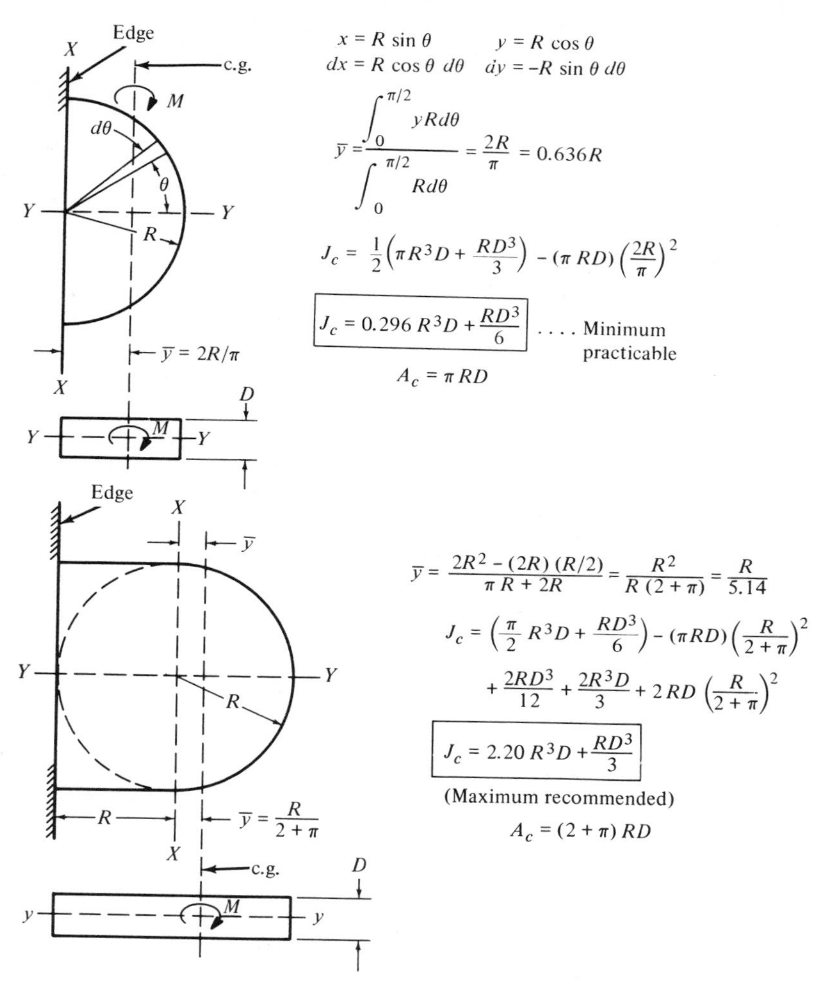

Fig. 6–23 Shear constants—derivation for round edge columns.

column cannot be considered fully effective. For these small cantilevers, the authors advise use of no larger value than that shown for a circular shear section tangent to the edge (labeled "Maximum Recommended" in Fig. 6-23). Similarly no larger value than that shown in Fig. 6-24 is recommended for round corner columns.

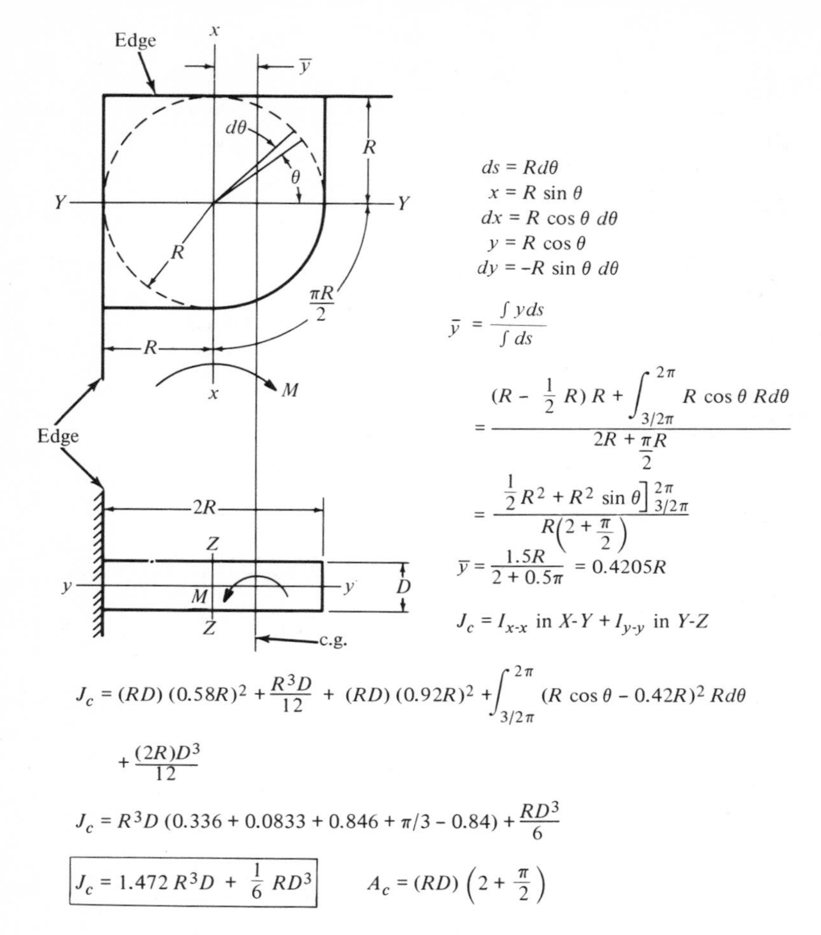

$$J_c = (RD)(0.58R)^2 + \frac{R^3D}{12} + (RD)(0.92R)^2 + \int_{3/2\pi}^{2\pi} (R \cos \theta - 0.42R)^2 \, R d\theta$$

$$+ \frac{(2R)D^3}{12}$$

$$J_c = R^3D\,(0.336 + 0.0833 + 0.846 + \pi/3 - 0.84) + \frac{RD^3}{6}$$

$$\boxed{J_c = 1.472\,R^3D + \frac{1}{6}\,RD^3} \qquad A_c = (RD)\left(2 + \frac{\pi}{2}\right)$$

Fig. 6–24 Shear constants—derivation for round corner columns.

For irregular column shapes, such as octagonal, hexagonal, etc., values for the shear constants can be interpolated between those for round and square shapes. Quick estimates, suitable for preliminary designs, can be similarly made for L- or T-shaped columns.

Preliminary Trial Proportions

For any practical application of flat slabs to be designed in conformance with the Code, the selection of preliminary sizes as near to final sizes as

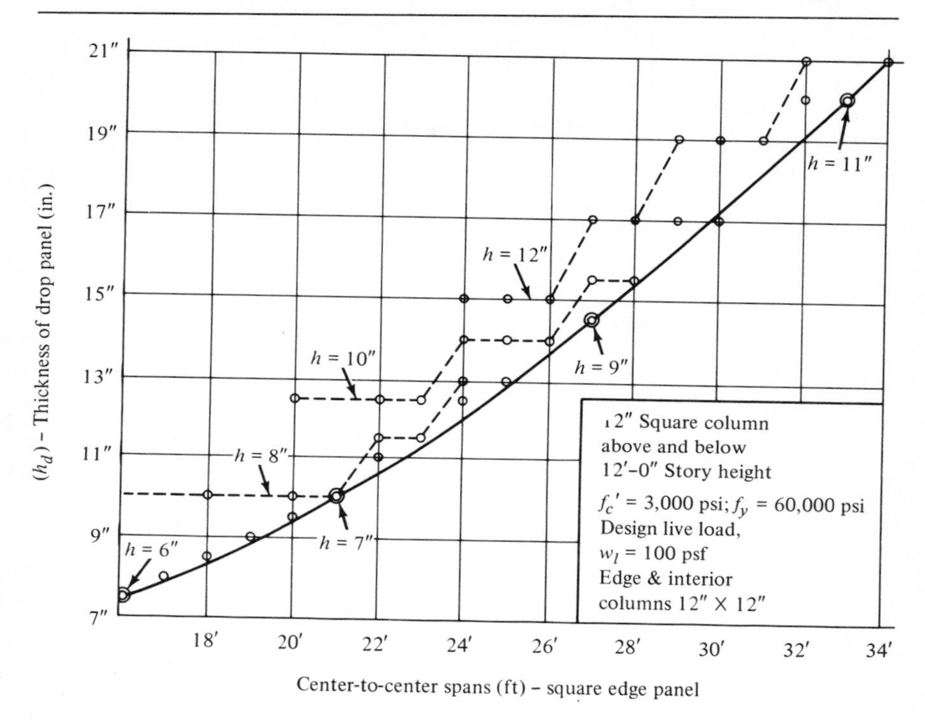

Fig. 6–25 Required drop panel thickness for 12 in. square columns.

possible is vital. Since the Code design procedures are essentially reviews, not necessarily convergent, a poor initial selection may require several adjustments each of which requires a complete review—only to prove inadequate or overdesigned.

Minimum values for the drop panel thickness (slab depth plus the projection of 25 to 100 percent of the slab depth) from a number of computer studies are shown in Figs. 6-25, -26, -27, -28, and -29 for *design* live loads of 100, 200, 300, and 400 psf. Column sizes were established as a minimum of 12 in. square, considered by the authors to be a minimum for loads sufficiently heavy and spans sufficiently long to require the flat slab with drop panel instead of a flat plate. As load and span increase, the column size was increased up to a maximum of 1/12 span, still using the minimum depth of the drop panel, 1.25 slab thickness. For heavier loading-longer span combinations, the drop panel depth was increased to a maximum of twice the slab thickness. When shear in the slab at the edge

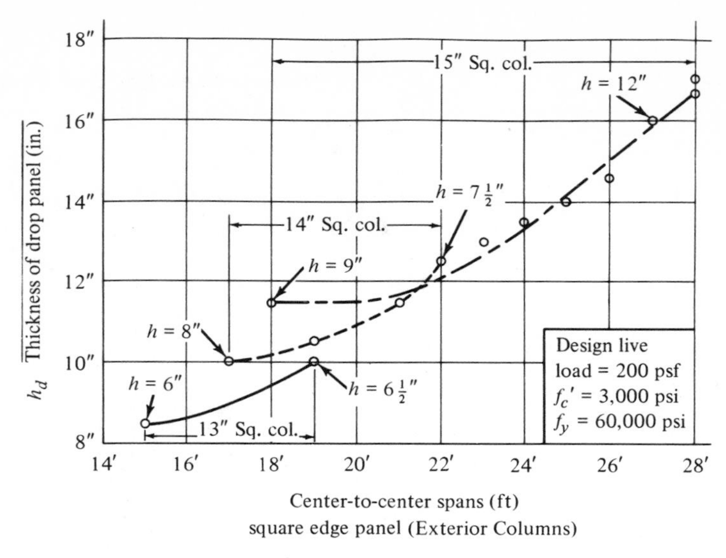

Fig. 6–26 Required drop panel thickness for minimum size square columns, exterior.

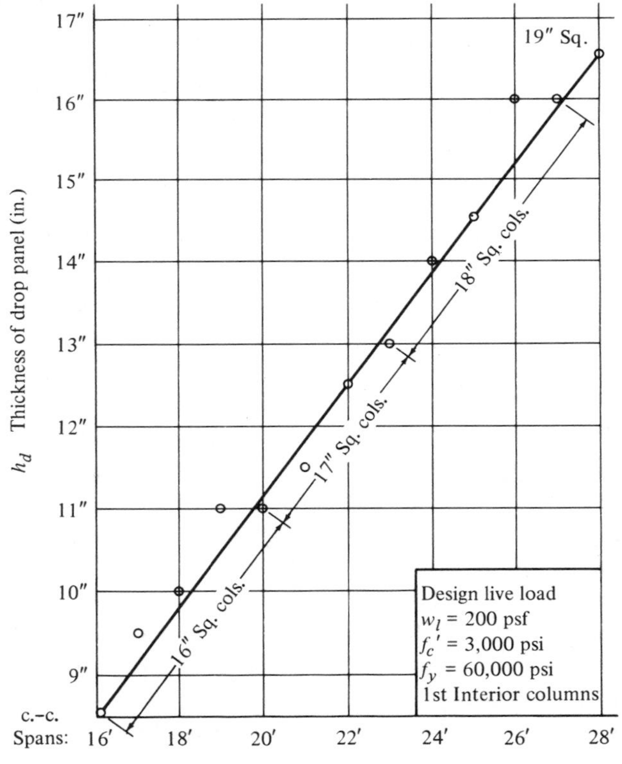

Fig. 6–27 $W_L = 200$ psf drop panel thickness for minimum size square columns.

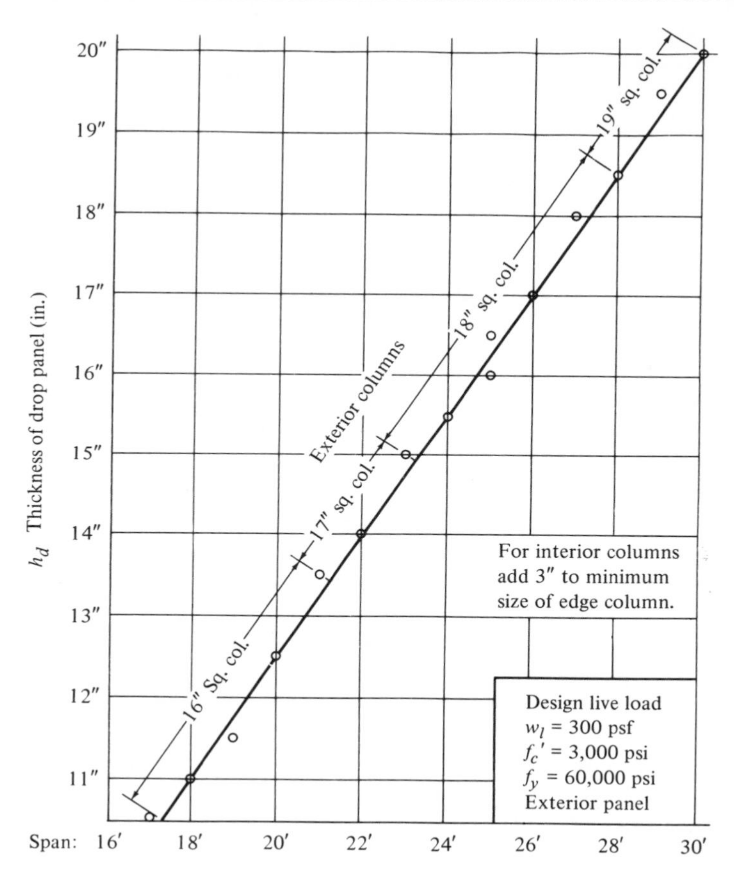

Fig. 6–28 W_L = 300 psf drop panel thickness for minimum size square columns.

of the drop panel became critical, the drop panel size was increased from the minimum of 0.33 span to 0.40 span. The preliminary trial proportion charts were prepared for square exterior panels, since the first interior column is more heavily loaded than other interior columns with all panels equal in size. The minimum column sizes shown are for flush edge exterior columns with no edge beams and accompanying minimum first interior columns. Story height was 12 ft with the same size columns assumed above and below the slab. All charts are for 3,000 psi concrete strength of concrete weighing 150 pcf and Grade 60 rebars. Spans shown are center-to-center of columns. The symbols used are h_D = drop panel thickness; h = slab thickness; c_1 and c_2 = square column cross-section dimensions.

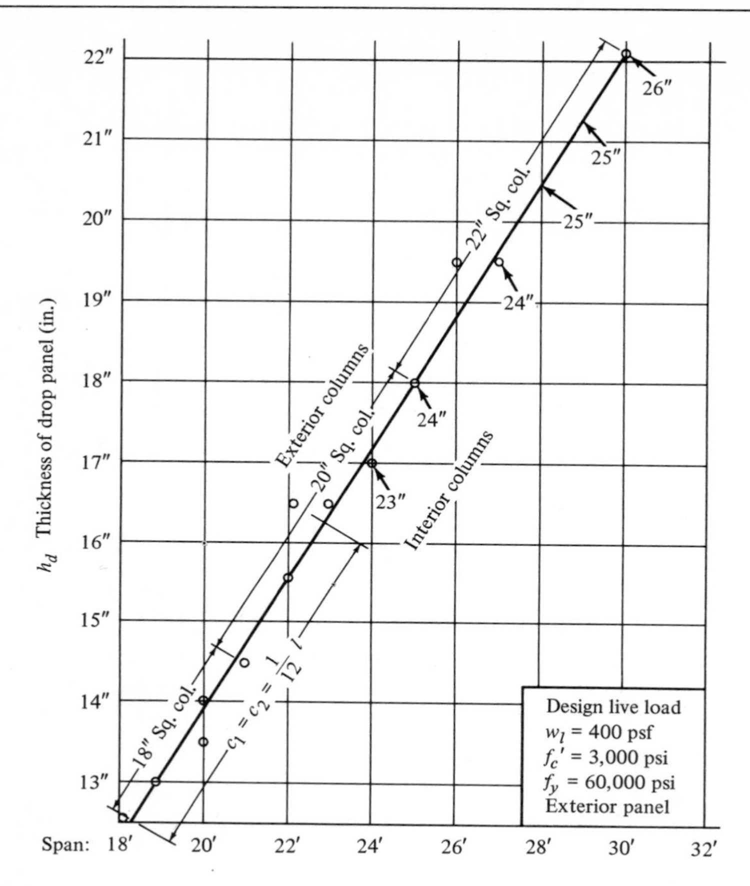

Fig. 6–29 W_L = 400 psf drop panel thickness for minimum size square columns.

7 two-way (waffle) flat slab design

GENERAL

A "waffle" flat slab is a two-way joist system. Under the general provisions of the 1971 Code, Chapter 13, the two-way joist portion may be combined with a solid column head or with solid wide beam sections on the column centerlines for uniform depth construction. These are the usual applications, favored for economy of formwork and the unobstructed level ceiling with a regular pattern of void spaces utilizable for lighting or ventilating fixtures. The arrangement of two-way joists with a solid head results in a slab stiffness distribution similar to that for a solid slab with a drop panel. The use of solid "beam" sections on the column centerlines also provides a solid full depth for shear transfer to the columns and permits concentration of moment reinforcement at the column. Other common variations are cantilevered edges or use of spandrel beams deeper than the uniform depth elsewhere.

By far the largest amount of waffle flat slab construction employs standard size reusable forms for economy. Standard form sizes, conforming to the Code limitations defining joist construction (Section 8.8.2), provide for ribs at least 4 in. wide, spaced not more than 30 in. clear, and for a depth not more than $3\frac{1}{2}$ times minimum width. Within these limitations it has been feasible to establish industry-wide standard sizes for reusable square void forms,* 19 × 19 and 30 × 30 in., in 6, 8, 10, and 12 in. depths and 8, 10, 12, 14, 16, and 20 in. depths, respectively. For nonstandard sizes, within the limitations of the Code or exceeding these limits, custom-made forms are required. The code permits use of 10 percent higher allowable shear capacity of concrete for joists within the established dimensional limits (Section 8.8.8). Waffle slab construction outside these limits is designed as multiple two-way slabs and beams.

The principal complications to two-way slab design introduced by the waffle slabs are the additional computations required for stiffness of the

* U.S. Department of Commerce, National Bureau of Standards, *Simplified Practice Recommendation.*"

solid and ribbed areas, and the additional critical shear sections in the ribs. These complications will be illustrated in the design examples following.

DIRECT DESIGN

EXAMPLE 1. *Waffle slab with solid head—square interior panel*

Spans $l_1 = l_2 = 39'\text{–}0''$ use standard dome void forms

Design live load 180 psf in a 36″ square module

30″ × 30″ × 20″ forms for 6″ ribs @ 36″; depth 20″ + 3″ top slab

Solid head (or drop panel) $D = 5 \times 36''$ module + 6″ outside rib = 15′–6″ square

For this example, an interior panel, the distribution of the panel moments under the direct design method is not made dependent upon the relative stiffness of the columns and slab unless the sum of the column stiffnesses above and below the slab are such that $\alpha_c < \alpha_{\min}$ (Section 13.3.6.1-a) (see Fig. 5-3, Chapter 5). $\alpha_1 = \alpha_2 = 0$ (Section 13.0); $l_1 = l_2$; $\alpha_1\ l_1/l_2 = 0$; $\alpha_2\ l_1/l_2 = 0$.

Panel moments (Sections 13.3.3.2, 13.3.4.1, and 13.3.4.3):

Column strip $-M = -(0.65)(0.75)\ M_o = -0.488\ M_o$

(6 ribs) $+M = +(0.35)(0.60)\ M_o = +0.210\ M_o$

Middle strip $-M = -(0.65)(0.25)\ M_o = -0.162\ M_o$

(7 ribs) $+M = +(0.35)(0.40)\ M_o = +0.140\ M_o$

Critical sections for shear:

1. Two-way shear at columns ("punching shear") (Sections 11.10.1-b, 11.10.2, and 11.10.3). Allowable shear on concrete, $v_c = 4\sqrt{f_c'}$, on the square section $(c + d)$ in the solid head, where d = average effective depth in the solid head.
2. One-way shear in the 24 joists (ribs) at a distance d from the edge of the solid head (Sections 11.2.2 and 8.8.8) $v_c = 2.2\sqrt{f_c'}$. Note that the code permits an increase in the allowable one-way shear carried by the concrete (Section 11.4.2) up to $3.5\sqrt{f_c'}$ upon a "detailed analysis" with allowable shear computed as

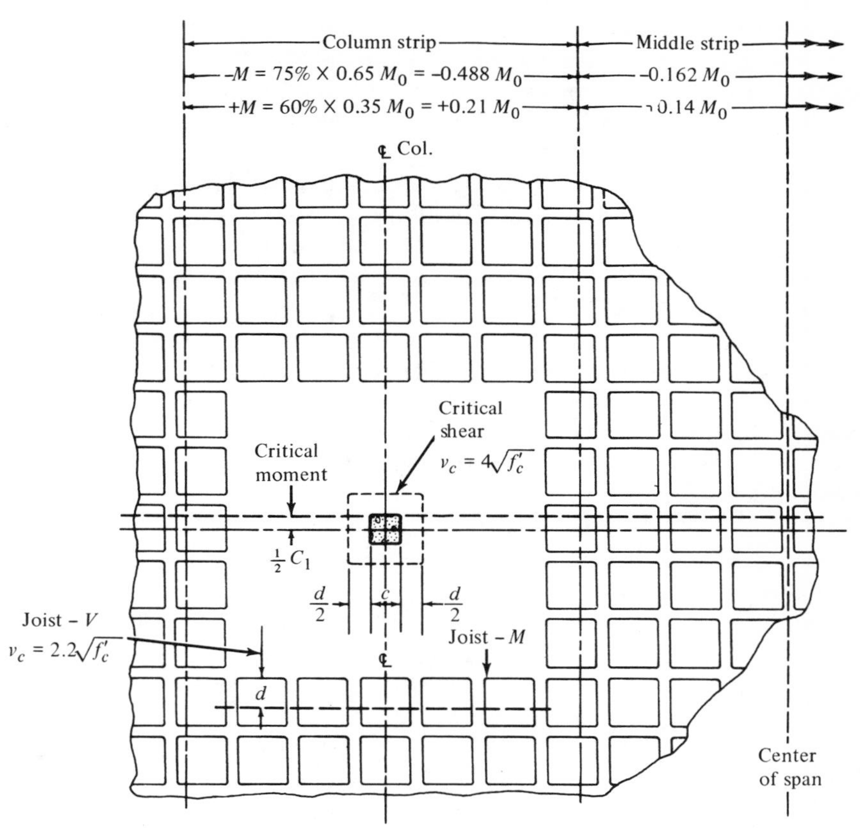

Fig. 7–1 Layout, panel moments, and critical sections—waffle slab with solid head.

$$v_c = 1.1(1.9\sqrt{f_c'} + 2500\ \rho_w V_u d/M_u)$$

The edge of the solid head will be at approximately one-sixth of the span, the applied moment, M_u, at this point will be negative but quite low, and the steel percentage, ρ_w, will be relatively high, so that the allowable shear on this critical section by the detailed analysis will usually be higher than the nominal value $2.2\sqrt{f_c'}$. For economy, stirrups in any type of joist ribs are to be avoided wherever possible, and so the authors advise use of the "long" formula here to determine the maximum shear capacity of the concrete without stirrups.

Critical sections for moment

1. The critical section for negative moment of the panel is at the face of the column as for flat plates (Section 13.3.3.1). If the required ratio of negative reinforcement in the joist ribs of the middle strip is excessive (Section 10.3.2) the bottom bars can be extended to serve as compressive reinforcement (Section 10.3.5). Practice recommended by the industry (CRSI) is to extend at least one of the two bottom bars in the ribs to overlap on the column centerlines in all full-length joists. This practice avoids an abrupt stress concentration resulting from cutting all bars at the column centerline and possible cracking at this point.

2. Column strip joists develop a critical section for negative moment at the face of the solid head. Again the bottom bars may be considered effective for compression since the code requires all straight bottom bars to extend 24 bar diameters or 12 in. minimum into the solid head (drop) (Section 13.5.6). This embedment is usually sufficient for compression development with most concrete strengths used. See Table 13-1 for compression development lengths.

EXAMPLE 2. *Waffle slab with solid beam sections on column centerlines —square interior panel*

Use the same dimensions as in previous example for comparison.

$l_1 = l_2 = 39'\text{–}0''$ 36″ module; 6″ ribs @ 36″; 20″ + 3″ depth

$w_l = 180$ psf Beams and joist-slab are monolithic, and so $E_{cB} = E_{cs}$

Omit one line of the dome forms on the column centerlines to form a solid beam section 42 in. wide, plus overhanging flanges of $4h_f = 12$ in. on each side (Section 13.1.5) (see Fig. 7-2).

Code Interpretations

The problems of determining the moments for this particular application are aggravated by incomplete, interdependent provisions in the Code, some ambiguous and some apparently conflicting. The location of critical moment sections and the determination of clear spans for the calculation of moments at these critical sections can be interpreted differently. The authors' interpretations are shown in Fig. 7-3.

"Supports" may be walls, columns, or beams (Section 13.1.6). Clear span, l_n, is measured between faces of supports (Section 13.0), or clear span, l_n, extends from the faces of columns, capitals, brackets, or walls (*note:* no beams) (Section 13.3.2.2). The width of a T-beam may include

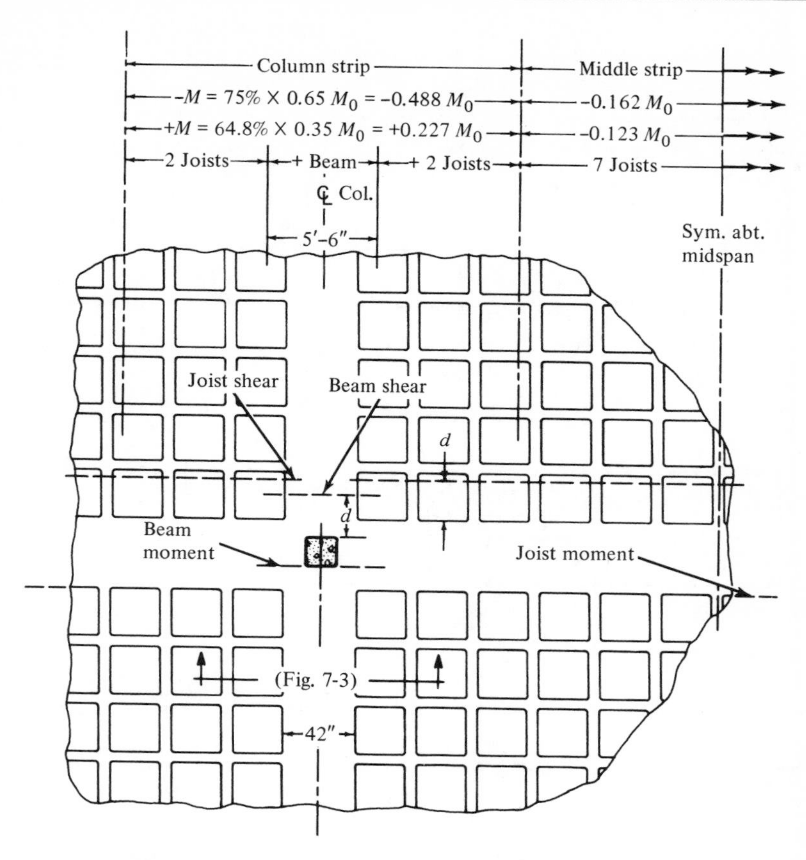

Fig. 7–2 Layout, panel moments, and critical sections—waffle slab with two-way beams.

a portion of the slab up to 4 h in width on each side of the beam (Section 13.1.5) (*note:* purpose is not stated; assume for assignment of direct loads and for calculation of moment resistance). The width of a T-beam (for torsion) can include an overhanging flange width not to exceed 3 h (Section 11.7.2).

The authors' interpretations of the above Code provisions are as follows:

1. The columns support the beams; clear span, l_n, for beams is the distance between faces of the columns (= 37′ –0″).

2. The beams support the "slab system," which in this case is the two-way joists and top slab. Clear span, l_n, for the joists is the distance between the solid faces of the beams (= 35′ –6″). Note that the authors here con-

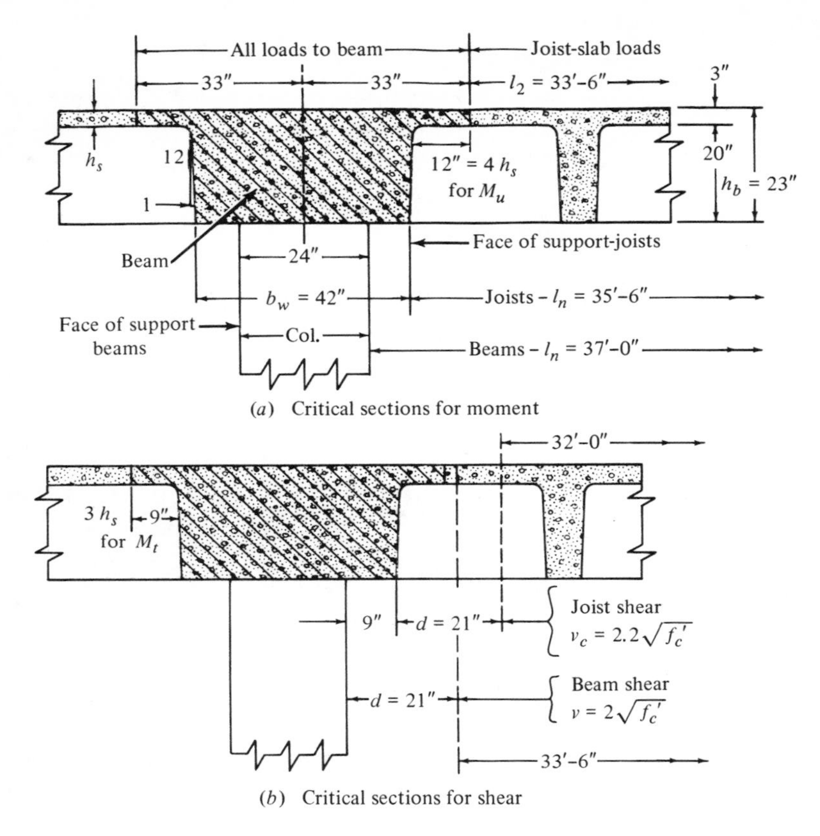

Fig. 7–3 Cross-section dimensions and critical sections.

servatively choose to disregard the overhanging flanges and use b_w as the beam width.

3. The "slab system" loads to the beams (M_o) are transmitted by the joist-slab portion having a transverse width between beam flanges (= 33′ –6″), and a clear span, l_n = 35′ –6″, which is loaded in the center 33′ –6″ portion.

Location of the critical shear sections and the distribution of loads causing the critical shear thereon are equally complex problems, particularly since they do not coincide with the critical sections for moment nor, therefore, the loads for moment (see Fig. 7-3). The author's interpretations are as follows:

1. Joists behave as beams in shear and so the one-way shear provisions

apply. Critical shear sections are at a distance, d, from the supports of the joists (joist d from the face of the beam webs). (Even though the joists are part of a two-way system and one Code provision (Section 13.2.2.), taken literally, would indicate that the face of the support for the joists is a line one-half c_2 from the column centerline, it does not seem realistic to accept this interpretation because the resulting "critical section" for joist shear could then lie inside a wide beam supported by slender columns.)

2. Beams behave in one-way shear since the solid area about the column is less than $d/2$ from its face. Critical shear sections in the beams, therefore, are at a distance, d, from the face of the beam supports (face of the columns).

3. Beams are to be designed for the loads applied directly upon them (Section 13.3.4.4). For this purpose the authors chose to use the full width of the beams, including the flanges ($4h$), consistent with a transverse width of 33′ –6″ for loads upon the joist-slab system. For convenience in the calculations, the authors recommend applying *all* uniform loads on the beam surface to the beam and using the same division into positive and negative moments as prescribed for the loads considered in the slab design —0.35 and 0.65, respectively.

PANEL MOMENTS. In order to determine the effect of the relative stiffness of the beams on the distribution of the panel moments, numerical values must be used. See Fig. 7-3 for cross-sectional dimensions. Cross-sectional properties:

$$I_s = 157{,}400 \text{ in.}^4 \text{ (for 13 joists @ 36 in., gross section)}$$

$$I_b = 50{,}000 \text{ in.}^4 \text{ (for 66 in.} \times 23 \text{ in. beam section)}$$

$$\alpha_1 = \alpha_2 = \frac{E_{cb}I_b}{E_{cs}I_s} = \frac{I_b}{I_s} = \frac{50{,}000}{157{,}400} = 0.318 \quad \text{(Section 13.0)}$$

$$\alpha_1 l_2/l_1 = \alpha_2 l_1/l_2 = (1.0)(0.318) = 0.318$$

Using the above constants, the moments are to be distributed among column strips, middle strips, and beams as prescribed by the Code (Sections 13.3.4.1, 13.3.4.3, and 13.3.4.4).

The Code prescribes interior panel negative and positive moments as 0.65 M_o and 0.35 M_o, respectively. The distribution to column and middle strips is prescribed for l_2/l_1 ratios of 0.5 and 2.0 and for $\alpha_1 l_2 l_1$ ratios of 0 and 1.0 or more. Linear interpolation between each of the two sets of parameters is required.

TABLE 7-1 Percentages of the Total Panel Moments to the Column Strip (Sections 13.3.4.1, 13.3.4.3)

RATIO	NEGATIVE MOMENT (%) (0.65 M_o)				POSITIVE MOMENT (%) (0.35 M_o)			
ℓ_2/ℓ_1	0.5	1.0	1.5	2.0	0.5	1.0	1.5	2.0
$\alpha_1 \ell_2/\ell_1 = 0$	75	75	75	75	60	60	60	60
$\alpha_1 \ell_2/\ell_1 \geqq 1.0$	90	75	60	45	90	75	60	45

Column strip and beam: $-M_u = -(0.65)(0.75)M_o = -0.488\ M_o$

Middle strip: $-M_u = -(0.65 - 0.488)M_o = -0.162\ M_o$

Column strip and beam: $+M_u = +(0.35)[0.60 + (0.75 - 0.60)(0.318)]M_o$

$= +(0.35)(0.648)M_o$

$= +0.227\ M_o$

Middle strip: $+M_u = +(0.35 - 0.227)M_o = +0.123\ M_o$

Since the ratios $\alpha_1 l_2/l_1 < 1.0$, the proportion of the column strip moment assigned to the beams must be determined by interpolation between 0 percent (for ratio = 0) to 85 per cent (for ratio = 1.0).

Moments to beam: $= (0.318)(0.85) \times$ column strip moments

$= 0.270 \times$ column strip moments

$-M_u = -0.488\ M_o(0.270)$
$= -0.1317\ M_o$

$+M_u = +0.227\ M_o(0.270)$
$= +0.0613\ M_o$

Column strip outside beam: $-M_u = (-0.488 + 0.1317)M_o$
$= -0.356\ M_o$

$+M_u = (+0.227 - 0.0613)M_o$
$= +0.1657\ M_o$

BEAM DESIGN.

Shear critical sections

Shear at a concentrated load or reaction is governed by the more severe of two conditions: (1) one-way (beam) shear at a critical section, d, from the face of the support, or (2) two-way (punching) shear at a critical section, $d/2$, from support (Section 11.10.1). Allowable two-way shear, $v_c = 4\sqrt{f_c'}$, is twice the one-way shear allowed by the "short" formula (Section 11.4.1). For the dimensions of beam, column, and slab depth used in this example (Fig. 7-3), the two-way shear critical section happens to be located practically at the face of the beam sections: $c_1 + (2)(d/2) = 45$ in. (see Fig. 7-4).

For two-way shear, the void areas must be regarded as partial depth holes which reduce the effective section falling between the radial lines to the center of the reaction (Section 11.12-a). In this case, the reduced section effective for two-way shear is approximately half the beam widths effective for one-way shear. If wider beam sections had been used, or smaller columns, two sets of shear calculations would have been required. Here, however, due to the essentially one-way behavior of these beams, one-way shear seems more appropriate and sufficiently conservative if the short formula, $v_c = 2\sqrt{f_c'}$, is used (Section 11.4.1).

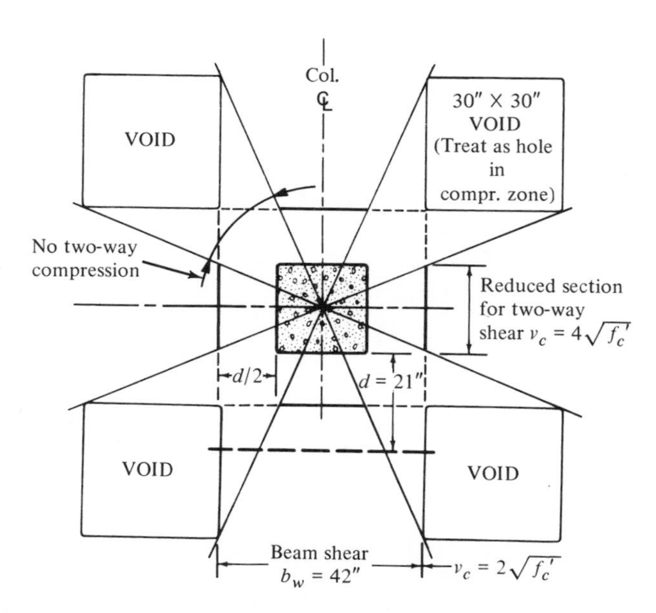

Fig. 7–4 Interior column shear sections—waffle slab with beams.

SHEAR STRESSES—REINFORCEMENT:

Use $f_c' = 4{,}000$ psi; $f_y = 60{,}000$ psi; concrete weight = 150 pcf.

Total load, $w = (0.135)(1.4) + 0.180 = 0.369$ ksf on the waffle portion.

$$w = [(2)(12)(3) + (42)(23)]\,\frac{0.150}{144}\,(1.4) + (0.180)\,\frac{(66)}{(12)} = 2.5 \text{ klf}$$

directly to the beam (Section 13.3.4.4) (see Fig. 7-5).

$$\text{Beam shear, } V_u = 1/4\,[(0.369)(33.5)^2] + (2.5)(33.5)(1/2)$$

$$= 103.5 + 41.9$$

$$= 145.4 \text{ kips}$$

In locating the critical sections an average value for the effective depth, $d = 21$ in., was estimated and used for convenience. In calculations for the shearing stress and for proportioning stirrups more conservatively, the variable actual value of d is used. Here $d = 20$ in. Average width, b_w (at mid-depth) is used.

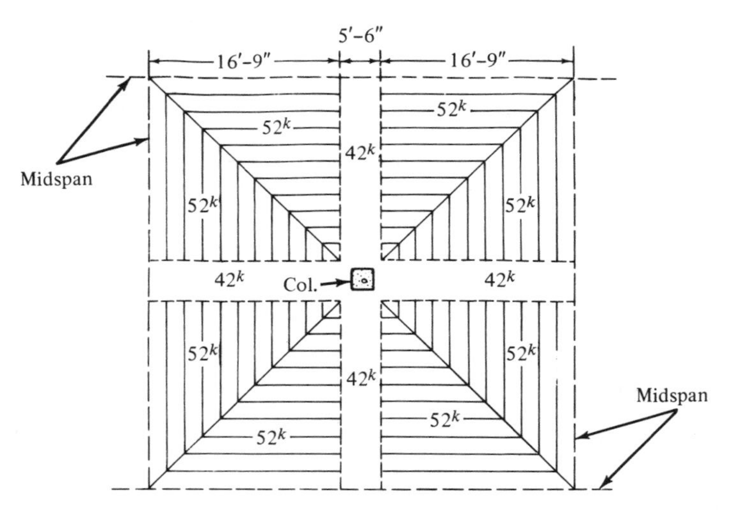

Fig. 7–5 Tributary areas and panel shears—waffle slab with beams.

$$v_u = \frac{145{,}400}{(0.85)(43.7)(20)} = 196 \text{ psi} > v_c = 2\sqrt{f_c'} = 126 \text{ psi}$$

Minimum shear reinforcement is required where $v_u > 1/2\ v_c$ in beams with a total depth greater than 10 in., 2 1/2 times the flange thickness (3 in.), or half the web thickness (43.7 in. average) (Section 11.1.1). In this case the beam depth is greater than the first two limits, either one of which would govern. Shear reinforcement is, therefore, required for $(v_u - v_c) = 196 - 126 = 70$ psi at the critical section, and minimum stirrups must be continued to the point where $v_u = 1/2\ v_c = 63$ psi.

$$\text{For vertical stirrups: } A_v = \frac{(70)(43.7)(10)}{60{,}000} = 0.51 \text{ in.}^2$$

(Section 11.6.1)

$$\text{Minimum stirrups:} \quad A_v = (50)(b_w)(s)/f_y = (50)(43.7)(10)/60{,}000$$

$$= 0.364 \text{ in.}^2 \quad \text{(Section 11.1.2).}$$

For both calculations, a spacing, $s = 10$ in. was used, as the maximum spacing allowed is $1/2\ d$ (Section 11.1.4). The minimum stirrups are sufficient for $v_u - v_c = 50$ psi. The point where $v_u = 50 + 126 = 176$ psi is approximately $\frac{196 - 176}{196} \times \frac{33.5 \times 12}{2} = 20$ in. from the critical section. Beyond this point minimum stirrups will be required to approximately $\frac{196 - 63}{196} \times \frac{33.5 \times 12}{2} = 136$ in. from the critical section. The critical section is 21 in. from the face of the column, approximately 12 in. from the face of the intersecting beam (see Fig. 7-6). Stirrups for $v_u = 196$ psi

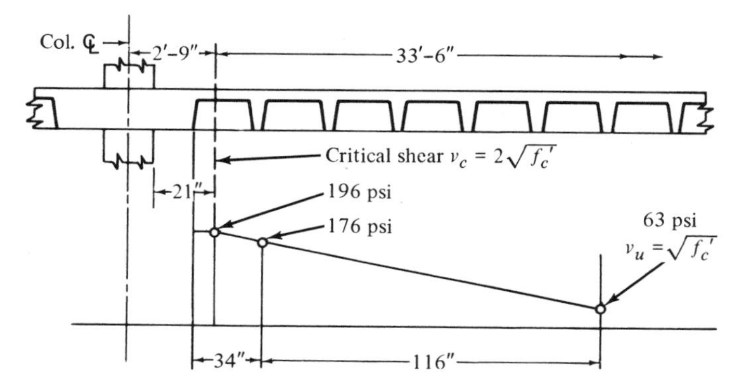

Fig. 7–6 Shear diagram beams.

will be used for the distance 12 in. + 21 in. beginning 1 in. from the face of the intersecting beam, with minimum stirrups for the remaining distance.

Use: 4–#4 ⌉⊔⊓⊔⌈ @ 10″

12–#3 ⌉⊔⊓⊔⌈ @ 10″

at each end of each beam.

TORSION. To complete this example, an investigation for the effect of torsion in the beams is required. Under balanced loadings or equal loadings in adjacent panels of the same size, the negative moments from any panel under consideration are balanced by identical negative moments from the adjacent panels. Columns then receive no moment from the floor system. When negative moments are unequal due to unequal loads or spans or both, the unbalanced difference between slab moments in adjacent panels is resisted by the columns. This unbalanced moment is transferred from the slab system to the columns principally by direct flexure from the slab system (in this case, from the beams) parallel to the direction of the unbalanced moments. The remainder of this unbalanced moment is transferred to the column by torsion between the slab system (in this example, the beams at right angles to the direction of the unbalanced moment) and the sides of the column parallel to the moment. With equal spans and loads as in this example, half of the remaining moment is transferred by torsion on each side of the column.

The Code prescribes an unbalanced moment, M_{un}, to columns for use in the direct design method (Section 13.3.5.2, Eq. 13-3). For flat plates, flat slabs, and waffle flat slabs without beams, the Code also prescribes the percentage of M_{un} which is assumed to be transferred by shear (Section 11.13.2). For two-way systems with beams, as in this example, the Code is less explicit and it becomes necessary to compute this percentage. An approximate procedure illustrated here consists of two steps:

(1) Compute M_{un} (Section 13.3.5.2, Eq. 13-3)

(2) Distribute the unbalanced moment, M_{un}, to the beam and panel joists on each side of the column, in accordance with Section 13.3.4.

The unbalanced moment resisted by the joists on each side of the column creates torsion within the beam supporting the joists, which is equal to the torsional moment, T_u. The torsional moment is assumed to be zero at the midspan and a maximum at the critical section located at a distance, d, from the face of the support (Section 11.7.4).

TORSION STRESS. To calculate the unbalanced moment transferred to an interior column, it will be necessary to determine the stiffness ratio,

α_{ec}, (Section 13.3.5.2) (see Fig. 5–4). Assume the story height, $l_c = 180$ in.; $h/l_c = 23/180 = 0.128$; $k_c = 1.41$. Compute I_c and K_c/E.

$$I_c = (24)(24)^3/12 = 27{,}700 \text{ in.}^4$$

$$K_c/E = (1.41)(4)(27{,}700)/180 = 868 \text{ in.}^3$$

Column sides, $c_1 = c_2 = 2$ ft; $l_1/l_2 = 39'\text{–}0''$; $c_1/l_1 = 0.05$ (see Fig. 5-5). $k_{AB} = 4.06$; compute K_s/E for the eleven joists plus one beam.

$$(K_s + K_b)/E = \frac{1}{(39)(12)}(4.06)\left[\frac{(11)}{13}(157{,}400) + (50{,}000)\right] = 1{,}589 \text{ in.}^3$$

Compute shear constants. $x = 23''$; $y = 42''$ (Section 13.4.1.5-c).

$$C = \left[1 - \frac{(0.63)(23)}{42}\right]\frac{(23)^3(42)}{3} = 111{,}600$$

$$K_t/E = \frac{(2)(9)(111{,}600)}{(39)(12)(1 - 0.05)^3} = 4{,}770 \quad \text{(Code Eq. 13-6)}$$

$$\frac{E}{K_{ec}} = \frac{E}{\Sigma K_c} + \frac{E}{K_t} \quad \text{(Section 13.4.1.5, Eq. 13-5)}$$

$$\frac{1}{K_{ec}} = \frac{1}{(2)(868)} + \frac{1}{4{,}980} \qquad K_{ec} = 1{,}285$$

$$\alpha_{ec} = \frac{K_{ec}}{(K_s + K_b)} = \frac{1{,}285}{(2)(1{,}589)} = 0.401 \quad \text{(Section 13.0)}$$

$$\alpha_c = \frac{\Sigma K_c}{\Sigma(K_s + K_b)} = \frac{(2)(868)}{(2)(1{,}589)} = 0.546 \quad \text{(Section 13.0)}$$

The ratio of the flexural stiffness of the columns to the slab, α_c, was determined at the same time for convenience. It is convenient to ensure that no adjustment is required in the positive moments before accepting column size selected as final. The minimum value of flexural stiffness ratio, α_c, to avoid increases in the positive moment is α_{min}. Interpolating for $\alpha_1 = \alpha_2 = 0.318$, and live/dead load ratio $= \frac{(180)/1.7}{135} = 0.784$, $\alpha_{min} = 0.370$ (Section 13.3.6.1), $\alpha_c = 0.546 > \alpha_{min} = 0.370$, and so no adjustment to the positive moment is required. See Fig. 5-3.

The unbalanced moment which must be transferred to the interior columns (Section 13.3.5.2) is

$$M_{un} = \frac{(0.08)(0.5)(0.180)(39)(37)^2}{\left(1 + \frac{1}{0.401}\right)} = 110 \text{ ft-kips}$$

The total unbalanced moment is transferred between the column and the floor by the beam in flexure and the joists on each side of the column to beams at right angles.

The total panel negative moment is 0.65 times the total static design moment equals (0.65) (M_o in the joists + $\mathbf{M}_o$ in the beams).

Joists, $M_o = (0.369)(33.5)(35.5)^2/8 = 1{,}947$ ft-kips

Beam, $M_o = (2)(41.9)(35.5)/8 = 372$ ft-kips due to direct loads on the beam (see Fig. 7-5).

The total $M_o = 1947 + 372 = 2{,}319$ ft-kips. The total negative panel moment, $-M_u = (0.65)(2{,}319) = 1{,}505$ ft-kips. The beam resists: (0.1317) (1947) + (0.65)(372) = 498 ft-kips. One-half of the remaining negative moment is resisted by the panel joists on each side of the column. The portion of the negative moment resisted by the joists on one side of the column is $\frac{(\frac{1}{2})(1505 - 498)}{1505} = 0.332$.

The portion of the total unbalanced moment to be transferred by torsion by the beams at right angles to the direction in which moments are being considered is in the same proportion.

$$T_u = (0.332)\ M_{un} = (0.332)(110) = 36.5 \text{ ft-kips}$$

$$\text{Torsional stress, } v_{tu} = \frac{(3)(36{,}500)(12)}{(0.85)(42)(23)^2} = 70 \text{ psi} < 1.5\sqrt{f_c'} = 95 \text{ psi}$$

and so the torsional effects may be neglected (Sections 11.7.2 and 11.7.1).

MOMENT REINFORCEMENT FOR BEAMS. Moment reinforcement for the beams can be selected as illustrated in Chapter 9. It should be noted, however, that the moment due to the loads applied directly to the beams must be added to the percentage of the two-way waffle slab moment resisted by the beams (Section 13.3.4.4). The total panel moment is distributed as follows :

Beams (Directly applied loads, 100%): $-0.1317\ M_o$; $+\ 0.0613\ M_o$

Middle strip joists (7 joists): $-0.162\ M_o$; $+\ 0.123\ M_o$

Column strip joists (4 joists): $-0.356\ M_o$; $+\ 0.1657\ M_o$

MOMENTS IN JOISTS.

$$M_o = (0.369)\tfrac{1}{2}(33.5)^2(\tfrac{1}{2})(39) - (0.369)\tfrac{1}{2}(33.5)^2(\tfrac{1}{4})(33.5)$$

$$= 2{,}310'k \text{ (see Fig. 7-7).}$$

Thus, $M_o = 2{,}310\ 'k$ is based upon a center-to-center span = 39′ –0″ between centers of supports (beams) (Section 13.1.6). This moment may be reduced to that at the face of the supporting beams taken as the critical section (Section 8.5.2.2) for moment in the individual joists. This reduction in negative moment becomes $-M_u = -M_{℄} + (0.369)(33.5)\tfrac{1}{2}(33.5)$ k (1.75) ft = $(-M_{℄}) + (364\ 'k)$. (Approximately the same result would be developed by use of Eq. 13-2, using $l_2 = 33.5'$ and $l_n = 35.5'$ in this example, where width of the supporting beams $\approx 1/10l_1$, but for wider beams the procedure in Fig. 7-7 seems more appropriate.) The reduction in the negative moment only is justified since the quantity $(d - \tfrac{1}{2}\,a)$ increases for negative moment bars passing from the joist to the solid beam section.

SHEAR IN BEAMS. The design of the two-way beams in this example follows the detailed procedures in Chapter 9 of this guide. The location of the critical section for one-way shear at $d = 21$ in. from the face of the column is shown in Fig. 7-2. The calculation of the total external shear force at this point is easily visualized as one-fourth of the sum of the total panel load minus the loads inside the square formed by the four critical shear sections.

One Code provision peculiar to two-way beams with two-way slabs is that the total external shear must be apportioned to these beams and slabs in the ratio of their relative stiffnesses when the quantity $(\alpha_1 l_2/l_1)$ is less than 1.0 (Section 13.3.4.7). In our example, with square panels $l_2/l_1 = 1.0$, and with the beams used, $\alpha_1 = \alpha_2 = 0.318$. The Code provides that 100 percent of the total shear due to loads in the tributary areas plus that directly applied to the beam (see Fig. 7-5) be assigned to the beams for values of $\alpha_1 l_2 l_1 \geq 1.0$ and that for values such that $0 < \alpha_1 < 1.0$, the percentage be determined by linear interpolation between zero and 100 percent (Section 13.3.4.7).

SHEAR IN JOISTS. Under the conditions of this example, the beams are wider than the columns and have an equal stiffness in both directions

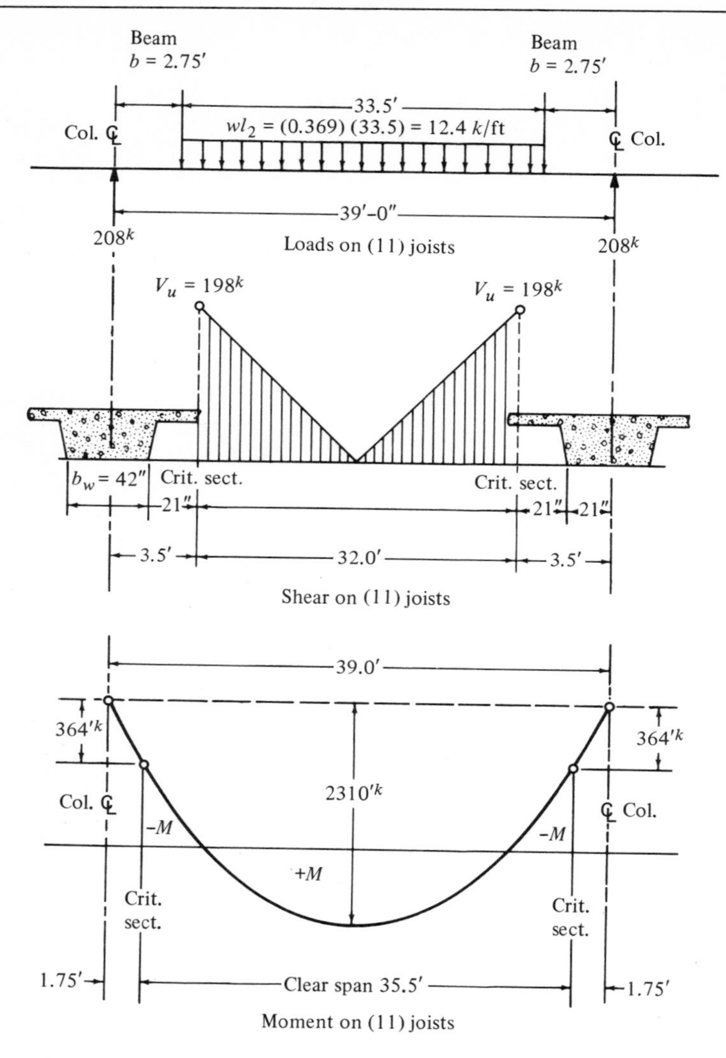

Fig. 7–7 Loads and critical sections for design of joists.

so that they obviously carry equal loads. The voids in the waffle system area become *holes* (in the compression flange) and both the beams and the joists behave as one-way members in shear. No two-way shear investigation is required. Beam shear is one-fourth of the loads outside the critical shear section. The proportion of the one-way shear assigned to the *joists,* however, can be reduced to (1.000 − 0.318) of the total shear on the waffle area (Section 13.3.4.7).

The Code specifically requires that the two-way beams be proportioned to resist the *shear* caused by 100 percent of the directly applied loads (Section 13.3.4.7) whereas, for moment, 100 percent of only those loads applied on the beam and *not considered in the slab design* need be considered (Section 13.3.4.4). A uniform live and dead load over the entire panel could, of course, be considered in the slab design. The practical result of a literal application of these requirements is to *reduce* the load for moment in the beam which would be resisted by fewer larger bars at less expense, and to *increase* the load for shear in the beam (consider stirrups to cost 50¢/lb in place). In our example, these peculiar requirements did not affect the beams since all shear is obviously carried to the column by the beams (see Fig. 7-8). As shown by the following calculations, the joists selected here did not require stirrups under either interpretation, but taking advantage of the reduction of shear to the joists could avoid the need for stirrups in a design where shear is larger. Stirrups in joists are not generally economical.

Total one-way (joist) shear at a constant distance, $d = 21$ in., from the face of the beams can be distributed as follows (see Fig. 7-7):

$$\text{Total } V_u = (\tfrac{1}{2}\,(12.4)[35.5 - (2)(1.75)] + 41.7 = 198 + 41.7 \text{ kips}$$

Using this maximum value, the average shear per joist is

$$V_u = \frac{1}{11}\,(198) = 18.1 \text{ kips per joist}$$

The average width of the joist web for shear, $b_w = 6 + \frac{1}{12}(20) = 7.67$ in.

$$v_u = \frac{18{,}100}{(0.85)(21)(7.67)} = 132 \text{ psi} < 2.2\,\sqrt{f_c'} = 139 \text{ psi} \quad \text{(Section 8.8.8)}$$

No stirrups are required in the joists (Section 11.2.1 and 11.2.2).

To take advantage of the reduction allowed for joists, the total shear could have been calculated for uniform loads at a section 21 in. from the face of the beam stem as

$$V_u = (0.369)(39)\left[\frac{39 - (2)(3.5)}{2}\right] = 230 \text{ kips}$$

plus the added dead load of the beam.

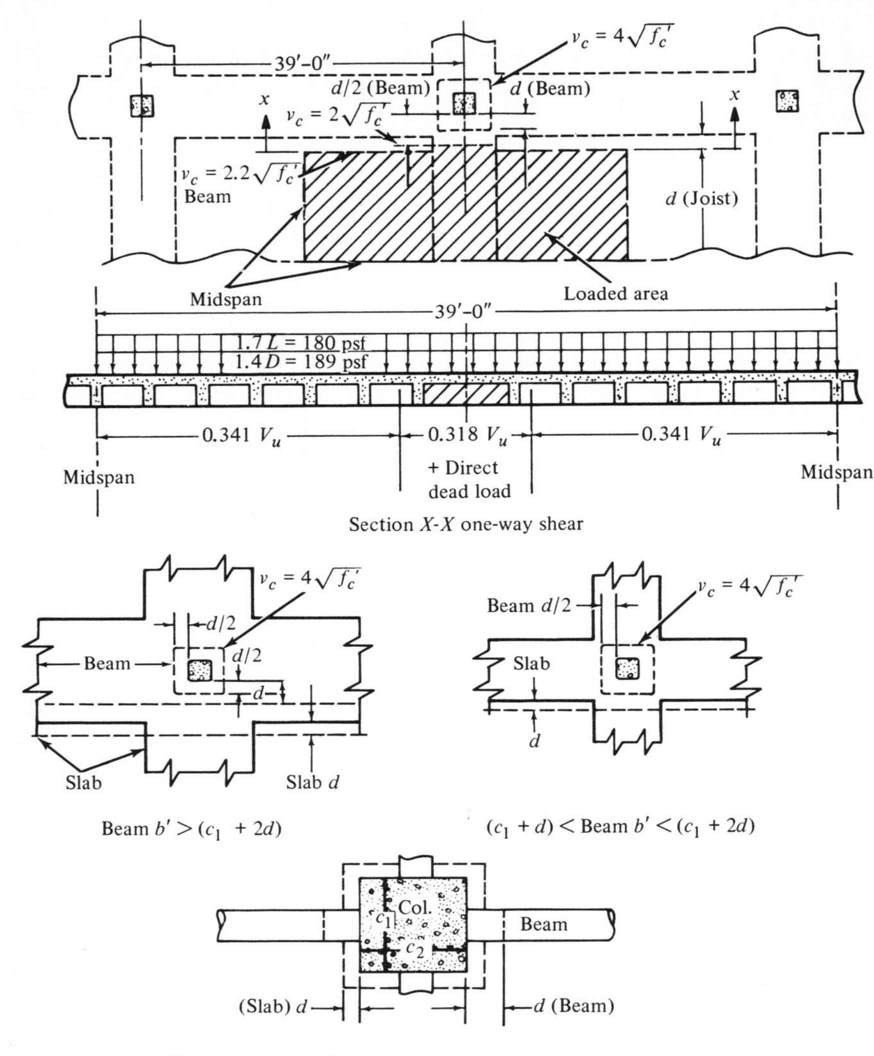

Fig. 7–8 Loads and critical sections for beams.

$$(20)(\tfrac{1}{2})(30 + 26.6)\,\frac{(150)}{144}\,\frac{(32)}{2}\,(1.4) = 13.6 \text{ kips}$$

Shear on the beam: $V_u = 13.6 + (0.318)(230) = 86.7$ kips

Shear on (11) joists: $V_u = (0.682)(230) = 156.8$ kips

To avoid these confusing double calculations for beam loads, one for moment and one for shear, the authors' recommendation is to consider all loads on the entire width of the beam, including the flanges, to be carried by the beam in both shear and moment and to add the different percentages prescribed for shear and moment from the remainder of the panel. Where beams are wider than the critical section for one-way shear ($c_1 + 2d$), investigate the two-way shear in the beam section, $(\frac{1}{2})d$ from the column face and one-way shear across the entire width of the panel at a distance equal to the *slab* depth from the face of the beams at right angles. Where beam width is less than ($c_1 + 2d$) but equal or greater than ($c_1 + d$), investigate one-way shear only and consider the beam section only. Where beams are less than the column width, an exact solution requires an investigation of both one-way and two-way shear at prescribed distances on the combined slab and beam section. Usually, it will be more practical to assign all shear to the beams. Regard void spaces (such as waffles) in the compression flange (bottom) as holes (see Fig. 7-8).

These recommendations will safely account for all shear to the column (Section 13.3.4.8), conform to the shear distribution provisions (Section 13.3.4.7), and be conservative for the moment in the beams and shear in the joists (Section 13.3.4.4). In addition to the practical reduction of design time (and possible error), these procedures tend to locate more steel in the beam in fewer, larger sizes.

MOMENTS FOR DESIGN OF JOISTS

Column strip joists (4): $+M = (0.167)(2310) = 386\ 'k$

$$-M = (0.356)(2310 - 364) = 693\ 'k$$

Middle strip joists (7): $+M = (0.123)(2310) = 284\ 'k$

$$-M = (0.162)(2310 - 364) = 315\ 'k$$

Reinforcement. $f_c' = 4{,}000$ psi; $f_y = 60{,}000$ psi

$$A_s = \frac{M_u}{\phi f_y(d - a/2)} = \frac{M_u}{(0.90)(60)(d - \frac{1}{2}a)} = \frac{M_u}{(54)(d - \frac{1}{2}a)}$$

$$a = (A_s f_y)/(0.85\ f_c' b) = 17.6\ A_s/b$$

Middle strip joists: $+M = 284/7 = 40\ 'k$; $b = 36$ in. $d = 21.1$ in.

$$a = 0.5\ A_s \qquad A_s = \frac{(40)(12)}{(54)(21.1 - 0.3)} = 0.43 \text{ in.}^2$$

Minimum positive moment, $A_s = \dfrac{(200)(6)(21)}{(60{,}000)} = 0.42$ in.^{2} (Section 10.5.1).

Use: 1–5#; 1–#4; straight bottom bars

Top bars: total width, $b_w = (7)(6) = 42.0$ in. for 7 joists

Assume $a = 0.4\ A_s$ $\qquad A_s = \dfrac{(315)(12)}{(54)(21.1 - 0.2)} = 3.34$ in.^{2}

Use: 18–#4 @ 12, straight top bars per middle strip (see Chapter 2, "Crack Control," maximum spacing for interior exposure).

Column strip joists (4), two on each side of the beam section.

$+M_u = 386\ 'k/4 = 96.5\ 'k$ per joist. $\qquad$ Assume $a = 0.5\ A_s$

$$A_s = \frac{(96.5)(12)}{(54)(21.3 - 0.3)} = 1.01 \text{ in.}^2 \text{ per joist rib}$$

Use: 1–#7; 1–#6; straight bottom bars

Top bars: $b_w = (4)(6.0) = 24.0$ in. $\qquad a = (17.6/24.0)\ A_s$
$A_s = 0.7\ A_s$

$$A_s = \frac{(693)(12)}{(54)(21.1 - 0.4)} = 7.4 \text{ in.}^2 \qquad \text{Use 24–\#5}$$

BAR LENGTHS. Two systems of design details for reinforcement of flat slabs are described and specifically permitted (Section 13.5 and Fig. 13.5.6). In order to satisfy maximum spacing limits for control of crack width (Commentary 10.6) the authors selected the system using all straight bars with the top bars uniformly spaced across the strip in which they are required without reference to the location of the joist ribs. See Fig. 7-9 for minimum bar lengths (Code Fig. 13.5.6).

WAFFLE SYSTEM—FLAT PLATE VERSUS TWO-WAY SLAB ON BEAMS. The comparison of design examples (1) two-way waffle flat plate with solid head about column and (2) two-way waffle slab flat plate with two-way solid beam sections is interesting. The first example with solid head designed under the 1963 Code in the CRSI handbook *Floor*

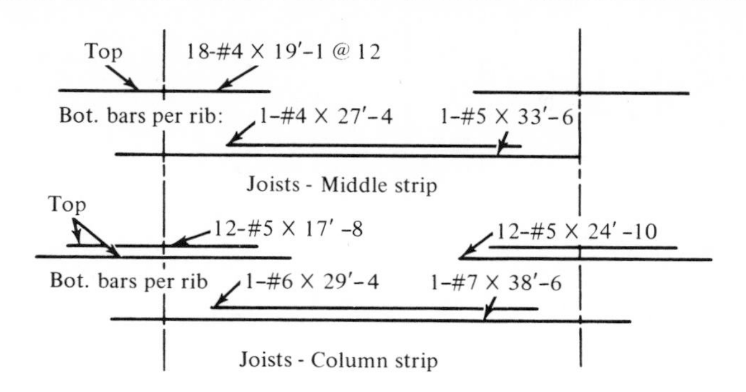

Fig. 7–9 Minimum bar lengths.

Systems, page 13–55, shows 3.04 psf reinforcement. This design utilized one straight and one truss bar per joist, with additional straight top bars added only in the column strips around the column. A column (or column capital) 2.1 ft in diameter was required. Table 7-2 compares reinforcing steel required for the various designs.

TABLE 7-2 Comparisons of Reinforcement Required; 1963 vs. 1971; Flat Plate vs. Two-Way Beams; and All Straight vs. Straight and Trussed Bars

Design	Bars	$C_1 = C_2$ Sq. Col.	% Total Stirrups	Total Steel
1963 Solid head	Truss-str	2.0 ft	0	3.04 psf[a]
1971 Solid head	Truss-str	2.0 ft	0	3.13 psf[b]
1971 Solid beams	Truss-str	2.0 ft	6.9%	3.48 psf[b]
1971 Solid beams	All str	2.0 ft	6.6%	3.64 psf

[a] 1971 max spacing limits not satisfied. Also add for splices if truss bars are limited to standard 60 ft length.

[b] Will require welded wire fabric or some additional bars for crack control.

Considering designs under the 1971 Code, it appears that:

1. The two-way flush soffit solid beams require about the same amount of reinforcement as the solid head system for square panels. The beams will have definite advantages for rectangular panels, varying spans and loads, and location of large openings.

2. The combination of truss and straight bars does not result in significant saving in weight of reinforcement if maximum bar spacing limits are satisfied. For strength requirements only, a saving of about 5 pecent was possible in this example with a 30 in. maximum spacing between top portions of truss bars in middle strips.

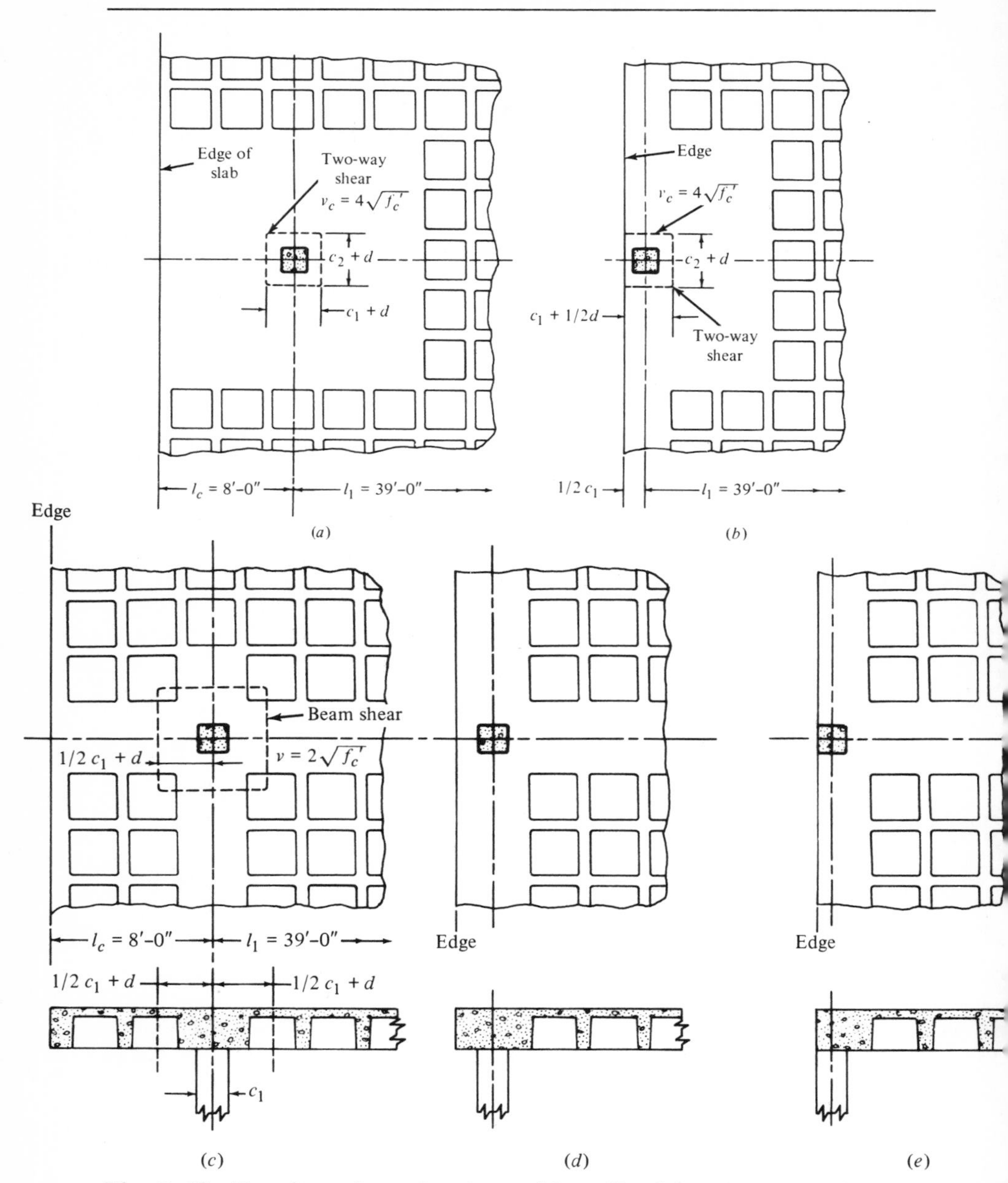

Fig. 7–10 Exterior column locations with waffle slabs: (*a*) waffle flat plate with solid head cantilever edge; (*b*) waffle flat plate with solid head flush edge; (*c*) two-way beam waffle flat plate with cantilever edge; (*d*) two-way beams edge column center of beam; (*e*) two-way beams edge flush with column.

EXTERIOR PANELS

The design of the exterior panels for waffle slabs involves the same considerations illustrated for solid two-way flat plates and flat slabs with drop panels (Chapters 5 and 6). The additional considerations necessary with the waffle system make the use of small sketches helpful for design, as well as for this explanation (see Fig. 7-10). The cantilever arrangements with edge columns set back from the building line as in (*a*) and (*c*) assist in developing the top bars for negative moment, reduce the unbalanced negative moments at the columns, and provide a symmetrical four-sided shear section which simplifies computations. The Code has no specific requirements for the consideration of the variable cross sections of spandrel beams (or slabs) illustrated. It specifically provides that the torsional member shall be assumed to have a constant cross section equal to the width of the column, c_1, and depth of the slab (Section 13.4.1.5-a). The next provision permits it to consist of a beam plus portions of the slab, again as if constant (Section 13.4.1.5-c).

In view of the variations possible within these definitions, some of which are shown in Fig. 7-10, it would seem more realistic to evaluate the torsional stiffness, K_t, by the procedure used for flat slabs with drop panels (see Fig. 6-6). This permissible application (Section 13.4.1.5-a) permits consideration of portions of the flange in the waffle slab up to 4 t on either side of a rib or solid beam section (Section 13.1.5).

For conditions (*b*), (*c*), and (*d*) it should be sufficiently accurate to consider the torsional member as a constant section with dimensions $x = 23$ in., $y = 42$ in. as for the interior beams. Similarly for the condition in (*e*), a constant cross section with dimensions $x = 23$ in. and $y = 33$ in. should suffice. For the common condition shown in (*a*), a precise evaluation would be complex as the actual torsion section is solid for one-sixth of the span at each end and either a channel, twin-tee, or thin slab in the center two-thirds of the span depending upon which definition one accepts. A simplified approximation with $x = 23$ in. and $y = 42$ in. at the ends and a channel 42 in. wide plus two 6 in. wide flanges for the center two-thirds of the span should be acceptable, if an average value is developed as in Fig. 6-6 (see Fig. 7-11).

Torsional Stiffness

Using the suggested approximations, computations for torsional stiffness, K_t, follow:

Conditions of Fig. 7-10 (*b*), (*c*), and (*d*): $K_t/E = 5{,}000$ in.^{3} (See

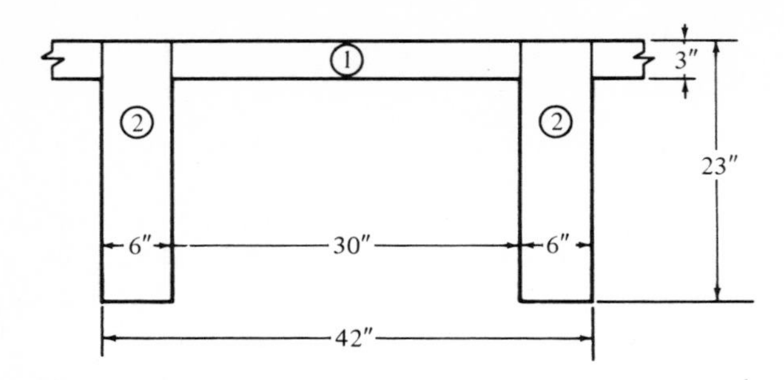

Fig. 7–11 Midspan section for torsional member of Fig. 7–10 (*a*).

example for square interior panel, = 39′ –0″, $c_1 = c_2 = 24$ in. waffle is 20 in. + 3 in. ribs @ 36 in. modules.)

Condition of Fig. 7-10 (e): $C = (1 - 0.63 \times 23/33)(1/3)(23)^3(33)$

$= 75{,}000$ (Section 13.4.1.5, Eq. 13-7)

$$K_t/E = \frac{(18)(75{,}000)}{(39 \times 12)(1 - 2/39)^3} = 3{,}380 \text{ in. (Eq. 13-6)}$$

Condition of Fig. 7-10 (a):

(1) $x = 3$ in., $y = 30$ in.

(2) $x = 6$ in., $y = 23$ in.

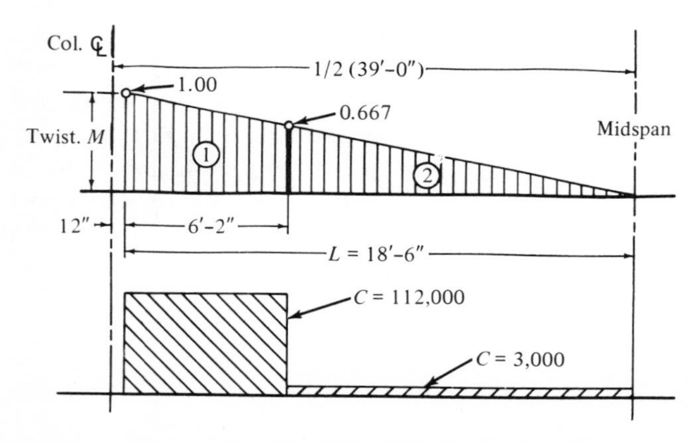

Fig. 7–12 Twisting moment distribution and C-values.

Waffle area:

$$C = \Sigma(2)(1 - 0.63 \times 6/23)(6)^3(23)/3 + (1 - 0.63 \times 3/30)(3)^3(30)/3$$

$$= 2{,}770 + 253$$

$$= 3{,}023$$

Solid head area, $x = 23$ in., $y = 42$ in. $C = 112{,}000$ (same as beam in (c)). Averaging these values according to the area of the torsion moment diagram (Fig. 7-12),

$$\text{total area } \tfrac{1}{2}(1)(\text{L}) = 0.5 \text{ L}$$

$$\text{area } (2) = \tfrac{1}{2}(2/3)(2\text{L}/3) = 2\text{L}/9 = 0.222 \text{ L} = 0.444 \text{ A}$$

$$\text{area } (1) = (0.5 - 0.222)\text{L} = 0.278 \text{ L} = 0.556 \text{ A}$$

$$\text{Average } C = (112{,}000)(0.556) + (3{,}000)(0.444) = 63{,}600$$

$$K_t/E = \frac{(2)(9)(63{,}600)}{(39)(12)(1 - 2/39)^3} = 2{,}850$$

It will be noted that the final value for K_t/E is not very sensitive to the assumption used for the cross section of the middle two-thirds of the span.

Exterior Panel Design Example

Application of direct design for case (*a*), Fig 7-10, for a square panel, $w_l = 180$ psf, for an edge panel, extends the previous design of an interior panel for the same conditions (see previous example, $K_c/E = 868$ in.). The solid head is 23″ deep × 15′ –6″ wide. Midspan I_s for 13 joists = 157,400 in.^{4} for the middle two-thirds of the span. At the face of column,

$$I_s = \frac{(15.5 \times 12)(23)^3}{12} + \frac{7}{13}(157{,}400) = 273{,}000 \text{ in.}^4$$

I_s = within the depth, c_1, of the column $= \dfrac{273{,}000}{(1 - 2/39)^2} = 303{,}000$ in.^{4} (Section 13.4.1.4).

Repeating the exact method for determination of K_s/E (Fig. 6-2) (see Fig. 7-13).

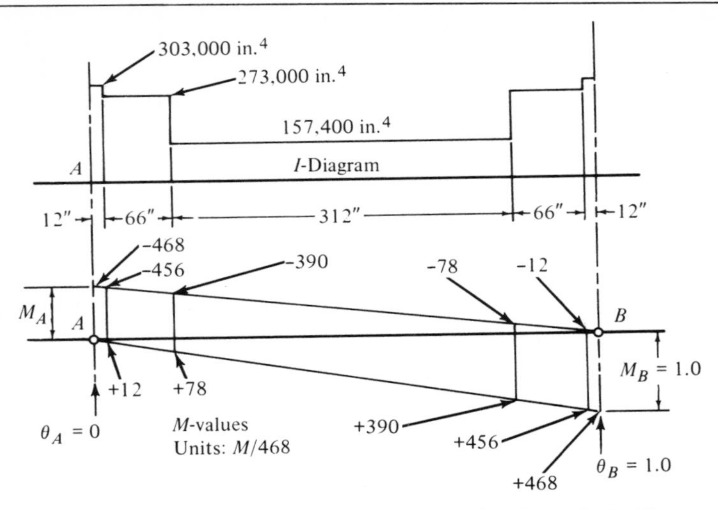

Fig. 7–13 Slab stiffnesses for determination of K_s/E.

$$0 = \left[\frac{-M_A\ 10^3}{468\ E}\right]\left[\frac{(\frac{1}{2})(12)(12)(464)}{303} + \frac{(12)(456)(462)}{303} + \frac{(\frac{1}{2})(66)(66)434)}{273} + \right.$$

$$+ \frac{(\frac{1}{2})(312)(312)(286)}{157.4} + \frac{(78)(312)(234)}{157.4} + \frac{(\frac{1}{2})(66)(66)(56)}{273} +$$

$$\left. + \frac{(12)(66)(45)}{273} + \frac{(390)(66)(423)}{273} + \frac{(\frac{1}{2})(12)(12)(8)}{303}\right] +$$

$$+ \left[\frac{M_B \times 10^3}{468\ E}\right]\left[(0.238)(4) + (18.05)(6) + (7.98)(34) + (94.3)(45) + \right.$$

$$+ (309)(182) + (154.8)(234) + (7.97)(412) + (2.9)(423) +$$

$$\left. + (0.191)(460)\right]$$

$$0 = -M_A(177{,}100) + M_B(101{,}700)$$

$$M_A = 0.574\ M_B \text{ (carry-over factor for moment distribution)}$$

$$\sum_B^0 \frac{M}{I} = \phi_B = 1$$

Note that symmetry permits subtraction of M_A-terms as a group.

$$1 = \frac{(1.00 - 0.574)}{468{,}000\ E} M_B\ 0.238 + 18.1 + 7.98 + 94.3 + 309 +$$

$$+ 154.8 + 7.97 + 2.90 + 0.238$$

$$\frac{M_B}{E} = \frac{(468{,}000)}{(0.426)(596)} = 1{,}842 \text{ in.}^3 = K_s/E$$

Note: As an approximation, a simple average I_s may be computed using the clear span, $l_n = 37'\text{–}0''$:

$$\frac{2}{39}(303{,}000) = 15{,}400 \text{ in.}^4$$

$$\frac{11}{39}(273{,}000) = 77{,}000 \text{ in.}^4$$

$$\frac{26}{39}(157{,}400) = 105{,}000 \text{ in.}^4$$

$$I_s = 197{,}400 \text{ in.}^4$$

$$K_s = \frac{4EI}{l_n};\ \frac{K_s}{} = \frac{(4)(197{,}400)}{(37 \times 12)} = 1{,}752 \text{ in.}^3\ (5\%\ \text{low})$$

$$\alpha_c = \Sigma K_c/\Sigma K_s = \frac{(2)(865)}{1{,}842} = 0.94 \text{ (Section 13.0)}$$

COLUMN STIFFNESS. Use $2 \times K_c$ for columns above and below the slab as in intermediate floors. Minimum $\alpha_{min} = 0.7$ (Section 13.3.6.1) (see Fig. 5-3) $\alpha_c > \alpha_{min}$ so that no additional positive moment steel is required in the exterior panel (Section 13.3.6.1-b, Eq. 13-4).

$$\frac{E}{K_{ec}} = \frac{E}{\Sigma K_c} + \frac{E}{K_t} = \frac{E}{1736} + \frac{E}{2850} = (0.000577 + 0.000351)E$$

$$K_{ec} = \frac{1}{0.009275} = 1078 \text{ in.}^3 \text{ (Sections 13.4.1.5, Eq. 13-5)}$$

$$\alpha_{ec} = K_{ec}/\Sigma K_s = \frac{1078}{1842} = 0.585 \text{ (slab on one side only)}$$

MOMENTS. Negative moment at exterior column (see chart, Fig. 5-7*a*):

$$-M_e = 0.240\ M_o \qquad M_o = \frac{(0.369)(39)(37)^2}{8} = 2{,}460 \text{ ft-kips}$$

$$+M = 0.525\ M_o \qquad \text{(Section 13.3.2.1, Eq. 13-2)}$$

$$-M_i = 0.710\ M_o$$

$$-M_e = (0.24)(2{,}460) = -592 \text{ ft-kips}$$

$$\text{Total cantilever moment, } M_c = \frac{(0.369)(39)(8)^2}{2} = 462 \text{ ft-kips}$$

$$\text{Dead load cantilever moment} = 189/369 \times 462 = 237 \text{ ft-kips.}$$

Unbalanced moment which must be resisted by the exterior column is reduced by the dead moment of the cantilever.

Shear

$M_{un} = -M_t$ (slab) + M_d (cantilever), for the case of maximum moment plus vertical shear with both cantilever and edge panel fully loaded plus moment (see Fig. 7-14, case 1). Edge shear = $\frac{1}{2}$ (0.369)(39)(37) = 270 kips. The negative design moment, M_e = 592 ft-kips at the face of the column (Section 13.3.3.1). At the centerline of the column, this moment will be increased approximately: $V \times \frac{1}{2}\ c_1 = 270 \times 1 = 270$ ft-kips. Similarly, at the first interior column the moment at the face (for design of the slab) will increase by 270 ft-kips. The transfer of shear from edge column to the first interior column due to the difference in end moments is $\frac{2015\ 'K - 862\ 'k}{39'} = 31$ kips.

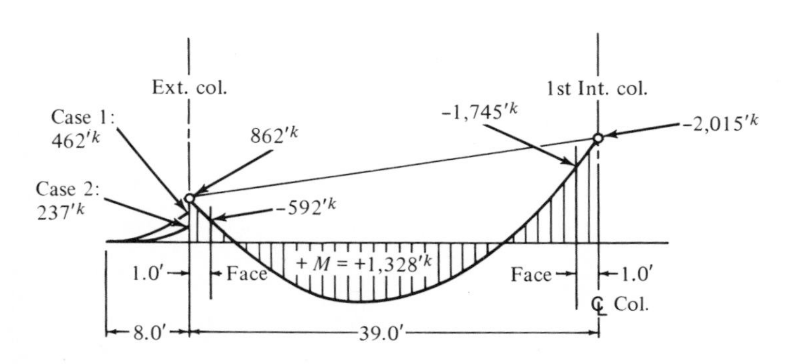

Fig. 7–14 End moments, exterior span.

Case 1. $V_u = 270^k + (189/369)(115)^k - 31^k = 298$ kips

$M_{un} = 862'^k - 237'^k = 625$ ft-kips to the column

Case 2. $V_u = 270k + 115k - 31k = 354$ kips

$M_{un} = 862'^k - 462'^k = 400$ ft-kips

Note that unbalanced moment is computed at the centerline of the column since it is the center of twist for the critical section shears (Section 11.13.2) (see Fig. 7-15). Moments for design of the slabs are those at the face of the columns (Section 13.3.2.2). The total unbalanced moment at the column is to be transferred partly by flexure, partly by shear.

CALCULATIONS FOR SHEAR SECTION CONSTANTS. (see Fig. 7-15).

Perimeter, $b_o = (4)(24 + 21) = 180$ in. (Section 11.10)
Shear section area, $b_o d = 180 \times 21 = 3{,}780$ in.^{2} (Section 11.10.2)
Moment of inertia of the shear section for resistance to torsion, J_c (Commentary Section 11.13.2):

$$J_c = (21)(45)^3/6 + (45)(21)^3/6 + (21)(45)(45)^2/2$$

$$= 318{,}000 + 69{,}500 + 957{,}000$$

$$= 1{,}346 \times 10^3$$

The fraction of the unbalanced moment which is considered to be transferred between the slab and the column by eccentricity of the shear about the centroid of the critical section (Section 11.13.2) $= F_s$.

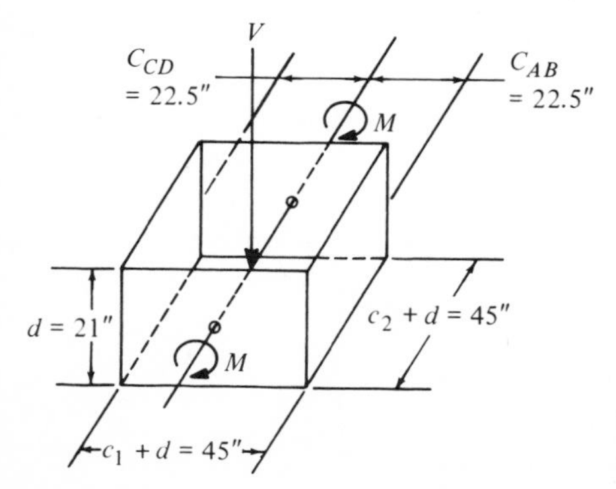

Fig. 7–15 Critical shear section dimensions.

Two-way (punching) shear at column

$$F_s = 1 - \frac{1}{1 + \frac{2}{3}\sqrt{\frac{c_1 + d}{c_2 + d}}} = 0.40 \text{ (Section 11.13.2)}$$

Case 1. $V_u = 298$ kips; $M_u = 625$ ft-kips

$$v_u = \frac{298{,}000}{(0.85)(3{,}780)} \pm \frac{(0.4)(625{,}000)(12)(22.5)}{(1346{,}000)(0.85)}$$

(Commentary 11.13.2)

$$= 93 \text{ psi} \pm 59 \text{ psi} = 152 \text{ psi} < 4\sqrt{f_c'} = 253 \text{ psi}$$

(Section 11.10.3)

Case 2. $V_u = 354k$; $M_u = 400$ ft-kips

$$v_u = \frac{354{,}000}{(0.85)(3{,}780)} + \frac{400}{625}(59) = 110 + 38 = 148 \text{ psi}$$

$$< 4\sqrt{f_c'} = 253 \text{ psi}$$

One-way (beam) shear

With a square edge panel the section for beam shear (Section 11.10.1-a) extends across the entire solid head at a distance, d, from the face of the column (Section 11.2.2). Since this section includes 15′ –6″ solid head plus 7 joist ribs, beam shear is not critical even at the allowable unit shear one-half that for the two-way shear (Section 11.4.1). In waffle flat slabs with solid heads, however, an additional critical section for one-way shear exists in the joists at a distance, d, from the face of the solid head. The total shear to the solid head is carried on 18 joists (see Fig. 7-10*a*). Average width of the compression area in one joist, $b_w = 6 + 21/12 = 7.7$ in. Total $b_w = (18)(7.7) = 139$ in. Take average $d = 21$ in. The total shear outside the critical section,

$$V_u = 350{,}000 - (369)(17.2 \times 19) = 230{,}000 \text{ lb}$$

$$v_u = \frac{230{,}000}{(0.85)(139)(21)} = 93 \text{ psi} < 2.2\sqrt{f_c'} = 139 \text{ psi}$$

(Sections 11.2.1, 11.4.1, and 8.8.8).

OBSERVATIONS ON EFFECT OF LOCATION FOR EXTERIOR COLUMNS. This example demonstrates the effect of the cantilever upon shear and moment at the exterior columns, for the arrangement shown in Fig. 7-10 (*a*). It should be expected that the shears and moments for the arrangement in Fig. 7-10(*c*) would be approximately the same, but would become increasingly critical for those in (*d*), (*b*), and (*e*). As the outside edge of the slab is located close to, or flush with, the outer face of the column, the critical shear section (Fig. 7-15) becomes three-sided (see Fig. 5-32). The moment of inertia of the flush edge three-sided section drops drastically, and the shear stress due to the unbalanced moment increases accordingly. For heavy loads or long spans, a larger column will be required, or a bracket or edge beam can be employed to utilize smaller columns.

8 two-way slab-beam design

GENERAL

Dissatisfaction with results of previously required procedures of analysis and design for two-way slab-beam systems was the original reason for the massive research upon two-way slab systems begun about 1955. All available load test results indicated that the flat plate or flat slab was a most efficient design with an adequate safety factor. The two-way slab on beams, analyzed as separate elements, appeared overdesigned with unnecessarily high safety factors. In the 1971 ACI Code, all two-way reinforced systems, with or without beams, are subject to the same integrated analysis and design procedures (Chapter 13). The requirements have been adjusted to provide nearly uniform strength design capacity for all two-way designs (Chapter 13).

DESIGN AIDS

Most of the charts presented in Chapter 5 for the determination of relative stiffnesses apply also to the two-way system with beams. In this chapter, only the additional procedures required for utilization of the beams will be presented. To avoid duplication, applications of Chapter 5 charts will be indicated by references.

ADVANTAGES

Practical economics indicate consideration of beams in a two-way system when the spans and/or loads are large enough so that required slab thickness and minimum column sizes for a slab without beams become large enough to offset the forming cost and added story height with the beams. The examples presented apply the two-way slab-beam to span and load combinations suitable to illustrate this effect. See Table 5-1 for comparative flat plate minimum size columns.

DIRECT DESIGN

All of the limitations upon the application of the direct design method (Section 13.3.1) previously discussed apply when beams are included. The one additional limitation applicable *only* if the slab panel is supported by beams *on all sides* is that the relative stiffness factors of the beams in the two perpendicular directions, $\alpha_1 l_2^2/\alpha_2 l_1^2$, shall not be less than 0.2 nor more than 5.0 (Section 13.3.1.6, Eq. 13-1). The requirement does not preclude the use of the direct design method for beams in one direction only or on one side only, such as spandrel beams. Note the use of the permissive "if."

INTERIOR PANELS

For the design of interior panels, the important effect is to assume more of the moment is resisted by the beams and less by the slabs. In square panels, the total negative panel moment assigned to the column strip (including the beam) is unchanged and total positive panel moment assigned is increased from 60 to 75 percent. For rectangular panels, however, as the

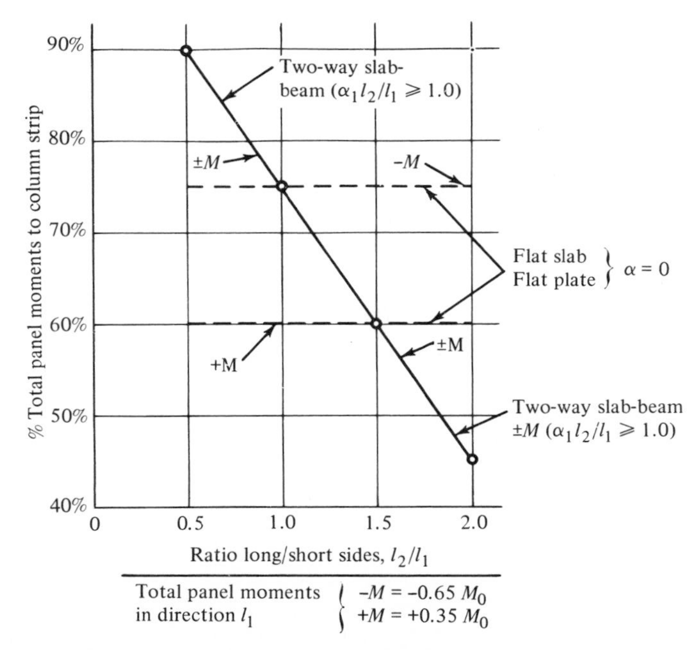

Fig. 8–1 Interior panel—column strip moments.

span ratio, l_2/l_1, approaches the limits, 0.5 and 2.0, the required distribution of the resisting moment per strip changes significantly due to the presence of beams (Sections 13.3.4.1 and 13.3.4.3) (see Fig. 8-1).

The following examples illustrate only the required distribution of the total panel moments for interior panels with beams. Design of the slabs and beams after the moments are determined follows the same procedures shown in Chapters 5, and 9.

EXAMPLE 1. *Square interior panel.* Assume the same beam section on all sides of the panel; same spans in adjacent interior panels. The limiting total width of top flange for this beam is 6 h (Section 13.1.5). Gross section properties are to be used (Section 13.0).

Determination of minimum slab thickness (Section 9.5.3.1)

Conformance to the limitations of the following three equations relieves the designer of deflection computations otherwise required by the code. A span must be assumed for the example in order to determine moment of inertia, I_s, for the slab and the ratio of beam/slab stiffnesses, α, since the first two equations are in review form. Try clear span $l_n = 40\ h$.

$l_1 = l_2$; $I_s = 3.33\ h^4$ (Section 13.0; $\alpha_1 = \alpha_2 = \alpha_m$; $E_{cb}I_b/E_{cs}I_s = 2.18$

(Section 9.0) $\beta = l_1/l_2 = 1.0$ for square panel

(Section 9.0) $\beta_s = 1.0$ for interior panel, beams on four sides

$$\text{Min } h = \frac{l_n(800 + 0.005\ f_y)}{36{,}000 + 5{,}000\ \beta[\alpha_m - 0.5(1 - \beta_s)(1 + 1/\beta)]}$$

$$= l_n/42.6 \quad \text{Eq. (9-6)}$$

$$h \geqq \frac{l_n(800 + 0.005\ f_y)}{36{,}000 + 5{,}000\ \beta(1 + \beta_s)} = l_n/41.8 \quad \text{Eq. (9-7)}$$

$$h \leqq \frac{l_n(800 + 0.005\ f_y)}{36{,}000} = l_n/32.7 \quad \text{Eq. (9-8)}$$

The controlling requirement (Eq. 9-7) is $l_n \leq 41.8\ h$. $l_n = 40\ h \leq 41.8\ h$. $\alpha_1 l_2/l_1 = \alpha_2 l_1/l_2 \geq 1.0$ and so beam takes 85 percent of the column strip moments (Section 13.3.4.4).

Distribution of positive and negative panel moments (Sections 13.3.4.1 and 13.3.4.3).

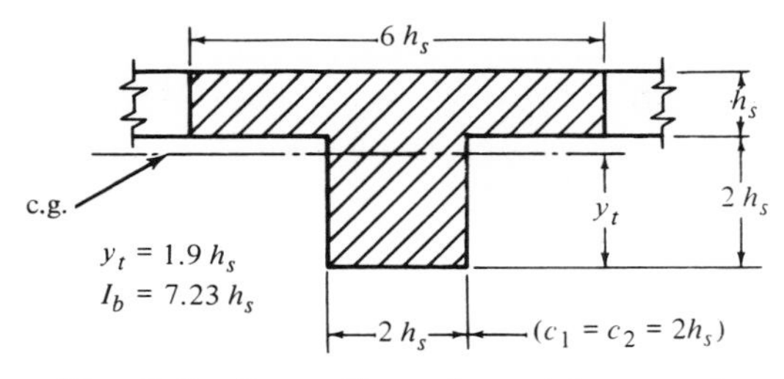

Fig. 8–2 Beam dimensions and properties.

$$-M_u = 0.65\ M_o;\ +M_u = +0.35\ M_o \text{ (Section 13.3.3.2)}$$

Total of beam and column strip: +75% (0.35 M_o); −75% (0.65 M_o)

Middle strip: +25% (0.35 M_o); −25% (0.65 M_o)

Beam: +63.8% (0.35 M_o); −63.8% (0.65 M_o)

Each side of the column strip outside of the beam:

+5.6% (0.35 M_o); −5.6% (0.65 M_o)

See Fig. 8-3 for summary layout.

EXAMPLE 2. *Rectangular interior panel*

Maximum $l_1/l_2 = 2.0$ (Section 13.3.1.2). Same beam proportions as in the previous example. Beams on all sides.

$$\beta = l_1/l_2 = 2.0; \quad \beta_s = 1.0 \quad \text{(Section 9.0)}$$

Try $h = 6''$; $l_1 = 27'\text{–}0''$ $l_2 = 13'\text{–}6''$ with beam web = 2 h

$I_{s1} = 2.25\ h^4$; $I_{s2} = 4.50\ h^4$; $\alpha_1 = 3.23$; $\alpha_2 = 1.61$; $\alpha_{av} = 2.42$ (Section 9.0).

$l_2 = 27\ h$; $l_1 = 54$ h; $l_n = 52\ h$ (Section 9.5.3.1).

Review for minimum thickness per Section 9.5.3.1:

Min $h = l_n/54.7$; $h \geqq l_n/51.0$; $h \leqq l_n/32.7$ Eqs. (9-6), (9-7), (9-8)

Check ratio of relative beam stiffnesses in the two directions:

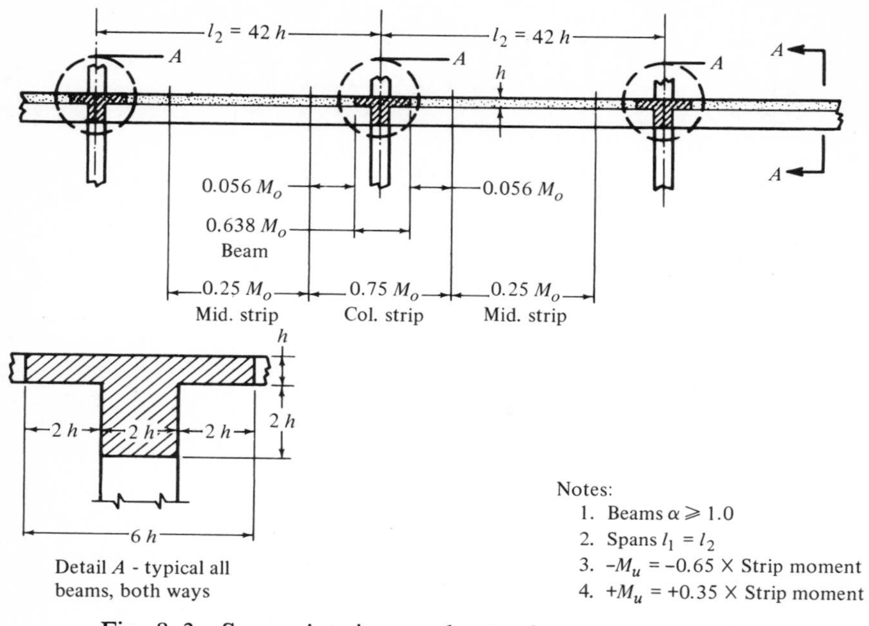

Fig. 8–3 Square interior panels—total moments per strip.

$$\alpha_1 l_2^2/\alpha_2 l_1^2 = 0.5; \quad \alpha_2 l_1^2/\alpha_1 l_2^2 = 2.0 \quad \text{(Section 13.3.1.6).}$$

$$0.2 < \text{Both} < 5.0$$

For use in assigning moments to each strip (Section 13.3.4):

$$\alpha_1 l_2/l_1 = 1.61 \quad \alpha_2 l_1/l_2 = 3.23$$

Moments in the short direction (l_2): Column Strip $\pm$ 45% total

Middle Strip $\pm$ 55%

Beam $\pm$ 38.2%

Column strip each side of beam $\pm$ 3.4%

Moments in the long direction (l_1): Column Strip + 90% total

Middle Strip $\pm$ 10%

Beam $\pm$ 76.5%

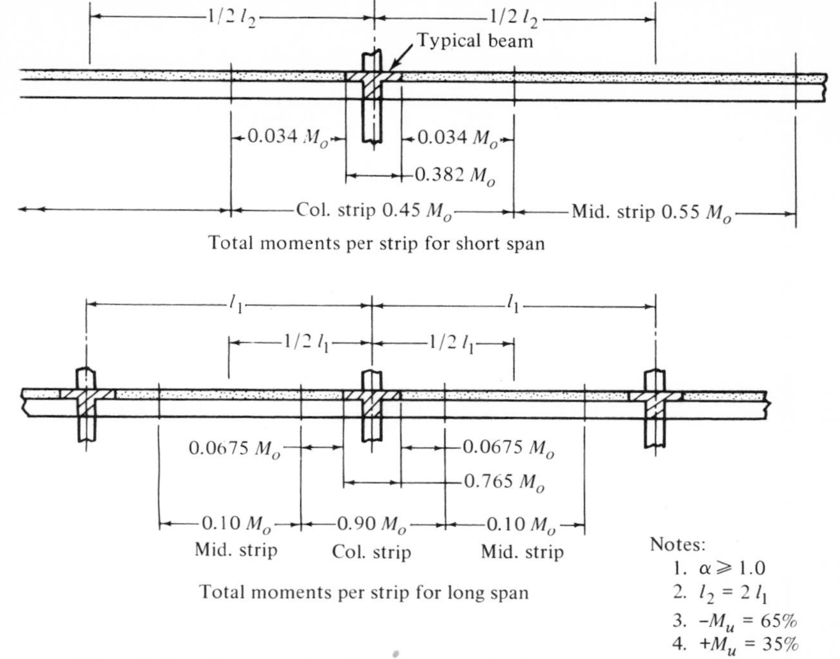

Fig. 8–4. Rectangular interior panel (2:1)—strip moments.

Column strip each side of beam ± 6.75%

See Fig. 8-4 for a summary layout.

SUCCESSIVE TRIALS

If the moment or shear is excessive in the beams for the dimensions assumed as a first trial, and any increase of the dimensions is desired, the distribution of the panel moments shown in Fig. 8-4 will be unchanged provided that the relative stiffnesses are not less than 0.2 nor greater than 5.0 and $\alpha_1\ l_2/l_1 \geqq 1.0$ (Section 13.3.1.6). Note that the expression, $\alpha_1 l_2/l_1 \geq 1.0$ determines the distribution to the column and middle strips as well as to the beams (Sections 13.3.4.1, 13.3.4.3, 13.3.4.4). Reduction of the beam sizes will not change the distribution unless the ratio, $\alpha_2 l_1/l_2$, or $\alpha_1 l_2/l_1$, falls below 1.0. It should also be noted that the tabulated percentages of moment for various values of α_1, β_t, l_2/l_1, and $\alpha_1 l_2/l_1$ are end points and that straight-line interpolation across and vertically in these tables is intended for intermediate values of these quantities.

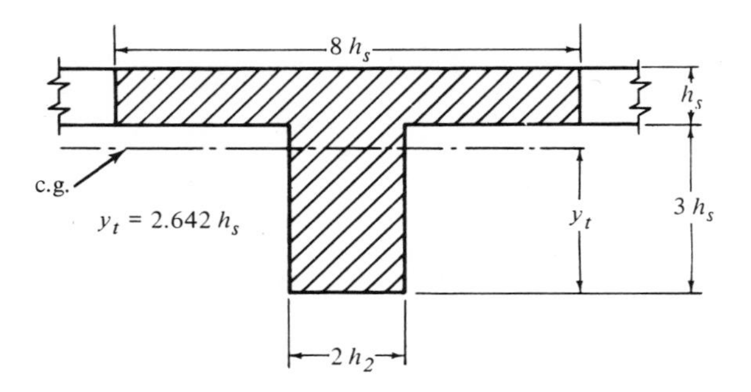

Fig. 8–5 Beam dimensions.

EXAMPLE 3. *Increase depth of beam in span l_1 (see Fig. 8-5).*

$$I_{b1} = 18.88\ t^4 \qquad I_{b2} = 7.23\ t^4$$

$$\alpha_1 = \frac{18.88}{2.25} = 8.4 \qquad \alpha_2 = \frac{7.23}{4.50} = 1.61$$

$$\alpha_1 l_2/l_1 = 4.19 \qquad \alpha_2 l_1/l_2 = 3.23$$

The distribution of the moments is unchanged since both ratios are greater than 1.0. Check relative beam stiffness factors,

$$\frac{\alpha_1 l_2^2}{\alpha_2 l_1^2} = 1.30; \quad 0.2 < 1.30 < 5.0 \quad \text{(Section 13.3.1.6)}$$

Check minimum thickness: $l_1/l_2 = s = 2.0$; $\beta_s = 1.0$ interior

$$\alpha_{av} = \frac{8.4 + 1.61}{2} = 5.0$$

Eq. 9-6: Min $h = l_n/78$

Eq. 9-7: $h = l_n/51$

Eq. 9-8: $h = l_n/32.7$

Eq. 9-7 controls, and so the min. $h \geq l_n/51 \geqq 3\frac{1}{2}$ in. (Section 9.5.3.1)

Design for Shear and Moment

Completion of any design after the determination of the applied bending moments for thicknesses within the above limits and actual loads follows the same procedures illustrated in Chapter 9. A number of specific minor questions of interpretation for the provisions of Chapter 13, however, will be encountered. For convenience, wide beams on column centerlines should be assumed to transmit all shear to the column (see pages 194-195). A number of common variations of slab, beam, and column proportions are shown in Fig. 8-6. In all cases, it will be necessary to check the beam (one-way) shear with allowable stress on concrete, $v_c = 2\sqrt{f_c'}$ (or the "long" formula) at a critical section a distance equal to the beam depth, *d,* from the column face (Sections 11.2.1 and 11.2.2). For cases where all beams are much wider or much narrower than the column, it may be necessary to check the two-way shear (Section 11.10.1-b) as for a slab, with $v_c = 4\sqrt{f_c'}$, at a critical section 0.5 d from the column face (Sections 11.10.2 and 11.10.3). For the small narrow beam the shear section will lie partly in the beams and partly in the slab (see Fig. 7-8). Loads within tributary areas inside lines at 45° between the column centers and the midspan comprise the maximum external shear load assumed per beam (Section 13.3.4.7). Note that the wording used in many general codes prescribing live load reductions for beams would preclude use of such reductions for beams which are designed as merely deeper but integral parts of a slab. Except as noted, the slab supported by beams on four sides will be subjected to one-way shear only, on all sides, at a critical section one slab depth from the face of the beam *web*. For this purpose, the overhanging flanges are not to be considered as "faces of supports."

Torsion

Where the beam is wider than the column (see Fig. 8-6) the beam section for torsional stiffness (Sections 13.4.1.5-c and 13.1.5) is slightly different than that defined for torsional strength (Section 11.7.2). For torsional

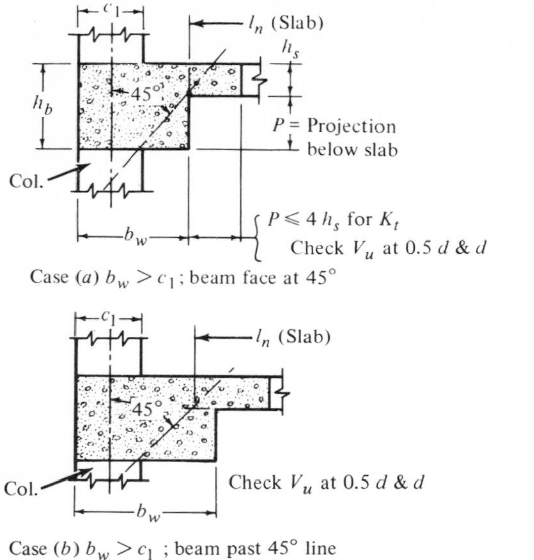

Case (a) $b_w > c_1$; beam face at 45°

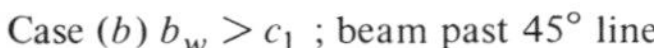

Case (b) $b_w > c_1$; beam past 45° line

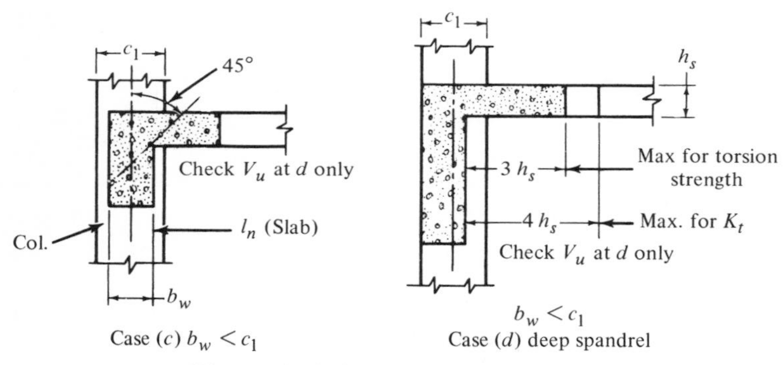

Case (c) $b_w < c_1$

Case (d) deep spandrel

Fig. 8–6 Edge beam variations.

stiffness, K_t, the beam section is assumed to consist of the sum of the rectangles, $(b_w)(h) + (h_s)$ (projection $\leq 4\ h_s$). The projection is taken as the larger above and below the slab where a spandrel beam projects thus. For torsional strength, the beam section is assumed to consist of the sum of the rectangles, $(b_w)(h) + (h_s)(3\ h_s)$ (Section 11.7.2).

Moment and Shear

The clear span of the slab is measured from the face of the column, capital, bracket, or wall (Section 13.3.2.2). When the slab is supported upon beams, the Code is ambiguous. The authors recommend the most conservative interpretation for each case (see Fig. 8-6). Where the beam is wider than the column, the clear span, l_n, can be taken from the face of the beam or the intersection of the extension of the slab soffit line with a 45° angle in a vertical plane from the center of the column, whichever is longer. This interpretation is justified because the portion of the beam within this angle satisfies the definition of a capital (Section 13.1.6) and presumably, a bracket, although no definition of a bracket is provided. For beams narrower than the column, the Code seems to permit the clear span for the slab to be measured from the face of the column (Section 13.3.2.2), but the more conservative interpretation is recommended, taking l_n from the face of the beam (Fig. 8-6).

EXTERIOR PANELS

The design of exterior panels supported by beams on all sides is somewhat more complicated than that for the interior panels considered in Examples 1, 2, and 3. Relative stiffnesses and ratios of relative stiffnesses, including the torsional stiffness of the spandrel beam, must be computed as in those Examples and Chapter 5, to determine the distribution of the flexural design moments. Due to the variety of combinations of beam depth, beam width, location of the beam with respect to the column centerline, and slab thickness possible, no simple design aids as in Chapter 5 for flat plates can be offered. In design of these beams for shear, the torsional moments which must be combined with the vertical shear for the edge beam become significant. See Chapter 9 for an example and design aids for stirrup reinforcement to resist the combined torsion and shear.

OTHER TWO-WAY SLABS

The 1971 Code has left the design requirements for two-way slabs supported by beams not on the column centerlines (see Fig. 8-7), as well as two-way slab design for walls with supports on all sides somewhat in limbo. The two-way slab provisions (Chapter 13) apply specifically to two-way floor slabs, uniformly loaded, with beams if any on column centerlines or not offset more than 10 percent of the span as part of the "column strip" (Sections 13.3.4.4 and 13.3.1.4). While the Code states that two-way slab systems may be supported by walls, columns, or beams (Section 13.1.6), most of the specific provisions, such as taking clear span between faces of the columns or walls, obviously intend that beams, if any, be on column centerlines.

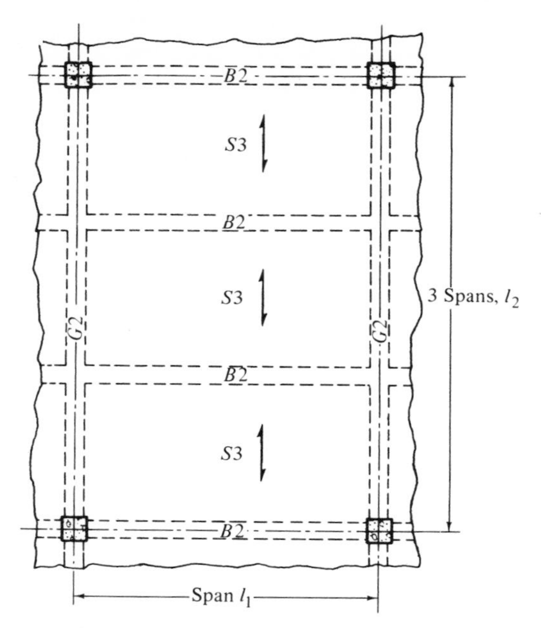

Fig. 8–7 Slab-beam-girder system.

Walls

Walls resisting earth or other lateral pressure and spanning vertically between floor and footing and horizontally between cross-walls or pilasters are seldom continuous over three or more spans vertically, and the direct design method for two-way behavior is not applicable. The equivalent frame analysis method including the concept of an "equivalent column" consisting of a column and attached torsional member is not easily extended to the common two-way reinforced wall slab. Such walls can be designed as two-way slabs to comply with the Code provisions for walls (Section 14.1.1), but for practical purposes this problem is outside the scope of the specific provisions for two-way slabs and must be designed under the permissive procedure clause (Section 13.2.1). For practical design of such walls the authors recommend the use of the 1963 ACI Code two-way slab provisions with appropriate adjustments for yielding or unyielding supports and nonuniform distributed loads. This procedure can qualify for approval under 1971 Code provisions (Sections 1.4, 13.2.1, 13.2.3, and 14.1.1).

For floor slabs supported on beams and girders as in Fig. 8-7, the authors recommend design as one-way slabs for practical reasons. When the ratio of sides is greater than 1:2, direct design as two-way slabs is not permitted. Even the ratio is expressed in terms of "spans" which could be interpreted as clear spans or as center-to-center of supports. In Fig. 8-7, the ratio of center-to-center spans is 2.0; if beams *B*2 were wider or if clear span were taken from column faces, the ratio of clear spans would be greater than 2.0. Equivalent frame analysis is permitted, but it is not easily extended to the conditions in Fig. 8-7. The use of one-way slab design will eliminate these and other ambiguities in the interpretation of the Code provisions for two-way slabs and thereby greatly speed up the design.

More importantly, it will eliminate ambiguities in the interpretation and application of the live load reductions permitted by the general codes, which will then become increasingly important for slab, beam, and girder as intended (see Chapter 9). The minimum shrinkage and temperature steel (Section 7.13) in the long span direction will suffice for actual two-way bending (not considered in the one-way design) if it is placed properly in the top and bottom of the slab at the supporting girders and midspan, respectively.

SUMMARY EVALUATION—TWO-WAY CONSTRUCTION

From numerical evaluations of the four principal variations of the two-way construction prescribed in the 1971 Code (Chapter 13), the authors offer the following general observations as to the practicable range for economical applications:

1. Flat plate with no beams—thicknesses 5 in. to 10 in.—spans 15′–0″ to 30′–0″ maximum—design live loads up to 300 psf maximum. All maximum limits subject to suitability of required minimum exterior column size.

2. Flat plate with shear heads to reduce to column size. Same range as flat plate, except column size can be reduced. Higher load/span combinations within above maximum limits can be utilized with reasonable size columns.

3. Flat plate with edge beam or cantilever. Same range as flat plate with shear heads at exterior columns; limited by interior column size.

4. Flat slab with drop panels and/or capitals. Slab thicknesses, 4 in. minimum to 8 in. minimum, spans 20′–0″ to 36′–0″ maximum. Design loads to 1000 psf.

5. Flat slab with drop panels and/or capitals and spandrel beams. Thicknesses 4 in. to 8 in. maximum.

6. Two-way slabs with beams on four sides on column centerlines. Thicknesses 3 1/2 in. to 12 in. maximum.

9 beams and girders

GENERAL

Although the terms "beam" and "girder" are frequently used in all codes, and, specifically, in the 1971 ACI Code, it was considered unnecessary to provide definitions for these terms in the Code. The common usage of these terms was evidently considered sufficiently definitive. For the purposes of this book, beams and girders are considered to consist of flexural members, generally horizontal, which act as a primary part of the structural framing system. A beam is herein considered to carry distributed loads such as parallel walls or tributary areas of the floor or roof systems; a girder, although it too may support these distributed loads, supports major concentrated loads such as columns or beam reactions. Most of the special provisions of the ACI Code will be applicable to both beams and girders under the authors' definitions of common usage, even where the Code refers merely to beams. Most of the requirements of the general codes in effect, such as fire protection and reduction of live loads, will also apply equally to both. For convenience in this chapter, all references to beams will apply equally to girders; provisions peculiar to girders will be identified.

CODE REQUIREMENTS

Code requirements applicable only to beams and girders may be grouped roughly into four categories, subdivided as follows:

1. *Analysis*—live load reductions (general code), Fig. 2-1; flange effects, "T-beams" (Section 8.7); approximate frame analysis (Section 8.4.2); elastic frame analysis (Section 8.5); limit analysis (Section 8.6); depth effect for depth/span ratios greater than 0.4 for continuous spans or 0.8 for simple spans (Section 10.7); lateral support effect (Section 10.4). Beams with two-way slabs, see Chapter 8.

2. *Design*—limiting maximum steel ratio, ρ, (Section 10.3.2); limiting minimum ρ (Section 10.5.1); torsion reinforcement (Section 11.8); mini-

mum stirrups (Section 11.1); crack control for "beam" cover (Section 10.6); deflection limits (Section 9.5); development of end anchorage (Section 12.2.2).

3. *Details*—ties for compression steel (Section 7.12.5); confinement within joints (Section 7.12.6); minimum cover (Section 7.14.1); bundled bars (Section 7.4.2); fireproofing cover (general codes).

4. *Construction*—casting concrete (Section 6.4.2); joints (Section 6.4.3).

Some general requirements will be discussed first, but interpretation of most of these Code requirements where several apparent interpretations are open is best undertaken by examples.

ANALYSIS

Loads

Design loads are required to be increased by load factors (Section 9.3) for the following three combinations of load effects:

1. $U = 1.4\,D + 1.7\,L$ (Code Eq. 9-1)
2. $U = 0.75\,(1.4\,D + 1.7\,L + 1.7\,W)$ (Code Eq. 9-2)
3. $U = 0.9\,D + 1.3\,W$ (Code Eq. 9-3)

where U = required strength; D = dead load effect; L = specified live load effect; and W = specified wind load effect. Equation (9-1) must always be investigated (Section 9.3.1). Equations (9-2) and (9-3) must be investigated when the beam is part of a structural frame expected to resist lateral loads and not fully braced against movement due to these (Section 9.3.2). Equation (9-3) will control design only in unusual cases where the magnitude of the wind force is relatively very large. Its use can usually be avoided by inspection. It can be shown that Equation (9-3) will control design with live load present only if: $0.15\,D + 1.275\,L < 0.025\,W$ (unlikely); or when live load is entirely absent if: $0.15\,D < 0.025\,W$ (true when D, L, and W have additive effects and $W > 6.0\,D$, or when D and L have effects opposite those of W, and $-W > 0.693\,D$). Only the last possibility need be considered in practice, and it is not likely to be encountered often.

Note that the three basic load combinations are to be employed to compute the required strength for design to resist a variety of load effects with various special load factors for each (Sections 9.3.3, 9.3.4, 9.3.5, and 9.3.6). The Code also requires consideration of differential settlement, creep, shrinkage, and temperature change effects "where they may be significant" at an effective load factor of $0.75 \times 1.4 = 1.05$ (Section 9.3.7). The measure of "significance" is obviously when the most adverse combination of these effects realistically expected to occur and realistically estimated exceeds $0.25\ (1.4\ D + 1.7\ L)$ or $1.7\ W$.

Frame Analysis Methods

Analysis for design moments and shears in beams may utilize the assumptions and procedures for an (exact) elastic frame analysis (Section 8.5) or may consist of the approximate frame analysis coefficients (Section 8.4). The reader is referred to Chapter 2 for a discussion of the limitations on the approximate method and an explanation of the "exact" elastic method. For economy in design time and in finished design, the authors recommend use of the approximate method where applicable, and further, that the provisions for redistribution of negative reinforcement (Section 8.6) be applied to the results of any elastic frame analysis undertaken. Redistribution of negative moment is necessary to equalize the economy of finished designs by the two methods; it is the only concession to principles of limit analysis specifically permitted.

Unbraced Frames with Slender Columns

In addition to the elastic frame analysis moments, including effects of both vertical and lateral loading conditions, the Code requires that the magnified end moments required for slender columns be used as the beam design moments only in frames not otherwise braced against sidesway (Section 10.11.7).

Slender Beams

A special provision must be observed for slender beams (Section 10.4). Slender beams are defined as those in which the spacing of lateral supports to the compression flange or face is more than 50 times the least width of same. Since lateral support equal to a minimum of 2 percent of the total (internal) compression in the compression flange is usually considered adequate, and may often be satisfied by simple friction of separate floor materials transmitting the load to the beam, the beam slenderness problem will be encountered only rarely in routine design. It may arise in design of long span upturned roof beams or in the design of thin deep panels used as spandrel beams. Such spandrels are often precast and added to the frame as simply supported beams with a span equal to the column spacing and width of the compression flange as little as 4 in.

When a slender beam condition arises, the simple solution is to provide lateral bracing within the spacing limits, thus eliminating the condition. Where lateral bracing is impracticable, the load capacity must be reduced.*

Deep Beams

The ordinary beam formulas are inadequate for the design of deep beams. Nonlinear stress distribution at service loads and the effect of vertical stresses must be taken into consideration for the analysis of beams with depths greater than 0.4 of the span for continuous spans and 0.8 for simple spans (Section 10.7). Special requirements for shear reinforcement are prescribed for deep beams loaded only at the top, although loads at or near the bottom flange causing vertical tension are more apt to create critical diagonal tension (Section 11.9). Deep beams are required to have minimum reinforcement as prescribed for shear (Section 11.9.5) or walls (Section 14.2) in both directions, whichever is larger, in addition to the minimum flexural reinforcement (Section 10.5).

Deep beams present a complex problem in analysis. As the depth-to-span ratio increases, the effect of the load and stress in the vertical direction cannot be neglected as in shallow beams. The assumption of straight line stress distribution at service loads does not apply to deep beams and the curve of stress distribution takes on radically different shapes.

For flexural design purposes, the Code defines deep beams as those with total depth greater than 0.4 clear span for continuous beams and 0.8 for simple beams (Section 10.7). No separate limit is provided for end spans. The Code wording is "depth-span" ratios; the authors' recommendation is that "depth" be taken as total depth and "span" be taken as clear span. It is difficult to determine the effective depth in deep girders and the width of the support can make the center-to-center span meaningless.

For shear design purposes a deep beam is defined in the Code as a beam in which the ratio of effective depth to clear span is greater than 0.2 (Section 11.9.1). Effective depth, d, is defined as the distance from the extreme compression fiber to the centroid of the tension reinforcement (Section 11.0.). Due to the nonlinear distribution of stress under elastic analysis there is often no extreme compression fiber; the maximum compression stress may lie inside the beam depth with tension areas above and below. For practical purposes the authors recommend application of the special provisions for shear in deep beams (Section 11.9) to beams in which the

* Hansell, William, and Winter, George, "Lateral Stability of Reinforced Concrete beams," *ACI Journal*, **56** (Sept. 1959), 193-214. Also, Sant, J., and Bletzacker, R. W. "Experimental Study of Lateral Stability of Reinforced Concrete Beams," *ACI Journal*, **58** (Dec. 1961), 713-736.

overall depth is greater than 0.2 the clear span for simple or continuous beams (Section 11.9.1). The special provisions for shear apply only to beams loaded at the top. Beams loaded otherwise are more complex to analyze.

Where very important deep beams are encountered or where the design involves sufficient repetition to become important, the use of models to determine the elastic stress distribution will obviously be helpful. Such use of models is now specifically recognized for use as a supplement to the calculations which may be required by a building official (Section 1.2.2). Such calculations must include consideration of the end support width; position and extent—top, bottom, or full depth; continuity, if any; position extent, and type of loads—uniform, concentrated, applied at the top, bottom, or between; holes; etc., for practical applications. Finite element solutions are often required.

The design of deep beams is a complex procedure outside the scope of this Guide, but the user will find the reference under Chapter 11, Code Commentary, useful.*

EXAMPLE 1: INTERIOR BEAM B1. Select beams suitable for the support of a joist-slab floor system of alternate spans 20′ –0″ and 24′ –0″ center-to-center of the supporting beams. The design loads for the joist system are w_d = 100 psf and w_l = 250 psf. The beams span 21′ –4″ between 16 in. square columns. Assume the lateral load resistance is provided by walls. The beams are to be 15″ deep to maintain a level ceiling with the 12 + 3 joist-slab system (see Fig. 9-1).

Beam Loads

Since the joist-slab spans alternate 20 ft and 24 ft, the beams on interior column lines (B-1) receive the following loads: Beam design live load $w_l = \frac{1}{2}(20 + 24)(0.250) = 5.5$ klf. Beam design dead load, for a beam 24 in. wide as a first trial, $w_d = [\frac{1}{2}(20 + 24) - 2](0.100) + (2)\frac{(15)}{12}(0.150)(1.4) = 2.5$ klf. The foregoing factors satisfy the load factor combination prescribed as: $U = 1.4\,D + 1.7\,L$ (Section 9.3.1, Eq. 9-1). $U = 2.5 + 5.5 = 8.0$ klf. Live load reductions are usually allowed in the general codes only for live loads less than 100 psf, and so such reductions will not be considered in this example (see Chapter 2, Fig. 2-1). The

* Additional references are: Design of Deep Girders, No. ST-66, Portland Cement Association. Shear Behavior of Deep Reinforced Concrete Beams, Vols. I-V (A continuing series), Dec., 1970, by Robert A. Crist, Eric C. Wang Civil Engineering Research Facility, University of New Mexico, Albuquerque, New Mexico.

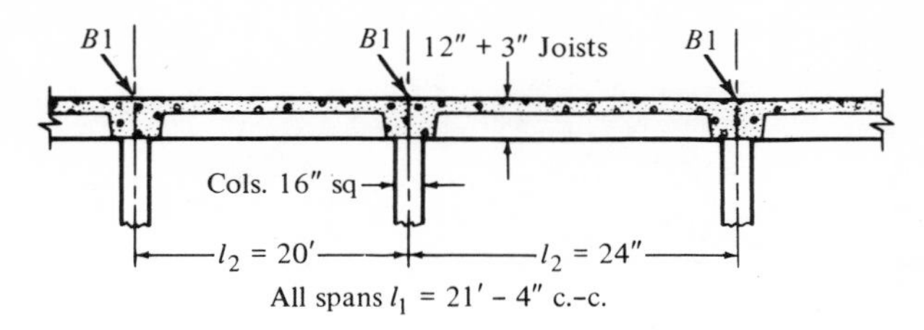

Fig. 9–1 Interior span beam *B*-1 and loads.

approximate analysis method will be used since this example is within the limitations thereon (Section 8.4.2).

Moments

Negative moment at faces of interior supports other than the first line, $-M_u = 1/11\ w(l_n)^2$; positive moment in interior spans, $+M_u = 1/16\ w(l_n)^2$ (Section 8.4.2).

$$-M_u = \frac{(8.0)(20)^2}{11} = -291 \text{ ft-kips}$$

$$+M_u = \frac{(8.0)(20)^2}{16} = +200 \text{ ft-kips}$$

Shear

Maximum shear at critical section, $d = 12$ in., from face of support,

$$V_u = \frac{(8.0)(18)}{2} = 72 \text{ kips at 12 in. from support}$$

Minimum shear at midspan for half-span load,

$$V_u = \frac{(5.5)(10)}{4} = 13.75 \text{ kips at midspan}$$

Required cover $= 1\frac{1}{2}$ in. min (Section 7.14.1.1); minimum stirrups are required since $h > \frac{1}{2}\ b_w$ (Section 11.1.1); effective depth to the top bars, $d = 12.4$ in.; to the bottom bars, $d = 12.6$ in (see Fig. 9-2).

STRENGTH DESIGN. Use $f_c' = 4{,}000$ psi; $f_y = 60{,}000$ psi. Moment reinforcement

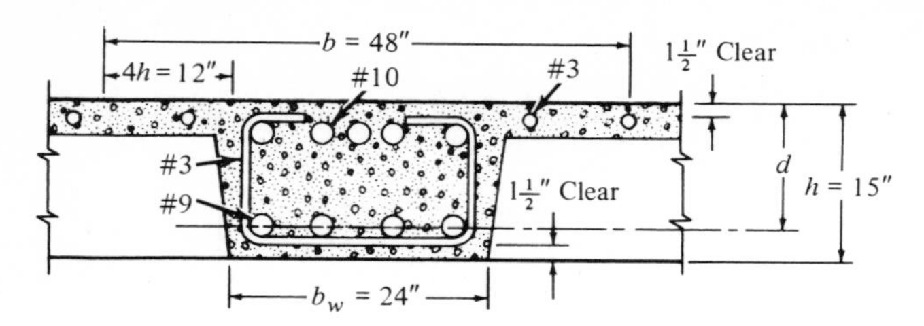

Fig. 9–2 Section dimensions—1st trial—beam B-1.

$$A_s = M_u/\phi f_y(d - \tfrac{1}{2}a) \qquad a = \frac{A_s f_y}{0.85\, f_c' b}$$

See Fig. 9-26(b).

$$0.75\,\rho_b = 0.0214$$

For top bars,

$$\frac{R}{1000} = \frac{M_u}{b_u d^2} = \frac{(291)(12)}{(24)(12.4)^2} = 946 > 930 \text{ for } 0.75\,\rho_b$$

Required $\rho = 0.0215$; top bars, $A_s = (0.0215)(24)(12.4) = 6.4$ in.^{2} Try 5–#10. Unless the web width, $b_w = 24$ in., is increased about 2 percent, the bottom bars must be extended for compression so that the net tensile steel ratio, $\rho - \rho \leq 0.75_b$ (Section 10.3.5) (see Fig. 2-8). For bottom bars,

$$\frac{R}{1000} = \frac{(200)(12)}{(48)(12.6)^2} = 315 < 930; \quad \rho = 0.0065$$

Bottom bars,

$$A_s = (0.0065)(48)(12.6) = 3.93 \text{ in.}^2$$

Try 4–#9.

Shear, $v_u =$

$$v_u = \frac{V_u}{\phi b_w d} \quad \text{(Section 11.2.1, Eq. 11-3)}$$

$$v_c = 2\sqrt{f_c'} \quad \text{(Section 11.4.1)}$$

12.5 in. from face of column,

$$v_u = \frac{(71.7)(1000)}{(0.85)(24)(12.4)} = 283 \text{ psi}$$

Allowable shear on the concrete, $v_c = 2\sqrt{f_c'} = 126$ psi. With bent-up bars for shear reinforcement, $(v_u - v_c) = 3\sqrt{f_c'} = 190$ psi is allowed (Section 11.6.2.2); with stirrups for shear reinforcement at a maximum spacing of 0.50 d for vertical legs (Section 11.1.4), allowable shear $(v_u - v_c) = 4\sqrt{f_c'} = 253$ psi (Section 11.6.3); for higher shear the spacing of vertical legs is 0.25 d maximum and the value of $(v_u - v_c) \leq 8\sqrt{f_c'} = 506$ psi (Section 11.6.4).

At this point the designer has several options:

1. Complete the first trial design with $b_w = 24$ in., providing stirrups at 0.25 $d = 3$ in. spacing.
2. Increase b_w, $24 \times \frac{283}{253} = +\ 27$ in. for stirrups at 0.50 d.
3. Increase b_w to avoid the minimum stirrup requirement (Section 11.1.1) so that $b_w \geq 2\ h = 30$ in. with $v_u \leq v_c$ (Section 11.1.1 c). Alternatively, increase depth since no minimum stirrups need be provided if the shear, v_u, is less than one-half v_c (Section 11.1.1 d). The required width to reduce v_u to $v_c = 126$ psi is $b_w = \frac{283}{126}$ (24) $= 52$ in. > 30 in.
4. Recalculate width for the first three options based upon allowable shear, $v_c = 1.9\sqrt{f_c'} + (2500)\ (\rho_w)\ \frac{V_u d}{M_u}$

(This last option involves tedius calculations for little gain, except at ends of simple spans where full bottom reinforcement is extended into the support. *Warning:* in cases such as tapered end joists with bent-up bars, the calculation will require use of a computer as it is impossible to predict critical sections.)

Open Stirrups Required for Vertical Shear, v_u

In this example, the width, b, will be increased from 24 in. to 28 in., reducing v_u to 243 psi $< 4\sqrt{f_c'}$ (option 2). The vertical stirrups may now be spaced at 0.50 d maximum. This width will eliminate need for compression steel. At the critical section, a distance d from the support:

$$A_v = \frac{(v_u - v_c)}{f_y}\ b_w s = \text{area of stirrups required}$$

(Section 11.6.1, Eq. 11-13),

where v_c = 126 psi; b_w = 28 in.; s = stirrup spacing; and f_y = 60,000 psi. For #3 ⊏⊐, $A_v = 2 \times 0.11 = 0.22$ in. Solving Eq. 11-13 for s, maximum spacing.

$$s = \frac{(0.22)(60{,}000)}{(243 - 126)(28)} = 4 \text{ in.}$$

At the maximum spacing, s = 6 in., the minimum area of stirrups required (Section 11.1.2, Eq. 11-1), A_v is

$$A_v = \frac{(50)(b_w)(s)}{f_y} = \frac{(50)(28)(6)}{60{,}000} = 0.14 \text{ in.}^2$$

Minimum design shear, at midspan for half-panel loads, as simple span:

$$V_u = \frac{(5.5)(10)}{4} = 13.75 \text{ kips}$$

$$v_u = \frac{(13.75)(243)}{71.7} = 47 \text{ psi}$$

(See Fig. 9-3.) If this value is taken as midspan shear, and a shear envelope is made, shear, v_u, decreases at 22 psi per foot. Space #3 ⊏⊐ @4 until

$$(v_u - v_c) = \frac{(2)(0.11)(60{,}000)}{(28)(6)} = 78.5 \text{ psi}$$

to permit use of the maximum spacing, 6 in.

$$v_u = 79 + 126 = 205 \text{ psi} \qquad \frac{(243 - 205)}{22} = 1.73 = 21 \text{ in.}$$

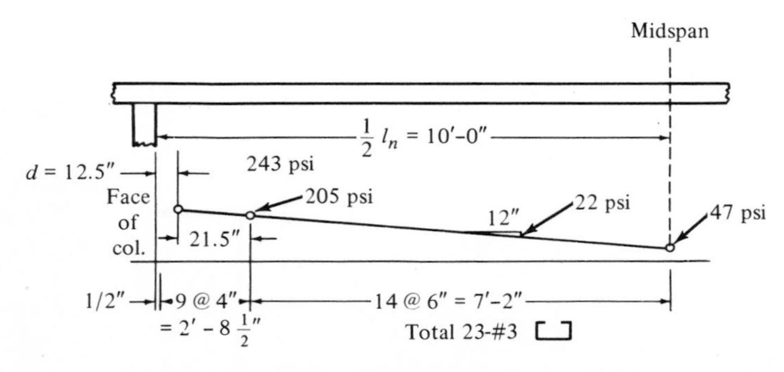

Fig. 9–3 Shear reinforcement—beam *B*-1.

Distance from face of support = 12.5 in. (d) + 21.5 = 34.0 in. Use 9 – #3 @4, 14 – #3 @6 [stirrup symbol]. These stirrups would suffice if torsion is negligible.

Torsion Requirements for Design

The design for shear must include an investigation of torsion (Section 11.7.1) due to unbalanced loading on any beam. The effect of torsion may be neglected in design provided that the torsional stress, v_{tu}, does not exceed $1.5\sqrt{f_c'}$ (Section 11.7.1), and so this calculation will seldom be required for interior beams supporting slabs with approximately equal spans and loads on either side. If torsion reinforcement is required ($v_{tu} > 1.5\sqrt{f_c'}$), it must be provided in addition to that required for shear and flexure (Section 11.8.1), but may be combined with the other reinforcement. In this example, if torsion reinforcement is required, closed stirrups only will be provided at a closer spacing or in larger sizes than required for shear alone.

ANALYSIS. Only the unbalanced joist-slab moment which will be transferred to the columns by the beam will create torsion in the beam. The total unbalanced moment at the centerline of the beam can be obtained from an elastic analysis or from the difference in the fixed-end moments from the short span with design dead load only and from the long span with design dead plus design live load (see Fig. 9-4).

Assume relative flexural stiffnesses as follows: $K_{ec} = 1.0$; $K_s = 1.64$ for the short span and $K_s = 1.36$ for the long span. The ratio of equivalent column stiffnesses to total stiffnesses at the joint

$$\frac{K_{ec}}{K_{ec} + \Sigma K_s} = \frac{1.0}{1.64 + 1.36 + 1.0} = 0.25$$

The total unbalanced fixed end moment, F.E.M._{un}, is:

$$\text{F.E.M}_{un} = (21.33)\left[\frac{(0.100 + 0.250)(24)^2}{12} - \frac{(0.100)(20)^2}{12}\right]$$

$$\text{F.E.M}_{un} = 287 \text{ ft-kips}$$

From Fig. 9-4, read total unbalanced moment to be transferred to an interior column equals 0.393 F.E.M._{un} = (0.393)(287) = 113 ft-kips. Note that the equivalent column stiffness, K_{ec}, was used to estimate the total unbalanced moment at the joint to avoid overestimating it by neglecting the torsional flexibility of the beam.

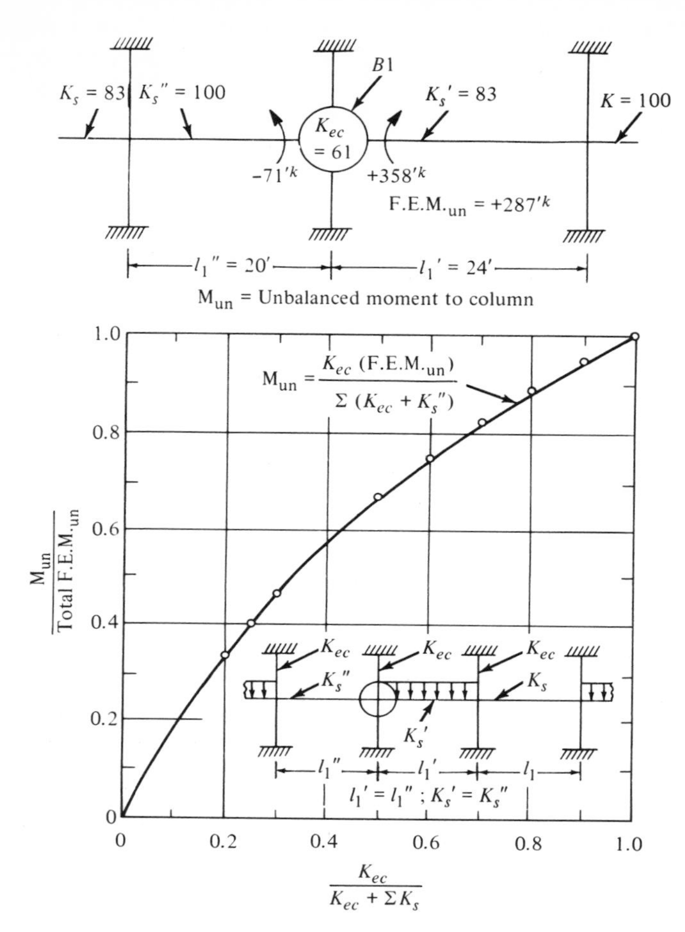

Fig. 9–4 Relative joint stiffnesses and unbalanced moment.

The next problem is to divide the unbalanced moment into portions which must be transferred to the column by direct flexure through the joists and by torsion through the beam. In this example, the direct flexural connection consists of one 10 in. wide joist on the column centerline since the span $l_1 = 21' -4''$, 4 in. more than an odd multiple of the joist module.

For two-way slab connections, the Code specifically prescribes a formula for separating the flexure and torsion (Section 11.13.2). Application of this formula using $c_1 = 10$ in. for the flexural connection and $c_2 = 16$ in. for the torsional connection results in assigning 37 percent of the unbalanced moment to torsion. The authors believe this Code provision is applicable only for two-way slabs or two-way beams wide enough to include a critical perimeter section for two-way shear, $(c_1 + d)$ $(c_2 + d)$. Since there is no specific Code requirement for the condition in this example, the authors believe that a conservative estimate of the unbalanced moment transferred by torsion must be made.

The critical sections for torsion are located in the beams on each side of the column at a distance $(\frac{1}{2}c_2 + d) = 1.71$ ft from the center of the column (Section 11.7.4). Assume all unbalanced moment from the joist-slab system outside the length between critical sections is transferred by the beams in torsion, half on each side.

$$T_u = \tfrac{1}{2}(M_{un})\left[\frac{(\frac{1}{2}\,l_2 - 1.71)}{\frac{1}{2}\,l_2}\right] = (\tfrac{1}{2})(113)\left[\frac{(10.67) - 1.71)}{10.67}\right] = 47.4 \text{ ft-kips}$$

Torsional Stress

Torsional stress,

$$v_{tu} = \frac{3\,T_u}{\phi\Sigma x^2\,y} \quad \text{(Section 11.7.2, Eq. 11-16)}$$

where x = the short dimension and y = the long dimension of rectangles contained in the beam cross section (Section 11.0) (see Fig. 9-5).

$$\Sigma x^2\,y = (28)(15)^2$$

$$= 6{,}300 \text{ in.}^3$$

(Neglect the overhanging flanges when stirrup reinforcement is required, as it is impracticable to reinforce the 3 in. flanges)

$$V_{tu} = \frac{(3)(47.4)(12{,}000)}{(0.85)(6{,}300)} = 318 \text{ psi} > 1.5\,\sqrt{f_c'} = 95 \text{ psi}$$

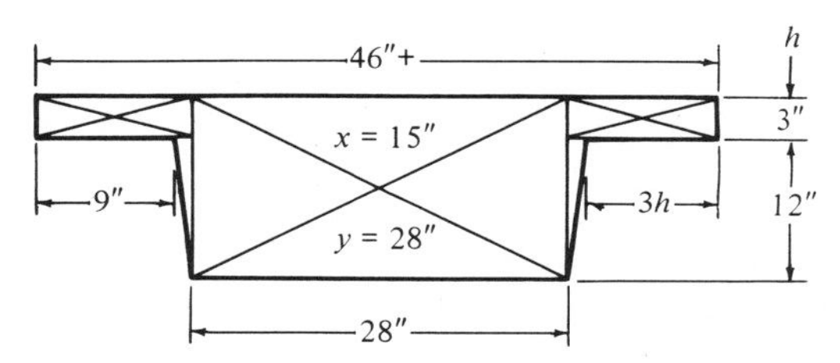

Fig. 9–5 Torsional section.

The torsional effects will require closed stirrups (Section 11.7.1). Maximum shear, v_c, allowed on concrete becomes:

$$v_c = \frac{2\sqrt{f_c'}}{\sqrt{1 + (v_{tu}/1.2\ v_u)^2}} = 86 \text{ psi} \quad \text{(Section 11.4.5)}$$

Selection of Stirrup Reinforcement

The torsion stress, $v_{tu} = 318$ psi, and the shear stress, $v_u = 243$ psi may be combined for the design of (closed) stirrups (Section 11.8.1). It will be found that application of the Code formulas is straightforward, but tedius (Sections 11.8 and 11.6). The charts in this chapter will eliminate much tedious calculation for reasonably regular sections in the usual range of sizes (see Fig. 9-25). Compute k_1 and k_2 as follows:

$$k_1 = \frac{v_u}{\sqrt{f_c'}} = \frac{243}{63.3} = 3.84 \qquad k_2 = \frac{v_{tu}}{\sqrt{f_c'}} = \frac{318}{63.3} = 5.03$$

Read: (1) that the combined torsion and shear does not exceed the maximum allowed, (2) that closed stirrups are required, and (3) that the maximum spacing of closed stirrups is $0.50\ d = 6.25$ in.

Compute the permissible torsional stress carried by the concrete, v_{tc}

$$v_{tc} = \frac{2.4\sqrt{f_c'}}{\sqrt{1 + (1.2\ v_u/v_{tu})^2}} = \frac{(2.4)(63.3)}{\sqrt{1 + \left(1.2\ \frac{243}{318}\right)^2}} = 112 \text{ psi}$$

(Section 11.7.5, Eq. 11-17).

Note that v_{tc} and v_c vary with the ratio of torsion/shear (see Fig. 9-25b). Assume the *ratio* can be considered constant for the following solution.

$(v_{tu} - v_{tc}) = 318 - 112 = 206$ psi. See Fig. 9-23 for $y_1 = 2\ x_1$, $(y - 3) = 2\ (x - 3)$.* For our example, $x = 15$ in. and $y = 28$ in., $(28 - 3) \approx (2)(15 - 3)$. Read for $(v_{tu} - v_{tc}) = 206$ psi, #4 stirrups @ 11 in.,

---

* For manual calculations to select stirrup size-spacing combinations, it is convenient to use the dimensions x_t and y_t as constants so that stirrup capacity is directly proportional to the size of the stirrup bars. The stirrup spacing design charts herein have been prepared upon the basis of over-all stirrup dimensions rather than from center-to-center of the stirrup legs (Section 11.0). Since the standard tolerance in fabrication of stirrups is $(\pm)\ \frac{1}{2}$ in. (CRSI) and the tolerances on cover in placing are $(\pm)\ \frac{3}{8}$ or $\frac{1}{2}$ in. (Section 7.3.2.1), the loss in precision in the use of overall dimensions is more theoretical than real.

maximum spacing for torsion. $(v_u - v_c) = 243 - 86 = 157$ psi (see Fig. 9-19). Read for $b_w = 28$ in. and $d = 12.5$ in., #4 stirrups @ 0.5 d provide capacity for $(v_u - v_c) = 145$ psi. 0.5 $d = 6.25$ in. $(6.25)\dfrac{(145)}{157} =$ 5.7 in. for the required maximum spacing for shear.

Combining, required spacing $s = 1/(1/5.7 + 1/11) \approx 3\frac{1}{2}$ in., beginning at the face of the column.

Compute the distance from the face of the column at which the stirrup spacing may be increased to the maximum of 0.5 $d = 6.25$ in. Shear capacity, $(v_u - v_c) = \dfrac{6.25}{12}(145) = 76$ psi at $s = 12$ in. Read Fig. 9-23 for $s = 12$ in., $(v_{tu} - v_{tc}) = 180$ psi torsion capacity. Change to the spacing $s = 6$ in. when $v_u = 76 + 86 = 162$ psi and when $v_{tu} = 180 + 112 = 192$ psi (see Fig. 9-3). These points are located 56.5 in. and 21.5 in. from the face of the column. At each end of the beam, use 17–#4 @ $3\frac{1}{2}$ ⎡⎤; 10–#4 @ 6 ⎡⎤ stirrups (see Fig. 9-6).

Longitudinal Bars for Torsion

The Code requires longitudinal bars distributed around the inside perimeter of closed stirrups at a spacing not to exceed 12 in., and not less than #3 bars in each corner of the stirrups (Section 11.8.5). In this example, perim-

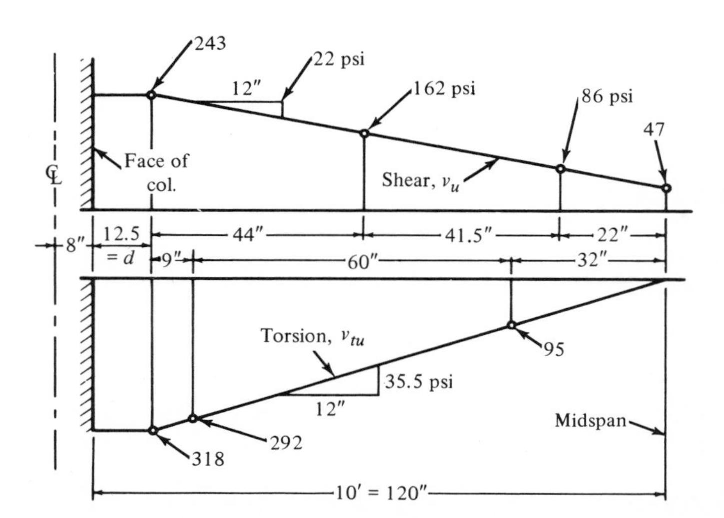

Fig. 9–6 Combined shear and torsion.

eter is 12 × 25, and so six bars will be required, three in the top and three in the bottom.

The minimum total area of the six bars, A_l, is the larger of

$$A_l = (2)(A_t)\,\frac{x_1 + y_1}{s} \quad \text{(Section 11.8.4, Eq. 11-20)}$$

$$A_l = \left[\frac{400\,(x)(s)}{f_y}\left(\frac{v_{tu}}{v_{tu} + v_u}\right) - (2)(A_t)\right]\frac{x_1 + y_1}{s} \quad \text{(Eq. 11-21)}$$

At the critical section for torsion, d from the face of the support, the top and bottom bars required for flexure will suffice. In this example, typically for most interior beams, loading for maximum flexure causes little torsion and loading for maximum torsion reduces both shear and flexure, and so the same bars will suffice for both loadings. This condition does *not* exist for edge or spandrel beams in torsion, and in such cases the torsional requirements for longitudinal bar areas (computed) are additive to flexural requirements (Section 11.8.1). In this example only the top bars required for torsion past the cutoff point permitted by flexural requirements need be computed.

Note that A_l is the value for six (equal) bars and that the spacing, s, at cutoff point is at the arbitrary maximum. In the second terms of the two expressions for A_l, use the *computed* spacing, s, at which the actual area of stirrups, A_t, would satisfy torsional stresses. (Otherwise, the designer providing stirrups larger than required, for simplicity of placing and detailing, may be computing heavier top than bottom bars at midspan where actual shear and torsion approach zero.)

At the cutoff point of the top bars for flexure, usually 0.30 l_n, $v_w = 136$ psi and $v_{tu} = 144$ psi. The *computed* spacing, s, required for torsion at this point $= \dfrac{180}{(144 - 112)}\,(12) = 67$ in.

$$A_l = (2)(0.20)\left(\frac{12 + 25}{67}\right) = 0.221 \text{ in.}^2$$

or

$$A_l = \left[\frac{(400)(15)(6)}{60{,}000}\left(\frac{144}{144 + 136}\right) - (2)(0.20)\right]\left(\frac{12 + 25}{67}\right) < 0 \text{ in.}^2$$

The required area for each of the six bars is (0.221/6 = 0.04 in.^{2} Use three #3, extend past the cutoff point of the top flexural bars $\left(\dfrac{0.04}{0.11}\right)$ (21)

$= 7$ in. < 12 in. minimum for a Class C tension lap splice of top bar. (three #3 × 10-0).

Note: The preceding example has been deliberately constructed to illustrate provisions new to the 1971 Code regarding shear and torsion. The maximum ratio of adjoining spans for use of approximate frame analysis and a relatively heavy live load to create torsion are used. Under the previous Code torsional stiffness equal or less than 20 percent of the flexural stiffness at a joint could be neglected in the frame analysis (1963 Section 905 c.3). This provision has been removed from the 1971 Code (Section 8.5.3). Torsion effects must be included with shear and bending where v_{tu} exceeds 1.5 $\sqrt{f_c'}$ (Section 11.7.1) and torsional stiffness, K_t, must be included in the equivalent column stiffness, K_{ec}, to determine K_{ec} for calculation of unbalanced moments to interior columns with two-way systems (13.3.5.2).

In the preceding example, the approximate frame analysis method was applied (Section 8.4.2). It will be noted that torsional stiffness per se is not involved in the analysis for moment or shear, but that unbalanced moment to be transferred to the column will be excessive unless the equivalent column stiffness, K_{ec}, is considered. Part of this unbalanced moment is transferred to the column by flexure and part by torsion. For two-way slabs, only the Code prescribes the fraction transferred by torsion (Sections 11.13.2 and 11.10.2). In this example, a very conservative interpretation has been employed to estimate the fraction transferred by flexure, leaving the remainder for transfer by shear.* It is further assumed that the entire live load is movable so that full live load and no live load can be present on adjoining long and short spans, respectively.

Under these conservative assumptions, it will be observed that torsion becomes very significant. With the beam dimensions which would usually be chosen for shear and moment only, the torsion stress developed requires closed stirrups at close spacings. To avoid cost of closed stirrups and labor to place them, a considerably wider beam would be required. The authors recommend that users of the Code assess the probability of pattern loading more closely than under previous codes to avoid overestimating unbalanced moments transferred through torsion to columns; and, where full pattern loading is realistically expected on unequal adjoining spans, that use of wider interior beams be considered to avoid the need of closed stirrups. Where closed stirrups must be used, of course, provision for use of two-piece closed stirrups (Section 7.12.7) will facilitate bar placing.

---

* K. Saether and N.M. Prachand, "Torsion in Spandrel Beams," *ACI Journal*, **66** (Jan. 1969), 24-30.

SELECTION AND DETAILS OF FLEXURAL BARS

Development (bond)

No design is complete until the proper selection of bar sizes and lengths is made. Determination of the division of total design moment into negative and positive bending moments and the areas of reinforcement for each is precise only for an assumed pattern of loads. As noted under "Redistribution of Negative Moments" (Section 8.6), reinforced concrete frames will adjust to various loading patterns provided the total area of steel top and bottom is sufficient and is extended through all possible critical tension sections for full development.

The Code provides an intricate set of overlapping requirements for minimum bar length (Sections 12.1, 12.2, and 12.3). The "Explanations of Revisions" first published with the proposed Code gives an elaborate explanation of "shifted" maximum moment diagrams. Only for an element repeated identically many times can the expense of design time to compute absolute minimum bar lengths for minimum steel weight be justified. The only practical approach for routine design is to establish arbitrary cutoff points or bend points as ratios of span in a "Typical Bending Detail." These "standard" bar lengths can be made applicable for all uniformly loaded beams in a structure; special details are then necessary only for unusually short or long beams, or girders with odd locations of concentrated load, etc.

With this approach, the provision of proper development lengths is reduced to a simple comparison of actual length available, computed as a ratio of the span to the cutoff point, to the length required for full development, l_d, for different bar sizes and selection of the largest size of bar usable. See Tables 13-4 and 5 for straight embedment, l_d, and hooked end embedment, E, respectively.

Typical cutoff or bend locations commonly used for approximately equal spans and uniform loads are available as ratios of span in several references* representative of commonly accepted practice. Some engineers prefer to reference these typical details to clear span and some c.-c. span. Clear span is more nearly correct for a variety of support widths, but the c.-c. span is more convenient for the engineer, detailer, estimator, and

* (1) *Manual of Standard Practice for Detailing Reinforced Concrete Structures,* ACI 315.
(2) *Manual of Standard Practice,* CRSI.
(3) *Building Code Requirements for Reinforced Concrete,* ACI 318-71, Fig. 13.5.6 (applicable to beams supporting two-way slabs only).

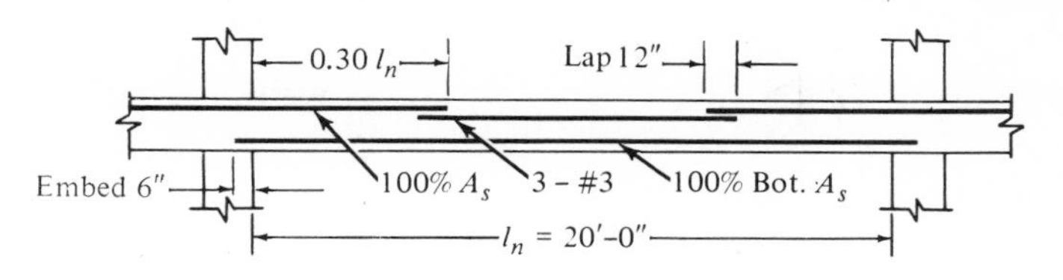

Fig. 9–7 Typical bar lengths—*A*.

placer. Whichever is used, the ratio employed should be adjusted to give approximately the same resulting length.

For this example, a typical bending diagram is shown in Fig. 9-7 for an interior span using straight bars only and vertical stirrups. The arrangement in Fig. 9-7 is perhaps the least sophisticated system and therefore the most foolproof from design board through construction. With it design time is reduced to a minimum. In this case the ratio to cutoff $0.30l_n$ barely sufficed; an optional equivalent more widely applicable would be 0.33 l_n. The available top bar development length is $0.30 \times 20'-0'' \times 12 = 72$ in. See Table 13-4, top bars, $f_c' = 4{,}000$ psi. The maximum size for these top bars is #10 ($l_d = 67$ in.). This bar selection now satisfies the general requirements of development on each side of the critical section, face of support (Sections 12.1, 12.1.3, 12.3.1, and 12.3.2.)

To complete the investigation for the top bar selection, the point of inflection for maximum negative moment must be located (see Fig. 9-8). The distance from the face of the support to the point of inflection (P.I.) is 4′–9″. The bar length provided extends past the point of inflection so that it is *not* cut off in a tension zone, avoiding elaborate alternative computations required to justify same (Section 12.1.6). Since none of the top bars are cut off short, the requirements of the Code regarding "continuing reinforcement" are satisfied (Section 12.1.5). The top bars must extend a distance equal to the largest value of: (1) *d*, (2) 12 bar diameters, or (3) $l_n/16$ (Sections 12.1.4, and 12.3.3).

Available length for development	= 72 in.
Distance to point of inflection	= 57 in.
Extension beyond point no longer required	= 15 in.

(1) d = 12.5 in.	$<$ 15 in.	Cutoff is sufficient length for #10 bars
(2) 12 d_b (#10): 12×1.27 in.	= 15 in.	
(3) $l_n/16$ 240/16	= 15 in.	

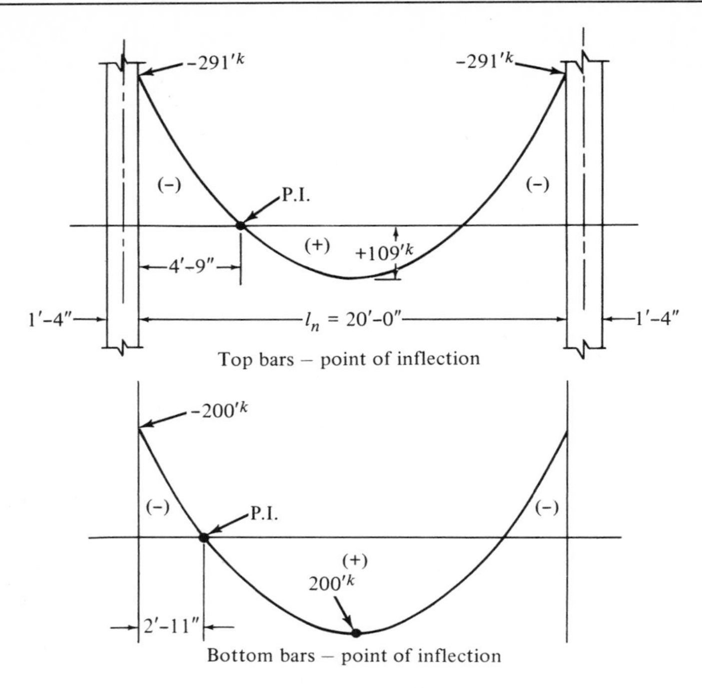

Fig. 9–8 Moment diagrams.

Top steel area required = 6.41 in.² Use five #10 in the beam and two #3, one in each flange for A_s = 6.57 in.² The two #3 are added to satisfy requirements for crack control (Section 10.6.1).

The area of bottom bars required = 4.00 in.² Four #9 bottom bars will be used. Since 100 percent of the bottom A_s is extended 6 in. into the support, assuming our example beam is part of a braced frame, no further development computations are required. For a beam which forms part of a

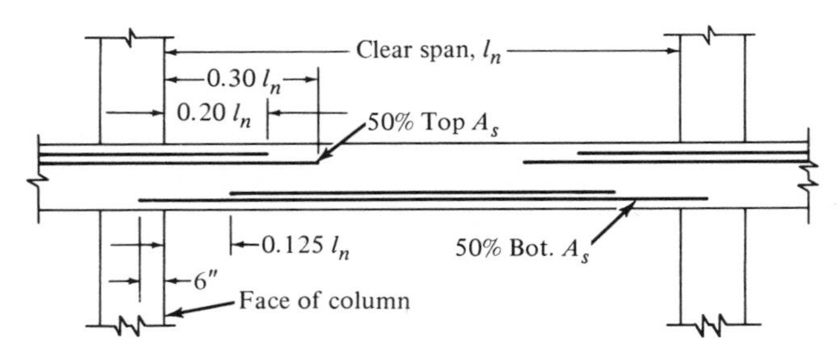

Fig. 9–9 Typical bar lengths—*B*.

primary lateral load resisting frame, 25 percent of the bottom steel must be extended into the support a distance l_d or an equivalent embedment for a larger percentage. See (Section 12.2.2) for tension.

Note that different sizes were selected for top and bottom bars as an aid to inspectors and placers to help avoid error. Preferably, these two sizes would be continued through all beams varying number but not size to suit spans and loads encountered. Since stirrups are supplied for the full length of the beam, some positive support for the stirrup cage and joist-slab top bars should be provided in the 8′–0″ gap between ends of the main top bars. The three #3 × 10′–0″ additional bars required for torsion with a lap of 12 in. at each end will provide sufficient support. See Table 13-8 for laps.

Cut-off Points for Reduced Reinforcement

For a structure with many repetitions of the same beam design, two commonly used refinements of the simple typical detail in Fig. 9-7 are shown in Fig. 9-9. The engineer must judge whether the indicated reductions in steel weight are repeated sufficiently to justify the cost of added design, inspection, and placing time. The additional computations required are shown for illustration below.

The short bottom bars are cut off in a compression zone, inside the point of inflection for positive moment (Fig. 9-8) thereby avoiding computations for cutoffs in tension zones (Sections 12.1.6 and 12.1.2). The Code requires that the following equation be satisfied at the point of inflection, for bars cut off in this region:

$$l_d \leq \frac{M_t}{V_u} + l_a$$

where l_d is based on f_y: M_t is the positive moment capacity at the section (P.I.): V_u is the applied (design) shear at the P.I.; and l_a is the embedment past the P.I., but not to exceed d or 12 bar diameters, whichever is greater for this calculation (Section 12.2.3). To save calculations, take M_t for 50 percent A_s as $\left(\frac{1}{2}\frac{M_u}{\phi}\right) = \frac{1}{2}\frac{(200)}{0.90} = 111$ ft-kips. At the P.I., $V_u = (8)(10 - 2.92) = 57$ kips; for the continuing #9 bottom bars, $l_d = 38$ in. in tension (Table 13-4); l_a 12 bar diam = 13.5 in.

$$\frac{(111)(12)}{(57)} + 13.5 = 23.5 + 13.5 = 37 \text{ in.} \approx 38 \text{ in.}$$

Development length from bottom bar cutoff to point of maximum positive

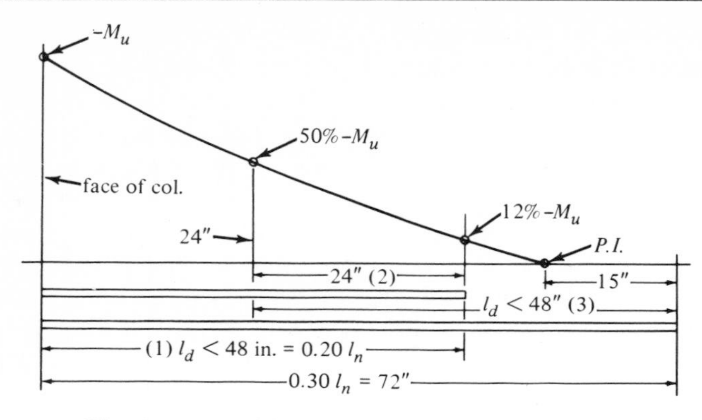

Fig. 9–10 Critical lengths for top bar cut-off.

moment, at mid-span, $= 7.50 \times 12 = 90$ in. > 38 in. The proposed cut-off point and selection of #9 bottom bars are satisfactory.

The cutoff point for the top bars involves a number of more rigorous conditions which must be satisfied (Sections 12.1.3, 12.1.4., 12.1.5, 12.1.6, 12.3.2, and 12.3.3). A complete redesign using smaller size bars (than #10, 100 percent at 0.30 l_n originally selected) will be required (see Fig. 9-10).

1. l_d for the cutoff bars cannot exceed 48 in. (Sections 12.1.1 and 12.3.2). From Table 13-4, $l_d = 42$ in. for #8, and 53 in. for #9, both "other" bars. #8 is the maximum size for this criterion.
2. Note that $d = 12$ in.; 12 bar diameters #8 $= 12$ in. < 24 in. (Section 12.1.3). This criterion is satisfied.
3. The long bars must be fully developed in 48 in. (Section 12.1.5). l_d for #8 $= 42$ in. < 48 in.
4. A final provision that must be met is that *one* of the following three conditions is satisfied at the cutoff point (Section 12.1.6): (a) Calculated $V_u \leq 2/3\ V_u$ allowed, (b) excess stirrups are present for 60 psi capacity, or (c) that the continuing bars are at half or less the allowable stress and that $V_u \leq 3/4\ V_u$ allowed (see Fig. 9-3). Bar stress in the contining bars is approximately 12 per cent $\times \dfrac{1}{0.5} =$ 24 percent.

Condition (c) is satisfied.

8–#8 top bars will satisfy the area required and meet development requirements for the typical detail of Fig. 9-9. In the example beam, these bars would be spaced approximately 3 in. center-to-center. This closer spacing than the design for the bar lengths in Fig. 9-7 will complicate placing to fit between column verticals and may preclude use of this more sophisticated typical detail at some supports in a practical situation. This consideration will also be involved in the decision for typical details on a specific project.

At this point, all the requirements for the first stage of design, design for strength, are complete. The Code requirements for serviceability constitute a second stage of design, usually performed simultaneously with the strength design. By use of design aids limiting the maximum spacing of bars for crack control and tabulations of maximum load for the usual range of span and depth combinations to control deflection in selecting depth, the second stage of design can be satisfied without any additional routine calculations. For illustrative purposes, it will be performed here as a review of the design just completed to determine whether any adjustments are required.

SERVICEABILITY DESIGN

Performance under service loads

DEFLECTION. The immediate deflection of a one-way beam must be computed if the beam thickness is less than the minimum thicknesses prescribed (Section 9.5.2.1, Table 9-5 a).

For interior spans, min $h = l_n/21 = 240/21 = 11.4 \text{ in.} < 15 \text{ in.}$

For end spans, min $h = l_n/18.5 = 240/18.5 = 13 \text{ in.} < 15 \text{ in.}$

Deflection need not be computed for this example. For a complete example of deflection calculations, see Chapter 4.

CRACK CONTROL. Determine the maximum spacing of the top reinforcement, #10 bars, with #4 stirrups and $1\frac{1}{2}$ in. clear cover, to satisfy crack control requirements for interior exposure. Part of the top steel will be placed in the 3 in. thick flanges (Section 10.6.2). The quantity, z, must not exceed 175 ksi for interior exposure (Section 10.6.3).

$$z = f_s\sqrt[3]{d_c\, A}$$

where

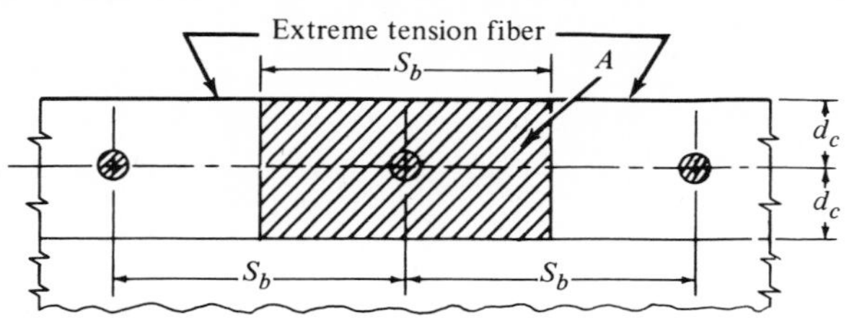

Fig. 9–11 Crack control.

A = the tributary area of concrete in tension per bar

f_s = tension in the bar at service load. If not calculated, f_s may be taken as 0.6 f_y.

d_c = the distance from the tensile face to the center of the bar.

Let s = the spacing of the bars, c. -c., then $A = (2)(d_c)(s)$. Service dead load = 2.5/1.4 = 1.78 k/ft; service live load = 5.5/1.7 = 3.23; total service load = 1.78 + 3.23 = 5.01 k/ft.

$$f_s = \frac{\phi(60{,}000)}{(8.0/5.01)} = \frac{(0.90)(60{,}000)}{1.596} = 33.8 \text{ ksi}$$

$$d_c = 1.500 + 0.500 + (\tfrac{1}{2})(1.27) = 2.63 \text{ in.}$$

Solving for the maximum spacing of the #10 bars, s:

$$z = (33.8)\sqrt[3]{(2)(d_c^2)(s)} = 175$$

$$2(d_c^2)(s) = \left(\frac{175}{33.8}\right)^3 = 138.7$$

$$\text{Maximum spacing, } s = \frac{138.7}{(2)(2.63)^2} = 10.03 \text{ in.} \quad \text{(See Table 3-2)}$$

Spacing for the #10 bars in the beam section (Fig. 9-2) is satisfactory, approximately 6 in. < 10 in. maximum. Note that it will not be possible to provide bars in the 3 in. thick flanges with the full required cover, $1\frac{1}{2}$ in., top and bottom (Section 7.14.1). In this case, #3 bars will be placed with $1\frac{1}{2}$ in. top cover so that they will fit under the top joist bars which require only $\frac{3}{4}$ in. cover (Section 7.14.2). Spacing of the #9 bottom bars is approximately $7\frac{1}{2}$ in., which is satisfactory without further check. These calculations complete the second stage of design, for serviceability, in Example 1.

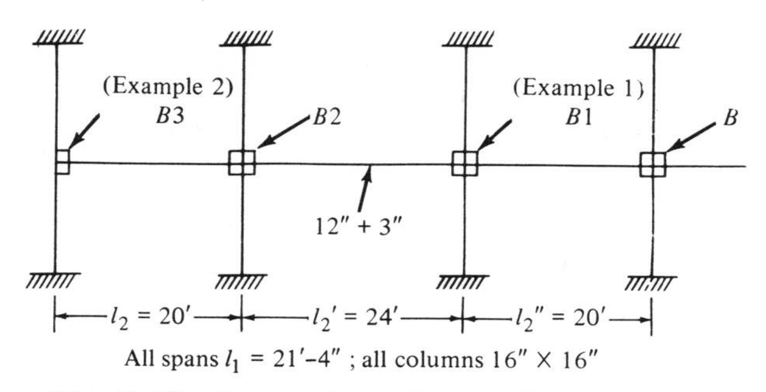

Fig. 9–12 Beam-column frame with beam *B*-3.

EXAMPLE 2. *Spandrel Beam—B-3*. Design for torsion beam *B*-3 (see Fig. 9-12): Continue Example 1 floor system loads, spans, etc. Consider beam *B*-3 as the spandrel beam between two interior columns in the direction of its span. Successive spans, l_1, are equal to 21′–4″ c.-c. of 16 in. × 16 in. edge columns. The floor system, span l_2 = 20′–0″ c.-c. between beams *B*-2 and *B*-3, is a one-way joist-slab, 12 + 3; with loads, w_d = 100 psf and w_l = 250 psf. Beam *B*-3 is to be 15 in. in total depth to maintain the level ceiling. Design for moment and servicability requirements will not be repeated because they will be similar to the design for *B*-1 in Example 1.

In example 1, the design charts of this chapter were employed to illustrate the short-cut approach to design for combined shear and torsion. In Example 2, each applicable Code formula will be used for purposes of comparison and to present the basis of the design charts. For users wishing to develop their own design charts for other stresses (lightweight aggregate, etc.), for an extended range, or simply to a larger scale for greater accuracy, a study of Example 2 should make such conversions simple.

Design for strength—shear and torsion

Total torsional moment through *B*3 $\leq -M_{ext}$ of the floor system. Use the approximate frame analysis (Section 8.4.2) to determine the end moment where the support is a spandrel beam. First trial width,

$$b = 20 \text{ in.} \quad -M_{u(\text{ext})} = \frac{1}{24}(w)(l_n)^2$$

$$w = 0.100 + 0.250 = 0.350 \text{ ksf on the floor area}$$

$$-M_{u(\text{ext})} = \frac{1}{24}(0.35)(20 - 16/12)^2 = 5.08 \text{ ft-kip/lf}$$

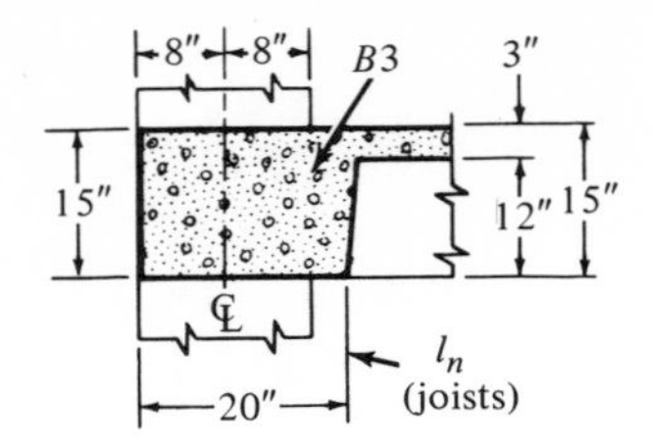

Fig. 9–13 1st trial size, *B*-3.

Torsion on one side of the column from midspan to the critical section at d = 12.5 in. from the face of the column (Section 11.7.4) = T_u.

$$T_u = (5.08)(10 - 1.04) = 45.5 \text{ ft-kips}$$

The resisting section for torsion (Section 11.7.2) = Σx^2 y (see Fig. 9-13). *Note:* neglect the small rectangle in the flange since it will not be reinforced. $\Sigma x^2 y = (15)^2(20) = 4{,}500$ in.^{3}

$$v_{tu} = \frac{(3)\ T_u}{\phi \Sigma x^2\ y} \quad \text{(Section 11.7.1, Eq. 11-16)}$$

$$v_{tu} = \frac{(3)(45{,}000)(12)}{(0.85)(4{,}500)} = 428 \text{ psi} > 1.5\sqrt{f_c'} = 95 \text{ psi}$$

allowable without torsion reinforcement, and so torsion reinforcement must be provided (Section 11.7.1).

Shear

If the torsion-shear reinforcement is to be combined, the shear stresses must now be determined (Sections 11.7.5, and 11.8.1). For the beam widths *B*3, 20 in.; and *B*2, 28 in., $l_n = [20 - (12 + 14)/12] = 17.83$

$$\text{Total shear, } V_u = \left[\frac{(0.35)(17.83)}{2} + \frac{(0.250)(20)}{12} \right.$$

$$\left. \frac{(20 \times 15)(0.150)(1.4)}{144} \right] (8.96)$$

$$= (3.12 + 0.417 + 0.437)(8.96)$$

$$= 35.7 \text{ kips}$$

Take $d = 12.5$ in.

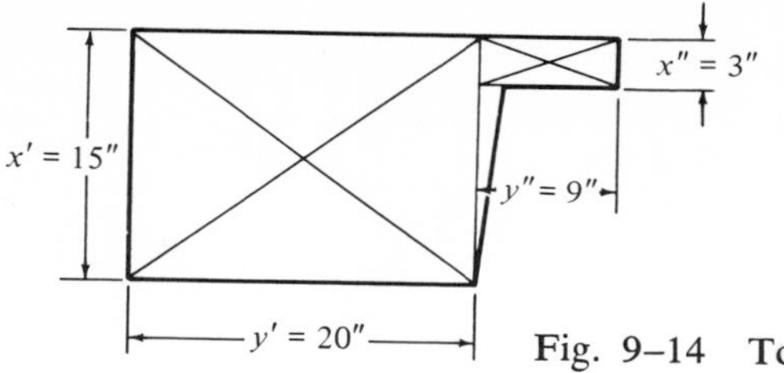

Fig. 9–14 Torsion section, *B*-3.

$$v_u = \frac{(35.7)(1000)}{(0.85)(20)(12.5)} \quad \text{(Section 11.2.1, Eq. 11-3)}$$

$$v_u = 167.5 \text{ psi} > v_c = 54 \text{ psi; see Fig. 9-25 b.}$$

Shear at midspan for partial panel loads (simple span basis):

$$V_u = \tfrac{1}{4}(8.96)(35.7)/10 = 8 \text{ kips}$$

Midspan $v_u = \dfrac{(8)(167.5)}{35.7} = 37$ psi (see Figs. 9-15 and 9-16 for stirrup dimensions and torsion-shear diagrams.

Maximum combined torsion-shear stress allowed
The maximum permitted,

$$v_{tu} = \frac{(12)\, f_c'}{\sqrt{1 + (1.2\; v_u/v_{tu})^2}} \quad \text{(Section 11.7.7, Eq. 11-18).}$$

$$\text{Allowable } v_{tu} = \frac{(12)(63.2)}{\sqrt{1 + (1.2 \times 0.392)^2}} = 686 \text{ psi} > v_{tu} = 428 \text{ psi}$$

Since the maximum torsional stress, v_{tu}, is less than that allowed, the

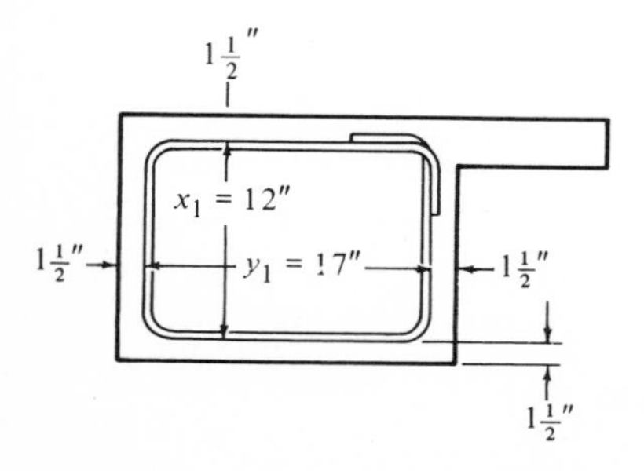

Fig. 9–15 Closed stirrup dimensions.

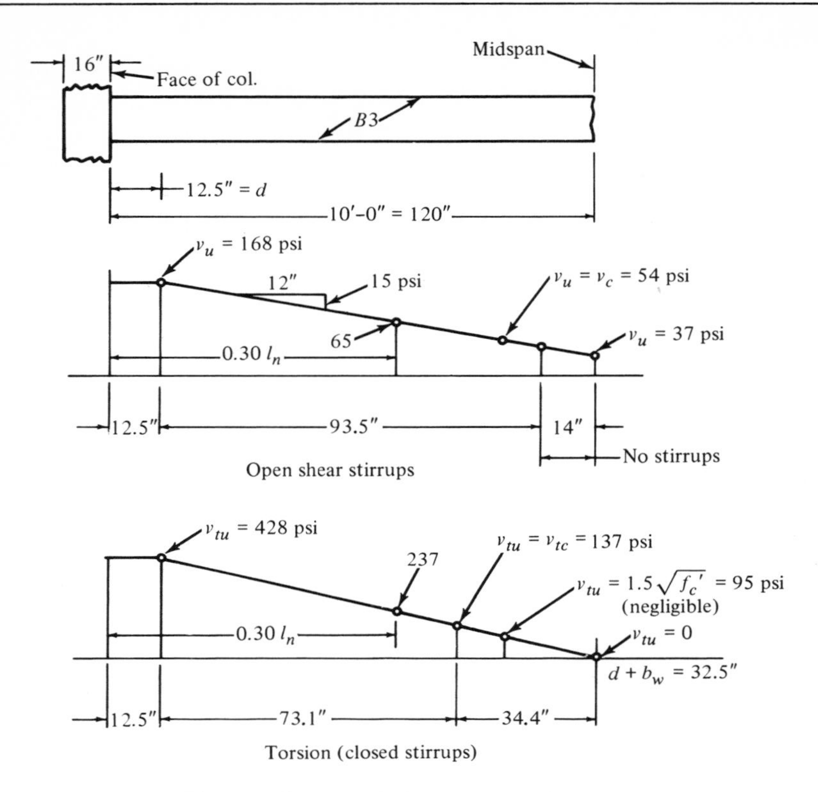

Fig. 9–16 Length for spacing stirrups.

trial width of 20 in. need not be increased. Additional trials are unnecessary (check solution above on chart, Fig. 9-25).

Reinforcement for the shear

The area of (open) stirrups required for shear,

$$A_v = \frac{(v_u - v_c)(b_w)(s)}{f_y} \quad \text{(Section 11.6.1, Eq. 11-13)}$$

$(v_u - v_c) = 167.5 - 54 = 114$ psi $< 4\sqrt{f_c'} = 253$ psi; maximum spacing for the vertical stirrups can be used (Section 11.6.3). Maximum spacing, $s = 0.5d$ (Section 11.1.4). $s = \frac{1}{2}(12.5) = 6.25$ in. max. At $s = 6$ in., $A_v = \dfrac{(114)(20)(6)}{(60{,}000)} = 0.228$ in.^{2} required.

Reinforcement for the torsion

The area of (closed) stirrups required for torsion,

$$\text{Area of one leg, } A_t = \frac{(v_{tu} - v_{tc})(s)(\Sigma x^2\, y)}{(3)(\alpha_t)x_1y_1f_y} \qquad \text{(Section 11.8.2, Eq. 11-19)}$$

where x_1 and y_1 are the short and long dimensions center-to-center of the legs of the closed rectangular stirrups, respectively (Section 11.0), and $\alpha_t = [0.66 + 0\ (0.33)y_1/x_1]$ but not more than 1.50 (Section 11.8.2) (see Fig. 9-15).

$$x_1 = 11.5 \text{ in.};\ y_1 = 16.5 \text{ in.};\ \alpha_t = 1.134$$

(*Note:* Charts are based on out-to-out dimensions of stirrups per standard industry practices; see footnote on page 232.)

Torsional stress carried by the concrete,

$$v_{tc} = \frac{2.4\sqrt{f_c'}}{\sqrt{1 + (1.2\ v_u/v_{tu})^2}} \qquad \text{(Section 11.7.5, Eq. 11-17)}$$

$$v_{tc} = \frac{(2.4)(63.2)}{\sqrt{1 + \left(1.2\ \dfrac{167.5}{428}\right)^2}} = 137 \text{ psi}$$

The minimum area required for one leg of the closed stirrups for torsion with stirrups spaced at 6 in. = A_t:

$$A_{tu} = \frac{(428 - 137)(6.00)(4{,}500)}{(3)(1.134)(11.5)(16.5)(60{,}000)} = 0.203 \text{ in.}^2$$

The maximum spacing of the closed stirrups, $s_{max} = \frac{1}{4}(x_1 + y_1) \leqq 12$ in. $\frac{1}{4}(x_1 + y_1) = (\frac{1}{4})\ (11.5 + 16.5) = 7.00$ in. $<$ 12 in. Since the maximum spacing for the closed stirrups is 7.00 in. and for open stirrups (for shear) is 0.5 d = 6.25 in., only the closed stirrups will be used, at a spacing of 6 in. or less, in sizes proportioned to resist combined shear and torsion. This arrangement will not only result in a minimum number of stirrups, but also will avoid the complications in design, detailing, and placing that would result from mixing two types of stirrups in tricky placing sequences.

Combine A_v and A_t

The combination $A_v + A_t$ will require more reinforcement than the minimum prescribed for shear alone (Sections 11.1.1 and 11.1.2). No minimum is prescribed for torsion. The computed A_v will therefore be used in the combination, $A_v + A_t$, rather than the minimum. Maximum reinforcement for combined shear and torsion at the critical section for a spacing, $s = 6.00$ in. $(2)(A_t) + A_v = (2)(0.203)$ in.^{2} + 0.228 in.^{2} = 0.634 in.^{2}

#5 closed stirrups @ 6 in. would provide an area = 0.62 in.^{2}

Compute the total number of stirrups required and any added longitudinal bars for torsion required in corners of stirrups. In Fig. 9-16 note that the torsion reinforcement is required for a distance of 12.5 + 75.2 + $(d + b_w) > 120$ in. from the face of the support. The Code requires that such reinforcement be provided for a distance $(d + b_w)$ beyond the point theoretically required (Section 11.8.6). $(d + b_w) = 12.5$ in. + 20 in. = 32.5 in. Use 19–#5 ☐ @ 6 at each end of the span.

Longitudinal Bars for Torsion

Longitudinal bars at least #3 in size must be provided in each corner of the closed stirrups. The spacing of these bars distributed inside the perimeter of the closed stirrups shall not exceed 12 in. (Section 11.8.5). In this example, three bars are required in both top and bottom. The bottom bars required for flexure will, of course, satisfy this requirement, and so only the minimum size of the top bars will be computed here. These top bars, which will also serve as part of the flexural reinforcement, must be extended to splice at midspan.

The required minimum area of all longitudinal bars, A_l, is the larger of the two following expression (Section 11.8.4):

$$(1)\ A_l = (2)(A_t)\left[\frac{x_1 + y_1}{s}\right] \quad \text{(Eq. 11-20)}$$

$$(2)\ A_l = \left[\frac{400\,xs}{f_y}\left(\frac{v_{tu}}{v_{tu} + v_u}\right) - 2A_t\right]\left[\frac{x_1 + y_1}{s}\right] \quad \text{(Eq. 11-21)}$$

At the usual cutoff point for top flexural bars, 0.30 l_c, $v_u = 65$ psi and $v_{tu} = 237$ psi. Computed spacing for #5 closed stirrups required for torsion is $s = \frac{428 - 137}{237 - 137}(6) = 17.5$ in. (see Fig. 9-15).

$$(1)\ A_l = (2)(0.31)\left[\frac{11.5 + 17.5}{6}\right] = 3.00 \text{ in.}^2 \text{ at } 12.5 \text{ in.}$$

from face of the support. At bar cutoff, $A_l = (3)(6/17.5) = 1.02$ in.^{2}; $\frac{1}{6}(1.02) = 0.17$ in.^{2}

$$(2)\ A_l = \left[\left(\frac{(400)(15)(6)}{60{,}000}\right)\left(\frac{237}{237 + 65}\right) - (2)(0.31)\right]\left(\frac{28.0}{16.5}\right) < 0$$

Use three #4 top bars continuous (lap spliced at midspan). These bars, or equivalent area, must be *added* to flexural requirements.

Shear and Torsion Design Notes

1. In Examples 1 and 2, two-piece closed stirrups for torsion were shown as the general design solution to satisfy the Code requirement of closed stirrups (Section 11.8.2). One-piece closed stirrups require either prefabrication of the entire beam cage and placing same as a unit—often impossible because top bars must pass between column verticals—or time-consuming assembly in place, perhaps by threading individual flexural bars through stirrups and between column verticals. For practical reasons, the one-piece closed stirrup, spliced by overlapping 90° hooks (Section 7.12.7) should be used only as a last resort. Two-piece closed stirrups should be used wherever possible.

(2) Two-piece closed stirrups must conform to specific code provisions for design (Section 7.12.7). Closure splices for one-piece and two-piece stirrups may consist of:

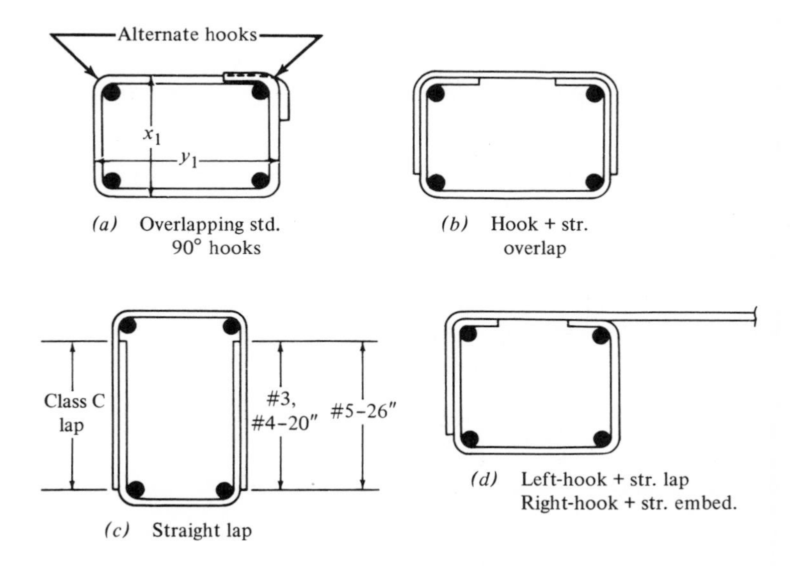

Fig. 9–17 Various closed stirrups.

(a) overlapping 90° end hooks around a longitudinal bar (Section 7.12.7 specifically for one-piece closed hooks and implicitly for two-piece closed hooks)
(b) straight lap lengths
(c) combinations of overlapping hook plus straight bar embedment

Where the straight bar embedment begins at the point of tangency for a hook at the top, the maximum length available to develop the remaining leg for 0.5 l_d, is the distance to the mid-depth of the beam, $\frac{1}{2}$ d, (Section 12.13.1.1). It will be noted that the Code requirements for the closure splices in two-piece stirrups (Section 12.13) are more severe than for the same types of closure splices in one-piece stirrups (Section 7.12.7) (see Fig. 9-17). The figure shows some of the special details for two-piece closed stirrups in beams of various shapes as suggested in the "Proposed Revision of ACI 315-65 Manual of Standard Practice for Detailing Reinforced Concrete Structures." Design limitations upon these arrangements imposed by the Code (Section 12.13) are also shown.

3. The simplest two-piece closed stirrup is shown in sketch (*c*) of Fig. 9-17. It requires a Class C tension lap splice (Sections 12.13.4 and 7.6.1.3) with lap length 1.7 l_d. See Table 13-4 for tension embedment

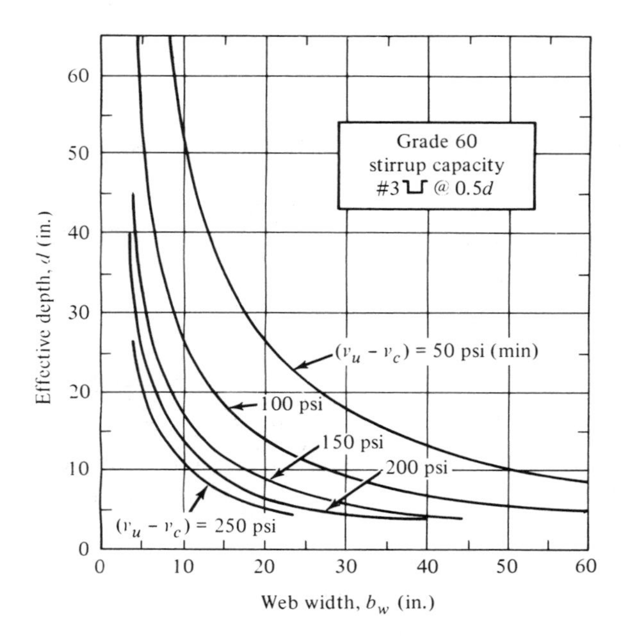

Fig. 9–18 #3 stirrup capacity.

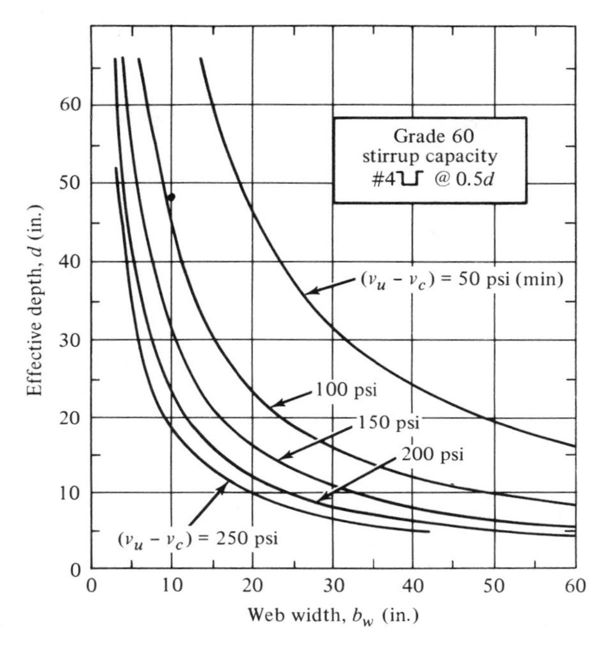

Fig. 9–19 #4 stirrup capacity.

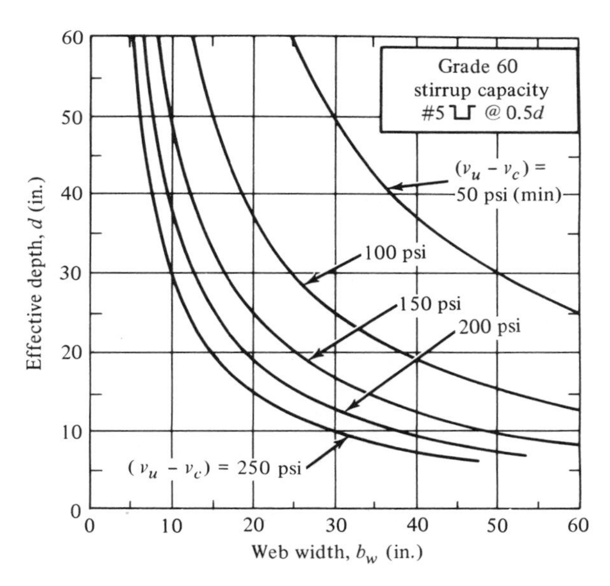

Fig. 9–20 #5 stirrup capacity.

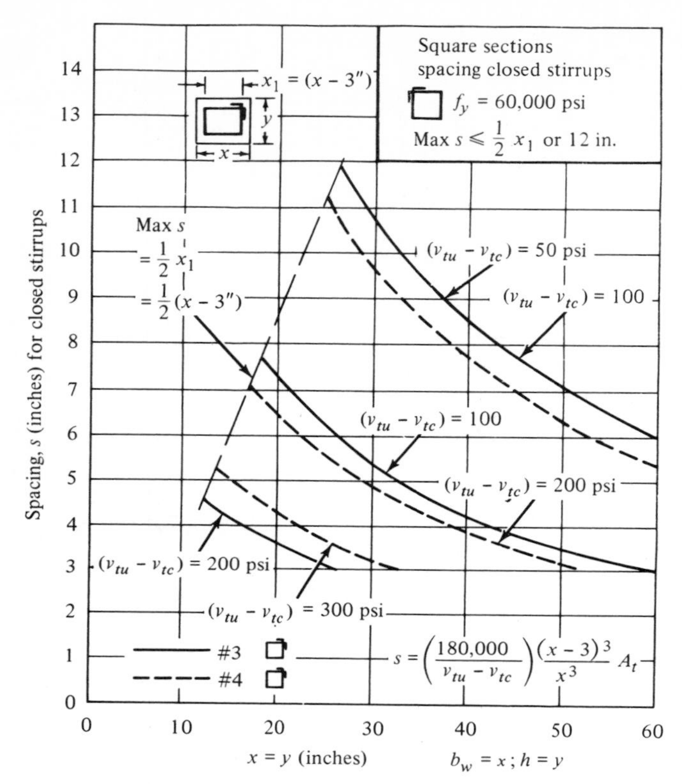

Fig. 9–21 Square sections—closed stirrup spacing chart.

lengths. For all concrete strengths 1.7 l_d = 20 in. for #3, 20 in. for #4, and 26 in. for #5. To determine minimum beam dimensions, add 3 in. for top and bottom cover of $1\frac{1}{2}$ in. for cast-in-place construction.

4. The sketch (*b*) in Fig. 9-17 shows a combination of overlapping between the mid-depth of the member and the start of the hook. For all concrete strengths, 0.5 *d* minimum for #3 = 6 in., for #4 = 6 in., and for #5 = 7.5 in. The resulting minimum beam depths are $17\frac{1}{4}$ in., 18 in., and $21\frac{3}{4}$ in. for #3, #4, and #5 stirrups, respectively.

5. The most efficient arrangement is that of sketch (*d*) since the second piece of the two-piece stirrup can be the negative moment bar developing the torsion which the stirrups must resist. Minimum beam depth is the same as for item 4.

6. To reduce design time for shear, simple charts of required stirrup size and spacing have been prepared for the equation $(v_u - v_c) =$

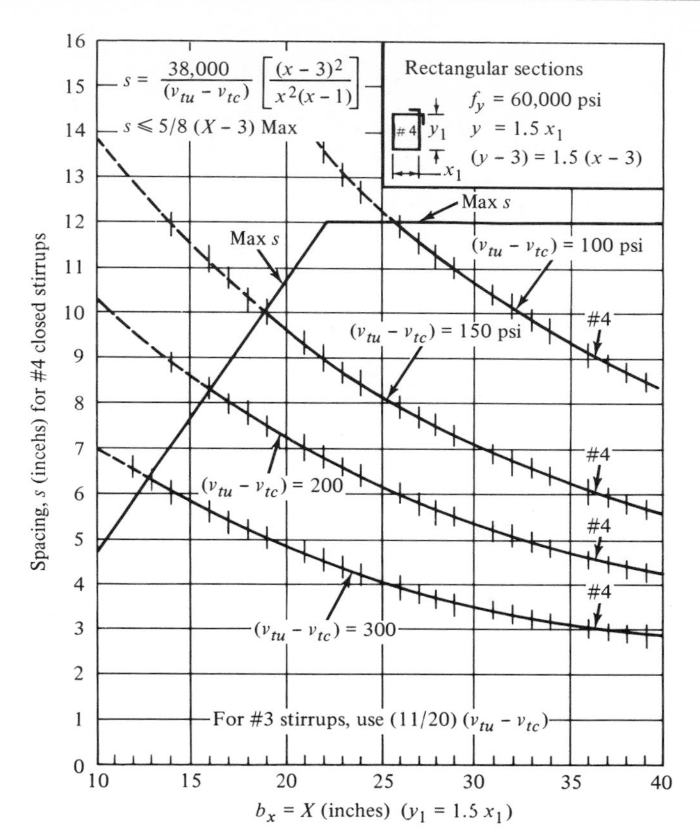

Fig. 9–22 Rectangular sections (1.5:1)—closed stirrup spacing chart.

$\frac{60,000A_v}{b_w s}$ (Section 11.6.1, Eq. 11-13). See Figs. 9-18, 9-19 and 9-20 for #3, #4, and #5 open ⅂⅃ stirrups (2 legs effective). The charts here present capacity for Grade 60 stirrups, in terms of excess shear $(v_u - v_c)$ to be carried by stirrups, applicable with any strength concrete. The charts were prepared for stirrups spaced at 0.50 *d*, the maximum spacing permitted (Section 11.1.4). The range of shear $(v_u - v_c)$ is 250 psi (maximum for $f_c' = 4{,}000$ psi) and spacing 0.50 *d* (Section 11.6.3) to the minimum, 50 psi, for any f_c (Section 11.1.2). The user can easily interpolate readings for any excess shear $(v_u - v_c)$ between values given, select appropriate size of *U*-stirrup, spaced at 0.50 *d*. Alternatively, the user can begin with a $(v_u - v_c)$ for any beam and a desired size of stirrup and reduce the spacing so that:

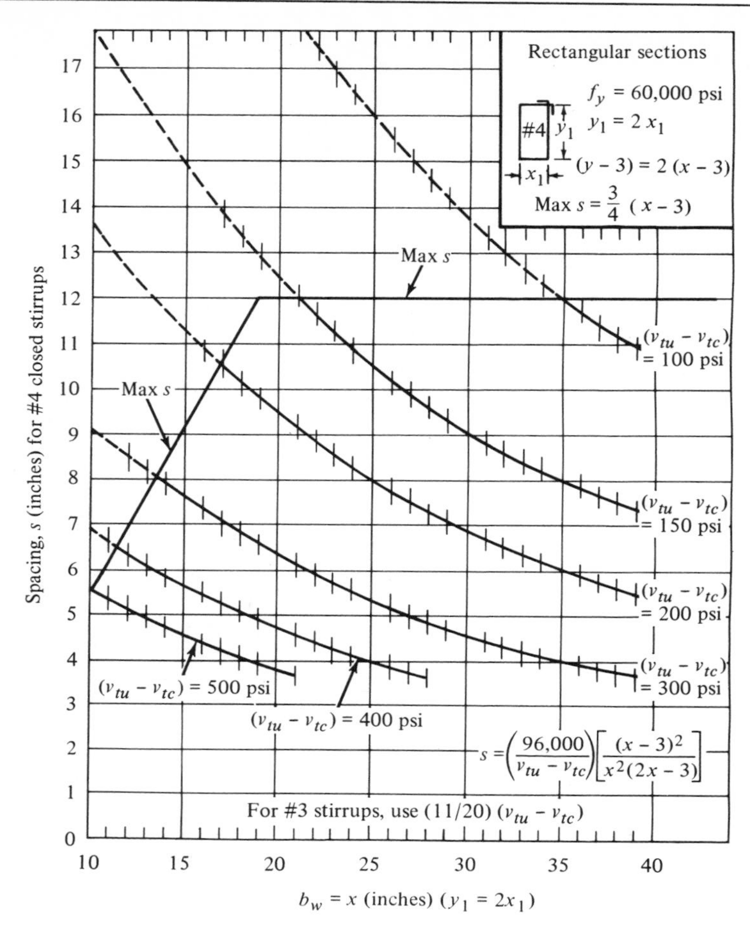

Fig. 9–23 Rectangular sections (2:1)—closed stirrup spacing chart.

$$s = \left[\frac{(v_u - v_c) \text{ tabulated}}{(v_u - v_c) \text{ actual}}\right] (0.50\ d)$$

If beam width makes multiple *U*-stirrups feasible (⅂⅃⅂⅃), allowing 8 in. width per ⅂⅃ in multiple stirrups, increase capacity proportionally as follows:

$$\text{Capacity } (v_u - v_c) = \left(\frac{\text{No. of stirrup legs furnished}}{2}\right) [(v_u - v_c) \text{ tabulated}]$$

Note that a longitudinal bar must be enclosed within each bend top and bottom in multiple stirrups (Section 12.13.2).

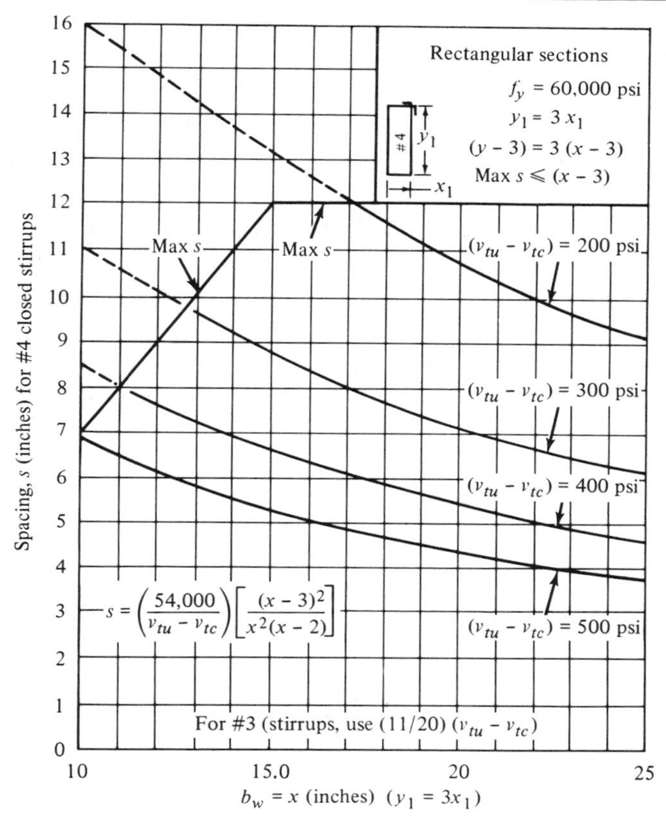

Fig. 9–24 Rectangular sections (3:1)—closed stirrup spacing chart.

7. Beams subject to torsion involve lengthy calculations if performed manually to satisfy a number of interdependent code requirements as in manual Example 2. Since all edge beams and many interior beams with possible heavy unbalanced live loadings on either side must be investigated, rapid answers to the following design questions in the order listed are essential to the practical designer:

(a) Whether the effect of torsion can be neglected (Section 11.7.1); is $v_{tu} \leqq 1.5\sqrt{f_c'}$;

(b) Whether the combined torsional stress plus shear stress is excessive requiring selection of a larger size section (Sections 11.7.7 and 11.6.4)

(c) Whether torsion or shear will control the maximum stirrup spacing and whether this spacing involves an impracticable number of stirrups, requiring in most cases selection of a larger size section.

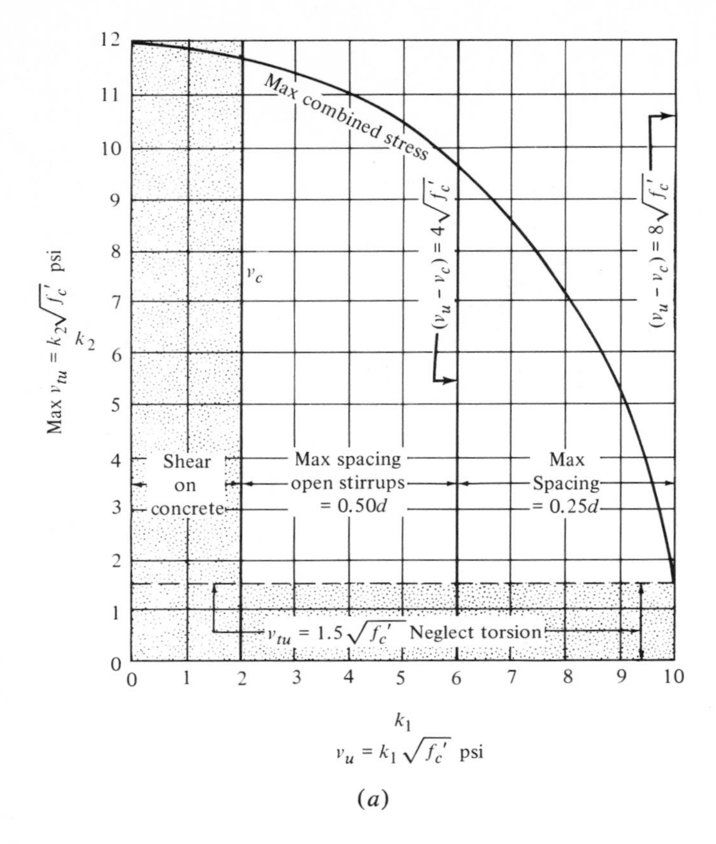

Fig. 9–25 (*a*) Maximum torsion with shear; (*b*) Torsion and shear carried by concrete.

Design aids to provide the answers to the above questions and to permit a rapid completion of the design where stirrups are required including the selection of stirrups have been included in this chapter (see Figs. 9-18 through 9-25). The procedure for the use of these aids is:

a. Compute the applied shear, V_u, and torsion, T_u, at the same criitcal section, a distance d from the face of the support.

b. Compute the unit shear and torsional stresses for the beam section selected as a first trial.

$$\text{Shear, } v_u = \frac{V_u}{\phi b_w d} \qquad \text{Torsion, } v_{tu} = \frac{3\,T_u}{\phi \Sigma x^2\, y}$$

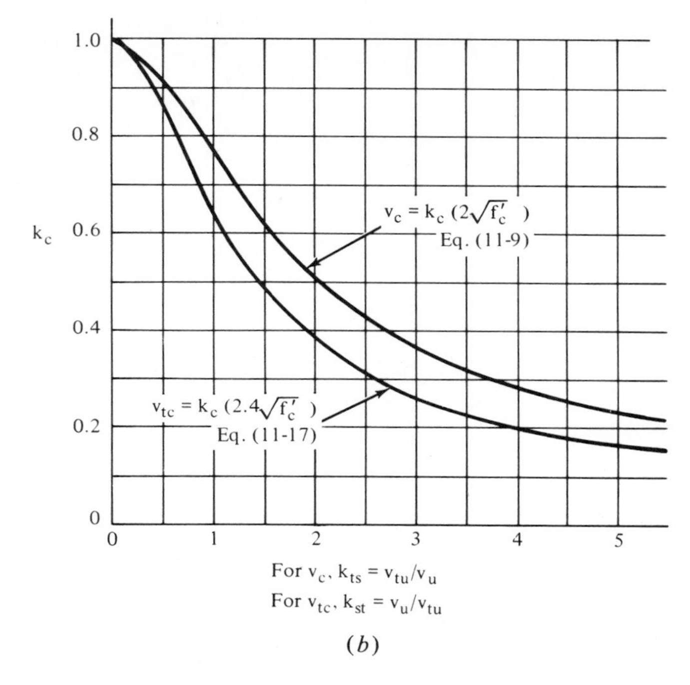

(*b*)

Fig. 9–25 *continued*
(*Courtesy of the Concrete Reinforcing Steel Institute*).

c. Enter Fig. 9-25 with values computed for $\sqrt{f_c'}$, v_u, and v_{tu}. Calculate values for k_1 and k_2. Read whether v_{tu} exceeds the maximum value allowable. Select a larger section and repeat if v_{tu} is above the curve.

d. Compute the allowable torsional stress that can be carried by the concrete, $v_{tc} = 1/5\ v_{tu}$, the maximum allowable total stress; compute $v_c = 2\sqrt{f_c'}$. Compute $(v_{tu} - v_{tc})$ and $(v_u - v_c)$, torsional and shear stresses, respectively, which must be carried by the stirrups. At this point, for discontinuous ends only, it will usually be possible to reduce the number of stirrups substantially by the use of the "long" formula for v_c.

$$v_c = 1.9\sqrt{f_c'} + 2{,}500\ \rho_w \frac{V_u d}{M_u} \quad \text{(Section 11.4.2, Eq. 11-4)}$$

Read the maximum spacing limits for open (shear) stirrups. Increase the

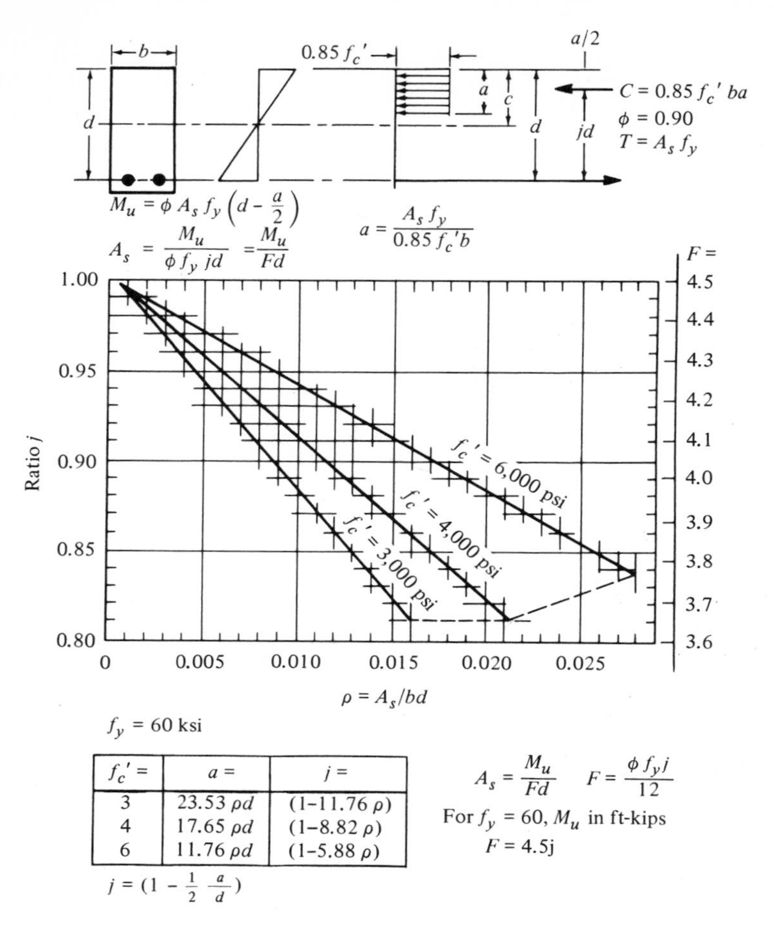

$f_c' =$	$a =$	$j =$
3	$23.53\,\rho d$	$(1-11.76\,\rho)$
4	$17.65\,\rho d$	$(1-8.82\,\rho)$
6	$11.76\,\rho d$	$(1-5.88\,\rho)$

Fig. 9–26 Rectangular section—flexural constants.

beam size and repeat if it is not desirable to reduce maximum stirrup spacing from 0.5 d to 0.25 d.

e. Enter Figs. 9-21 to 9-24 with the value $(v_{tu} - v_{tc})$. Read maximum spacing of closed stirrups for torsion, in inches for #3 and #4 stirrup sizes. For very large beams only, it may be desirable to compute spacings for #5.

f. Enter Figs. 9-18 to 9-20 with the value $(v_u - v_c)$. Read size of stirrups to permit spacing at 0.50 d. Compute size for spacing at 0.25 d.

g. Use judgment to select the simplest practicable arrangement of stirrups in minimum numbers to satisfy the Code. A sketch similar to that in

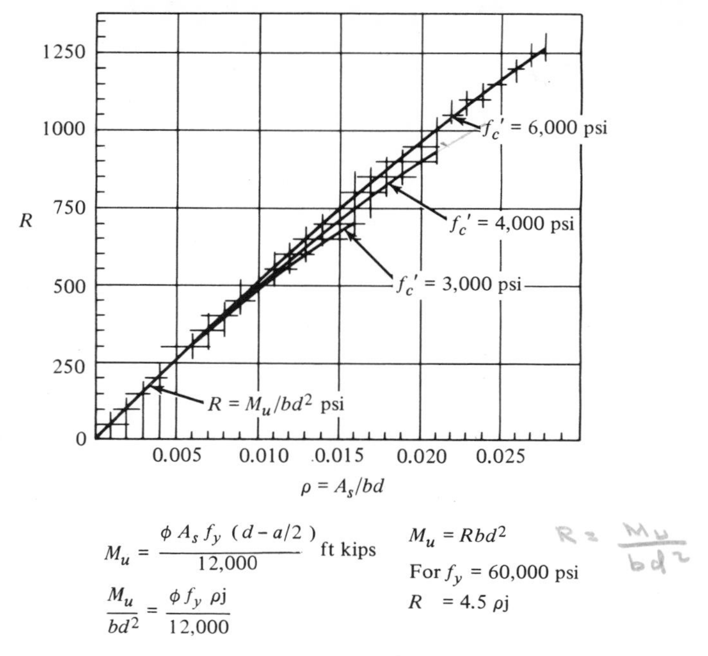

Fig. 9–26 *continued*

Fig. 9-16 will usually save time. The authors recommend that the user *not mix open and closed stirrups in an alternate placing sequence.* If torsion controls the spacing throughout, use only closed stirrups; if torsion controls for only a short distance at each end, it is better to use closed stirrups until torsion can be neglected, and then open stirrups beyond. The simple procedure illustrated in Example 1 can be adopted to determine various stirrup size-spacing combinations acceptable under the Code for comparison.

h. If the final design requires closed stirrups for torsion, longitudinal bars are also required as part of the torsional reinforcement. At the critical section for torsion, the longitudinal bars required for flexure can be utilized. These bars will usually suffice for interior beams, but must be increased in spandrel beams where the maximum flexure and torsion occur under the same loads.

The design charts (Figs. 9-27 and 9-28) will permit a rapid selection of the minimum size longitudinal bars for extension beyond the normal cutoff point of the top bars for flexure. Shear and torsion at this point are usually both low, and both charts are, therefore, based upon $(v_{tu} - v_{tc}) =$

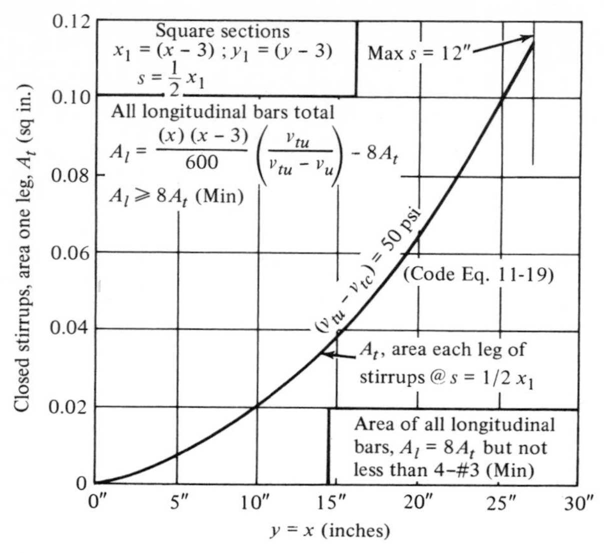

For values of shear to be carried by the reinforcement, $(v_{tu} - v_{tc})$, larger than 50, increase the area per leg of the closed stirrups, A_t, proportionately to the ratio: $\left[\frac{v_{tu} - v_{tc}}{50}\right]$

Fig. 9–27 Minimum torsional steel—square sections.

50 psi. Enter Fig. 9-27 with the actual $(v_{tu} - v_{tc})$; read A_t = the minimum area of one leg of closed stirrups at maximum spacing, $s = \frac{1}{2} x_1$; increase A_t proportionately if $(v_{tu} - v_{tc}) > 50$ psi. The total area of all longitudinal bars, $A_l \geq 8\, A_t$, but not less than four #3. Enter Fig. 9-28 with the torsion ratio $\left(\frac{v_{tu}}{v_{tu} + v_u}\right)$; read A_l/A_t. If the actual value $(v_{tu} - v_{tc}) < 50$, reduce the difference $(A_l/A_t - 8)$ in proportion to $\left(\frac{50}{v_{tu} - v_{tc}}\right)$. Calculate the number of bars required, one per corner and spaced not more than 12 in. around the inside perimeter of the closed stirrups. The area of each bar is *not* required to be equal, which will aid matching the total to that required.

Figures 9-27 and 9-28 were prepared for square sections. For office use, the authors suggest that readers prepare similar charts for various ratios of x_1/y_1 to cover the usual range of rectangular sections. As noted thereon the sample charts are merely solutions of the applicable Code equations (Eqs. 11-1, 11-20, and 11-21).

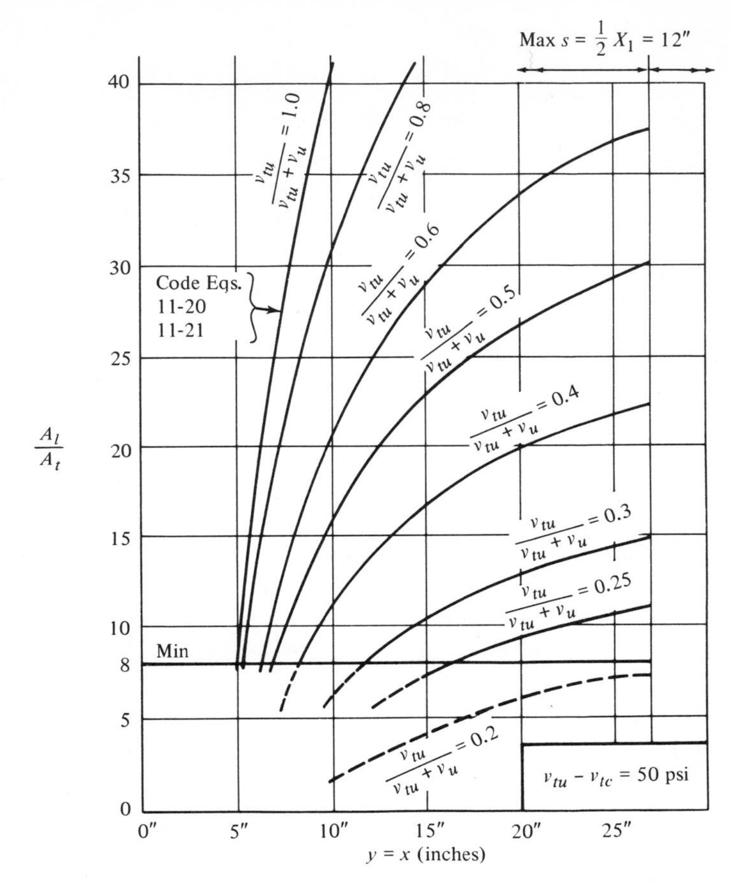

For values of $(v_{tu} - v_{tc})$ greater than 50, the *difference* $(A_l/A_t - 8)$ is reduced in proportion to: $\dfrac{50}{v_{tu} - v_{tc}}$

Fig. 9–28 Longitudinal torsion steel for high torsion rations, $v_{tu}/(v_{tu} + v_u)$, in square sections.

Flexural Design Notes

(1) In most of the manual examples, the flexural computations for the required areas of tension steel, A_s, were carried out by trial and review. The flexural design equations (1963 ACI Code) are not prescribed directly in the 1971 Code, which merely lists the necessary assumptions for design, It has been assumed that the necessary design equations resulting from

these assumptions are well known and accepted in the form shown in the 1963 Code.

These equations are established in simple terms for *review* of trial design sections, but the exact solution for *direct design* can become extremely complicated.

$$A_s = 1.7\ f_c'\ bd\ A_s/f_y - \tfrac{1}{2}\sqrt{\frac{2.89\ f_c'b^2d^2}{f_y^2} - \frac{6.8\ f_c'bM_u}{\phi f_y^2}}$$

Figure 9-26 solves this equation for the flexural constants in the familiar forms used in handbooks for the straight-line flexural design theory for many years. These constants provide quick *direct* solutions for steel area, A_s, or size of section, *bd,* to satisfy the 1971 Code.

$$(1)\ A_s = \frac{M_u}{Fd} \qquad (2)\ M_u = R\ bd^2$$

in which M_u is in ft-kips; A_s in sq in.; and d in in.

(2) In Fig. 9-26 (*a*) derivations are given for quick adjustments of the curves for any concrete strength, f_c' or yield point, f_y. Note that the relations are linear. Three curves are given for the tension steel area constant, *F,* with concrete strengths, $f_c' = 3{,}000$ psi, 4,000 psi, and 6,000 psi and the standard grade $f_y = 60{,}000$ psi.

(3) In Fig. 9-26 (*b*), three curves are presented for the compression constant, *R,* with concrete strengths, $f_c' = 3{,}000$ psi, 4,000 psi, and 6,000 psi using standard grade $f_y = 60{,}000$ psi. For other values of f_y, multiply the steel ratio, ρ, by the ratio $60{,}000/f_y$. It will be noted that the value of R is not sensitive to f_c', except as to the limiting maximum values of steel ratio, ρ. For most designs where 3,000 psi $\leq f_c' \leq$ 6,000 psi, the values of R can be read directly with sufficient accuracy for practical design.

10 columns

DEFINITIONS: COLUMN, WALL, AND PEDESTAL

One of the ambiguous areas in the 1971 Code is the distinction between a load bearing wall (see Chapter 11) and a column. The legal definitions emphasize that a column is "used *primarily* to support axial compressive loads" and that a wall is "used *primarily* to enclose or separate spaces" (Section 2.1). Columns and pedestals are distinguished by the ratio of their height, l_c, to their least dimension, c_1 or c_2; for columns, which must be reinforced (Section 10.9), the minimum ratio is 3; for pedestals, which may be plain concrete, the maximum ratio is 3 (Section 2.1). When a pedestal of any height is loaded beyond the capacity of plain concrete, it must be reinforced *and* designed as a column, and so the overlap of these definitions at the height-to-least dimension ratio equals 3 will seldom cause confusion (Section 15.7.1).

MINIMUM SIZES

There are no limits on the ratio of column cross-section dimensions, c_1/c_2; minimum column dimensions, c_1, c_2; or minimum column cross-sectional areas, A_g; specifically prescribed in the 1971 Code. Arbitrary minimum limits on column cross sections in all previous codes have been eliminated. This change is not drastic, however, as the indirect limits, minimum eccentricity $e = 0.5\ h$ for spiral and $0.10\ h$ for tied columns, but not less than 1 in. (Section 10.3.6) and the reductions prescribed for slender columns (Sections 10.10 and 10.11) effectively limit the minimum size of practicable columns.

DESIGN-CODE REQUIREMENTS

Reinforced concrete column design can be performed manually since the prescribed design assumptions result in simple geometric relationships for strain compatibility, stress in concrete, stress in steel, etc., but practical economics make the cost of design prohibitive for large numbers of columns as in routine work. Since the basic assumptions for design (Section 10.2) are unchanged from previous codes, and the minimum design requirements (Section 10.3) are but slightly revised, the engineer can still use design aids based on the last code, at least for "short" column capacity. These design aids will be applicable through the full range of the load-moment interaction curve except for the portion where $P_u < 0.10\ f_c'A_g$ (Section 9.2.1.2-c and -d). In this range, the column is approaching a condition of pure flexure which is unusual in practice. Although the capacity for this area is prescribed more simply in the new Code, it varies slightly above or below that previously allowed and so will require separate computations.

DESIGN AIDS

Load capacities for short columns with specific sizes, shapes, reinforcement, concrete strengths, and eccentricities are tabulated in the two CRSI Handbooks for both WSD and USD. Column interaction curves showing the full range of load and moment capacities for a variety of concrete-steel strength combinations are available in the ACI Design Handbooks. Computer programs for column design are available from the PCA.

SLENDERNESS EFFECT

New design aids should be developed for slender column design to conform to the new requirements (Sections 10.10 and 10.11). Since the new requirements are somewhat less restrictive on the capactiy of slender columns, use of existing design aids in the interim until new aids are available will usually satisfy the Code. As a quick check on the slenderness effect, note that its effect may be neglected when $kl_u/r < (34 - 12\ M_1/M_2)$ for columns braced against sidesway and $kl_u/r < 22$ for columns not braced against sidesway (Section 10.11.4). Where columns are braced, the effective length ratio, k, is 1.0; for columns not braced against sidesway, $k \geqq 1.0$ (Section 10.11.3). Other terms are defined as follows: l_u = unsupported length; r = radius of gyration for cross-sections (0.30 h for rectangular shapes and 0.25 h for round columns); M_1 and M_2 are the end moments of the column, where $M_1 < M_2$. M_2 is always positive, and M_1 is positive for single curvature, negative for double curvature (Section 10.0).

ALTERNATE DESIGN METHOD

For design by the alternate method (Section 8.10) using unity load and capacity reduction (ϕ) factors, column design capacity is taken as simply 40 percent of that computed by the strength design method.

DESIGN ASSUMPTIONS FOR COLUMNS (Section 10.2)

1. Strain compatibility and equilibrium of forces must be satisfied.
2. Strains in the steel and the concrete are proportional to the distance from the neutral axis.
3. The maximum compressive strain in the concrete is 0.003.
4. The stress in the steel is equal to the strain times E_s which must be equal to or less than f_y.
5. The tensile strength of concrete equals zero.
6. The concrete stress block may be taken as any shape that can be justified by tests (i.e., give the same answer as assumption 7.)
7. A rectangular stress block with concrete stress equal to 0.85 f_c' and extending from the extreme fiber in compression to a straight line parallel to the neutral axis, located at a distance of $a = \beta_1 c$, may be used; c = distance from the outer fiber to the neutral axis, $\beta_1 = 0.85$ when $f_c' \leq 4{,}000$ psi and 0.05 less for each 1,000 psi over 4,000 psi (Standard Whitney method) (see Fig. 2-7).

DESIGN REQUIREMENTS FOR COLUMNS (Section 10.3)

The basic minimum requirements for the design of short columns are (1) the tensile reinforcement ratio, $\rho = A_s/bd$, shall not exceed 0.75 of the ratio producing balanced conditions for the section under pure flexure (Section 10.3.2) and (2) the eccentricity for design, e, must not be less than 0.05 h for spiral or 0.10 h for tied columns, nor less than 1 in. (Section 10.3.6) (see Figs. 2-8 and 9-26).

APPLICATIONS

Consider designs of short tied columns with $e = 0.10\ h$. Use longhand application of the design assumptions and design requirements listed. Compare wall-like columns and walls. Use $f_c' = 4{,}000$ psi; $f_y = 60{,}000$ psi; try minimum steel ratios, $(\rho + \rho') = 0.005$, (Section 10.8.4) and maximum, $(\rho + \rho') = 0.08$ (Section 10.9.1).

Direct design using the basic assumptions is almost impossible by longhand methods of calculation for a given load, P_u, and moment, M_u, but an indirect design by computations of capacity for assumed *positions of the neutral axis* is simple. Computation of several control points for a load-moment interaction curve and constructing such a curve will permit the capacity for any given eccentricity to be read with an accuracy appropriate for design. The upper portion of the interaction curve, where compression controls the design, is very nearly a straight line. Straight line interpolation

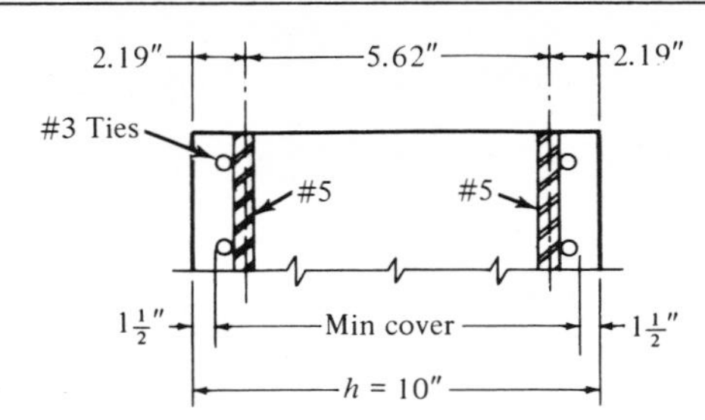

Fig. 10–1 Column cover.

gives a very close approximation for the maximum load capacity with minimum permitted eccentricity, $e = 0.10\ h$. See the typical interaction curve for the full range of e (Fig. 10-16).

EXAMPLE 1: Consider a wall-like column with interior exposure, 10 in. × 60 in., with minimum steel. Use 10–#5 vertical bars placed equally in the long faces. $A_g = 10 \times 60 = 600$ in.^{2} $(\rho + \rho') = 3.10/600 \approx 0.005$. Since the same basic assumptions and requirements for design apply for design of wall-like columns and for design of walls as columns (Section 10.16), the design procedures will be identical, but the capacities computed will vary slightly depending upon the cover of the bars. (With the steel ratio used in this example and the bar sizes indicated, the choice of terminology is the designer's option.)

For columns the minimum cover required is 1½ in. to the *ties* for cast-in-place construction (Section 7.14.1.1); the minimum size of ties is #3 (Section 7.12.3) (see Fig. 10-1).

For walls designed as columns, if ties are provided, minimum size #3, the minimum cover required is ¾ in. to the *ties* (Section 7.14.1.2) (see Fig. 10-2).

For walls designed as columns, lateral ties may be omitted (1) if the area of the vertical bars is 0.01 A_g or less, or (2) where the vertical bars are not required as compression reinforcement (Section 10.16.4) (see

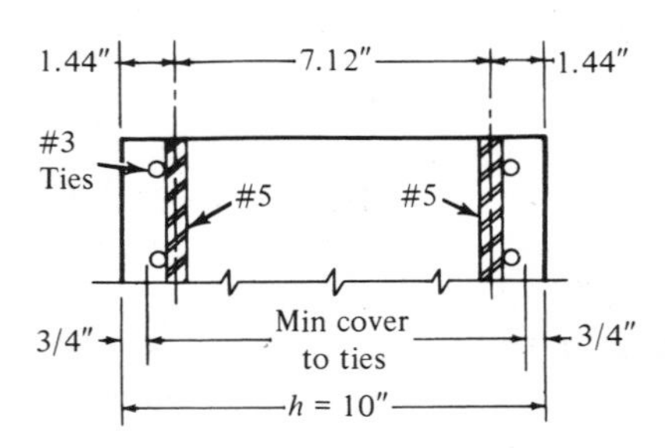

Fig. 10–2 Cover for wall with ties.

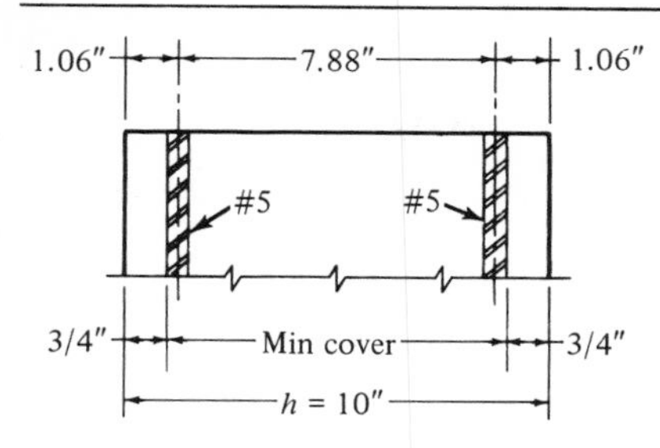

Fig. 10–3 Doubly reinforced wall with ties.

Fig. 10-3). The first provision permits the use of vertical bars in one or two faces without ties for tension or compression stress, provided that the area is less than 0.01 A_g. It would permit the vertical reinforcement to be concentrated in the compression face of the wall in amounts between 0.0012 A_g minimum (Section 10.16.2-a) and 0.01 A_g maximum without ties (Section 10.16.4). If the wall were subjected to loadings causing tension, additional vertical bars as required can be provided in the tension face without ties. *Ties are required only where vertical bars in compression exceed in area 0.01 Ag.* The above interpretation is the most liberal and is recommended as such by the authors, although the Code provisions are sufficiently ambiguous to permit other interpretations.

The previous three arrangements are for a doubly reinforced cross-section. Walls may also be singly reinforced if not more than 10 in. thick (Section 14.2-g) and designed by the "wall formula" (Code Eq. 14-1) under the empirical method, provided the eccentricity, $e \leq h/6$ (Section 14.1.2) (see Chapter 11). For larger eccentricities, such walls must be designed as columns (Section 14.1.1). See Fig. 10-4 for an equivalent single reinforced simple arrangement of 5–#7 verticals without ties, $(\rho + \rho') = 0.005$.

For the wall-like column of Fig. 10-1, compute the maximum capacity allowed ($e = 0.10\ h = 1$ in.) by constructing the upper portion of its load-moment interaction curve and interpolating along the curve to the point where $e = 1$ in. (see Fig. 10-5).

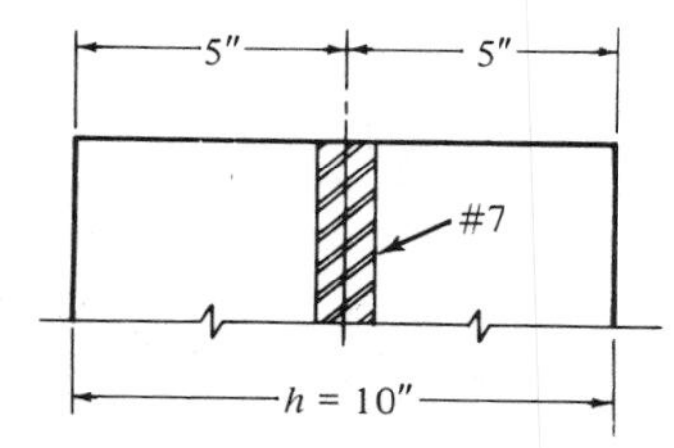

Fig. 10–4 Singly reinforced wall with ties.

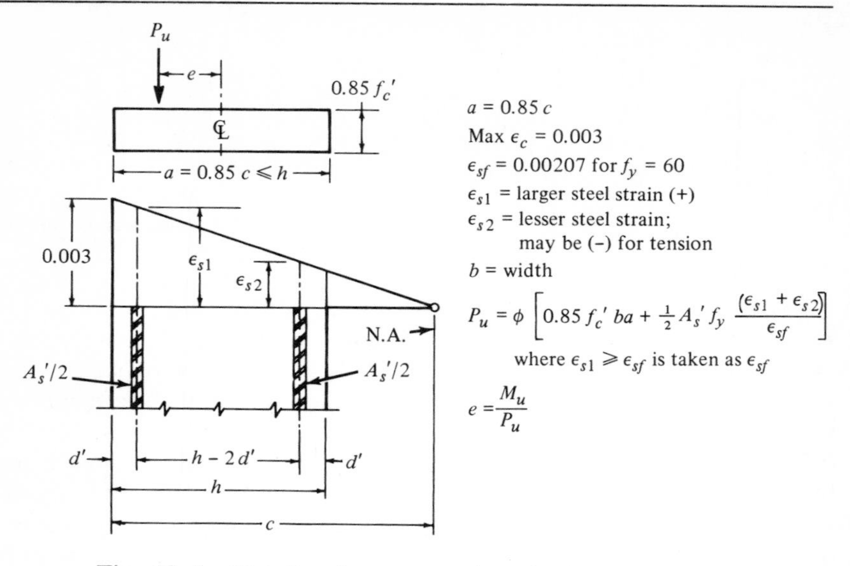

Fig. 10–5 Notation for stress and strain (example 1).

Case (a)

$e = 0$; $c = \infty$; $a = 10$ in.; $\xi_c = \xi_{s2} = \xi_{s1} = 0.003$; 10-#5, $A_s' = 3.10$ in.^{2}

$$P_u = (0.70)[(0.85)(4)(10)(60) + (3.10)(60)]$$

$$= (0.70)(2040 + 186)$$

$$P_u = 1{,}558\ k \qquad M_u = 0$$

Case (b)

(See Fig. 10-6.) Take $c = 10/0.85 = 11.76$ in.; $a = 10$ in.

$$\xi_{s1} > 0.00207,\ f_{y1} = 60 \text{ ksi};\ \xi_{s2} = \left[\frac{3.95}{11.96}\right](0.003) = 0.000991$$

$$P_u = 0.70\left[2040 + \frac{3.10}{2}(60) + \frac{3.10}{2}\left(60 \times \frac{0.000991}{0.00207}\right)\right]$$

$$= (0.70)(2040 + 93 + 44.5)$$

$$P_u = 1{,}524\ k \qquad Mu = P_u\,e = (0.70)(93 - 44.5)\,\tfrac{1}{2}\,(5.62) = 95.4''k$$

$$e = M_u/P_u = 0.0625 \text{ in.}$$

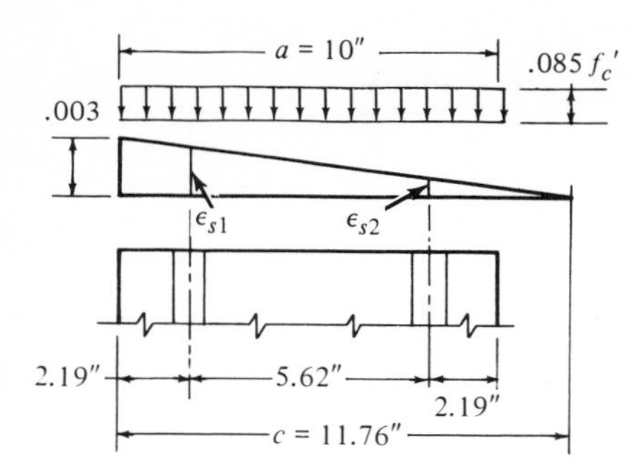

Fig. 10–6 Case (*b*).

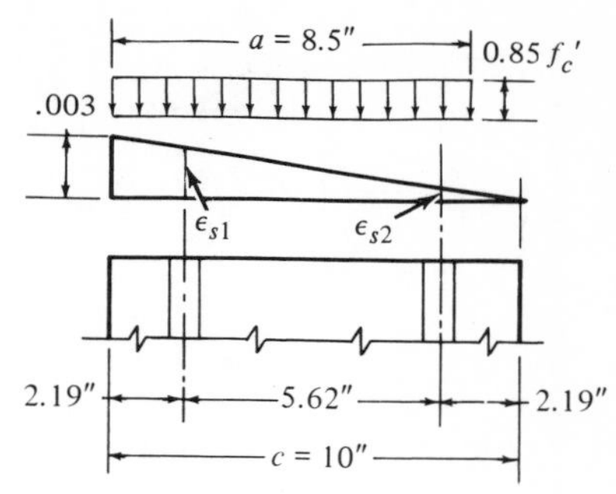

Fig. 10–7 Case (*c*).

Case (c)
(See Fig. 10-7.) Take $c = 10$ in.; $a = 8.5$ in.

$$\xi_{s1} = \frac{7.81}{10.00}(0.003) > 0.00207; \quad f_{y1} = 60 \text{ ksi}$$

$$\xi_{s2} = \frac{2.19}{10.00}(0.003) = 0.000657;$$

$$P_u = 0.70\left[(0.85)(4)(8.5)(60) + (1.55)(60) + (1.55)(60)\left(\frac{657}{2070}\right)\right]$$

$$= 0.70\,(1734 + 93 + 29.5) = 1214 + 65 + 20.7$$

$$P_u = 1300\ k \qquad M_u = (1214)(5.00 - 4.25) + (65.0 - 20.7)(2.81)$$

$$M_u = 1035''k;\ e = 1035/1300 = 0.796 \text{ in.}$$

Case (d)
(See Fig. 10-8.) Take $c = 9.4$ in.; $a = 8.0$ in.

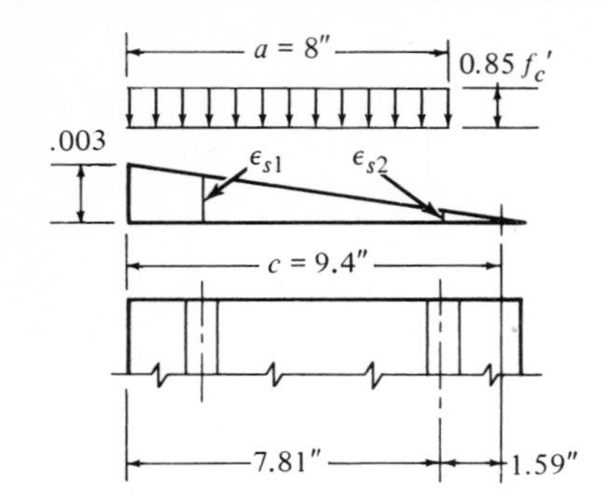

Fig. 10–8 Case (*d*).

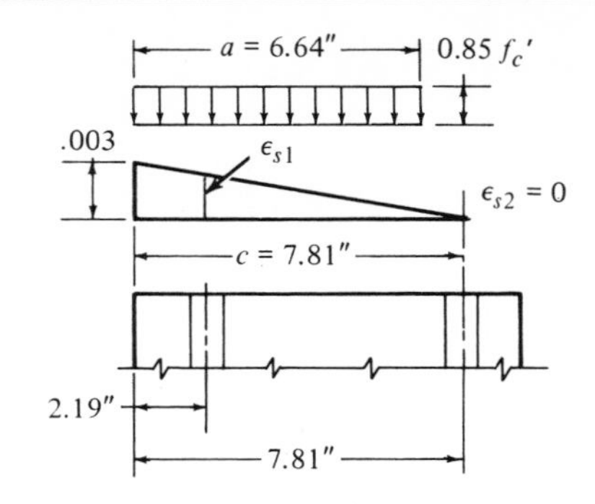

Fig. 10–9 Case (*e*).

$$\xi_{s1} = \frac{7.21}{9.40}(0.003) > 0.00207,\ f_y = 60 \text{ ksi}$$

$$\xi_{s2} = \frac{1.59}{9.40}(0.003) = 0.000507,\ f_{y2} = \frac{507}{2070}(60) = 14.7 \text{ ksi}$$

$$P_u = (0.70)[(3.4)(8)(60) + 1.55\ (60 + 14.7)]$$

$$P_u = (0.70)(1632 + 93 + 23) = 1142 + 65 + 16 = 1223\ k$$

$$M_u = (1140)(5 - 4) + (65 - 16)(2.81) = 1278 \text{ in.-kips}$$

$$e = M_u/P_u = 1278/1223 = 1.045 \text{ in.}$$

Case (e)
(See Fig. 10-9.) Take $c = 7.81$ in.; $a = 6.64$ in.

$$\xi_{s1} = \frac{5.62}{7.81}(0.003) > 0.00207,\ f_{y1} = 60 \text{ ksi}$$

$$\xi_{s2} = 0, \qquad f_{y2} = 0$$

$$P_u = (0.70)[(3.4)(6.64)(60) + (1.55)(60)]$$

$$= (0.70)(1355 + 93) = 948 + 65.1$$

$$P_u = 1013\ k$$

$$M_u = (948)\left(5.00 - \frac{6.64}{2}\right) + (65.1)(2.81) = 1{,}593 + 183$$

$$M_u = 1776 \text{ in.-kips} \qquad e = 1776/1013 = 1.75 \text{ in.}$$

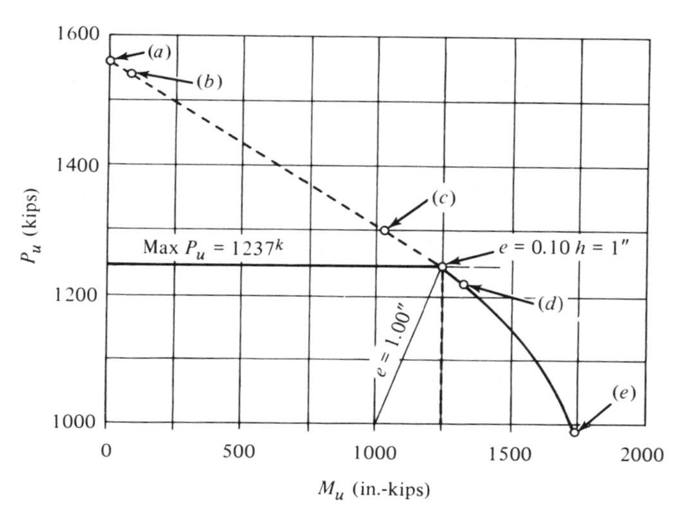

Fig. 10–10 Capacity at $e = 0.10\ h$ about minor axis.

If the five points for (P_u, M_u) computed in (a), (b), (c), (d), and (e) are plotted as the upper portion of a load-moment interaction curve, the capacity at $e = 1.0$ in. can be determined by simple straight line interpolation since this portion of the curve is very nearly a straight line (see Fig. 10-10). Interpolating between (c), $e = 0.796$ in. and (d), $e = 1.045$ in., for $e = 1.0$ in., $P_u = 1237$ kips and $M_u = 1237$ in.-kips.

The requirement for a minimum eccentricity has the effect of cutting off the top of the interaction curve since all usable combinations of load and moment lie below this line. Note that the column of Example 1 has been investigated for short column capacity with bending about the minor principal axis, that is, parallel to the short dimension of the cross section. From Fig. 10-10 it can be seen that columns less than 10 in. thick would be controlled by the minimum eccentricity of 1 in., since capacity is reduced more rapidly thereafter for the 10 in. example column.

CAPACITY ABOUT MAJOR AXIS. To complete the investigation of the maximum short-column capacity for the column of Example 1, it is necessary to determine whether the capacity at minimum eccentricity about the major or minor axis controls. The Code requires that the minimum eccentricity be considered separately about "*either* principal axis" (Section 10.3.6). It was not intended that biaxial bending at the minimum eccentricity in both directions be considered simultaneously (Commentary 10.3.6). Where a known eccentricity exists in both directions under any loading condition, a biaxial analysis is, of course, required (Commentary 10.3.6).

For the wall-like column of example 1, the minimum eccentricity, $e = 0.10\ h = 6$ in. about the major axis. The usual practice in detailing a 10 in. × 60 in. column for 10–#5 bars in two faces would be to space them equally, beginning with a half-space, unless the designer indicated some other specific arrangement desired (see Fig. 10-11). The capacity at $e = 0$ for bending in the longer direction is the same as case (a) previously calculated. $P_u = 1558$ kips; $M_u = 0$; $e = 0$.

EXAMPLE 1 (Cont.)

Case (f)
(See Fig. 10-12). Take $c = 54$ in. so that $f_{s5} = 0$.

$$f_{s1} = (48)\frac{(0.003)}{(54)}\frac{(60)}{(0.00207)} = (48)(1.61) > 60; \quad \text{use } f_y = 60 \text{ ksi}$$

$$f_{s2} = (36)(1.61) = 58 \text{ ksi}$$

$$f_{s3} = (24)(1.61) = 38.6 \text{ ksi}$$

$$f_{s4} = (12)(1.61) = 19.3 \text{ ksi}$$

$$f_{s5} = (0)(1.61) = 0 \text{ ksi}$$

$$P_u = (0.70)[(0.85)(4)(45.9)(10) + (0.62)(60 + 58 + 38.6 + 19.3)]$$

$$= (0.70)(1560 + 37.2 + 35.9 + 23.9 + 12.0 + 0)$$

$$= (1093 + 26.0 + 25.2 + 16.7 + 8.4)$$

$$= 1169 \text{ kips}$$

$$M_u = (1093)(30.00 - 23.00) + (26.0)(24) + (25.2)(12) - (8.4)(12)$$

$$= 8476 \text{ in.-kips}$$

$$e = 8476/1169 = 7.25 \text{ in.}$$

Case (g)
(See Fig. 10-13). Take $c = 60$ in.; $a = 51$ in.

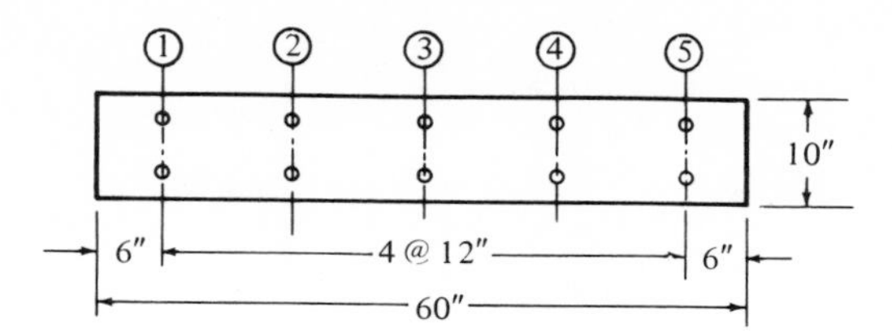

Fig. 10–11 Bar layout plan (Example 1).

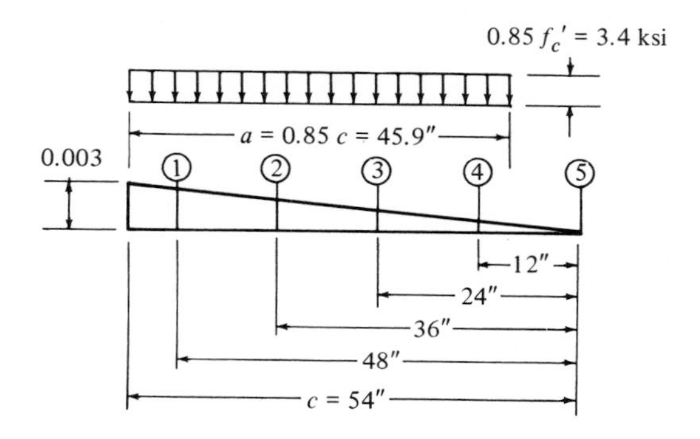

Fig. 10–12 Case (*f*).

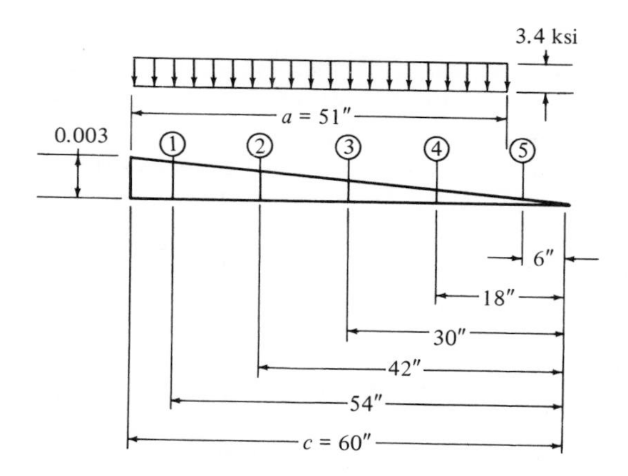

Fig. 10–13 Case (*g*).

$$f_{s1} = (54)\frac{(0.003)}{(60)}\frac{(60)}{(0.00207)} > 60; \quad \text{use } f_{s1} = 60 \text{ ksi}$$

$$f_{s2} = (42)(1.45) > 60; \quad \text{use } f_{s2} = 60 \text{ ksi}$$

$$f_{s3} = (30)(1.45) = 43.5 \text{ ksi}$$

$$f_{s4} = (18)(1.45) = 26.1 \text{ ksi}$$

$$f_{s5} = (6)(1.45) = 8.7 \text{ ksi}$$

$$P_u = (0.70)[(3.4)(52)(10) + (0.62)(60 + 60 + 43.5 + 26.1 + 8.7)]$$

$$P_u = (1214 + 26.0 + 18.9 + 11.3 + 3.8)$$

$$= 1302 \text{ kips}$$

$$M_u = (1214)(30 - 25.5) + (26)(24) + (26)(12) + (18.9)(0) -$$

$$(11.3)(12) - (3.8)(24) = 6172 \text{ in.-kips}$$

$$e = 6172/1300 = 4.75 \text{ in.}$$

Plotting the values for capacity from cases (a), (f), and (g) develops the straight line top portion of the interaction curve for short-column capacity bending about the major axis. Interpolation shows approximately $P_u = 1230$ kips and $M_u = 7380$ in.-kips at $e = 6$ in. (see Fig. 10-14). Note that even with this rather inefficient arrangement of steel for bending in this direction, the limiting capacities are close. In this case, bending about the major axis in the long or strong direction controls the maximum load, P_u, at minimum eccentricity. When the column is not braced against sidesway in the weak direction, this conclusion is of purely academic interest. The reductions of capacity prescribed (Sections 10.10 and 10.11) for slenderness will reduce the capacity for bending in the direction of the 10 in. dimension drastically for any practicable column not braced against sidesway. See the discussion on long columns following in this chapter.

SHORT CUT TO DIRECT DESIGN OF COLUMNS

The capacity of columns at minimum eccentricity allowed by the Code is a most important practical consideration because most routine interior building columns are in this class. Where walls or other bracing provide

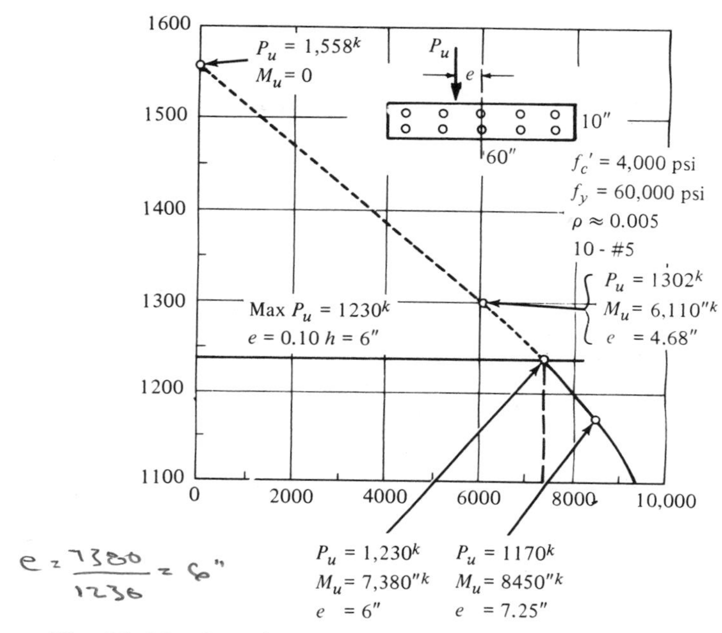

Fig. 10–14 Capacity at $e = 0.10\,h$ about major axis (example 1).

lateral resistance to sidesway of the structure, the design of most if not all columns will be controlled by the minimum eccentricity specified in the Code.

An examination of column capacity tables in various handbooks and computer analyses to determine the average compressive stress on the gross column area with load at minimum eccentricity reveals little variation with shape of the column cross-section, the bar layout or sizes of bars, or details of cover and ties. The major variables are $\phi = 0.70$ for walls and tied columns, f_c', f_y (Grade 60 is standard), and the total steel ratio, $\rho = A_s/A_g$. This reduction of the number of significant variables makes possible a very simple design aid for columns (see Fig. 10-15). The chart will enable the designer to make a fast selection of a preliminary size column with a low steel ratio that can be used for elastic analyses and easily adjusted if the design moment exceeds that corresponding to a minimum eccentricity by adding steel, usually without changing the column size, which would require a second trial analysis. The results are usually accurate within acceptable limits for final design where the actual eccentricity computed is 0.10 h or less. Note the narrow range of the band for each f_c' to allow for the lesser variables such as the size, ratio of cover

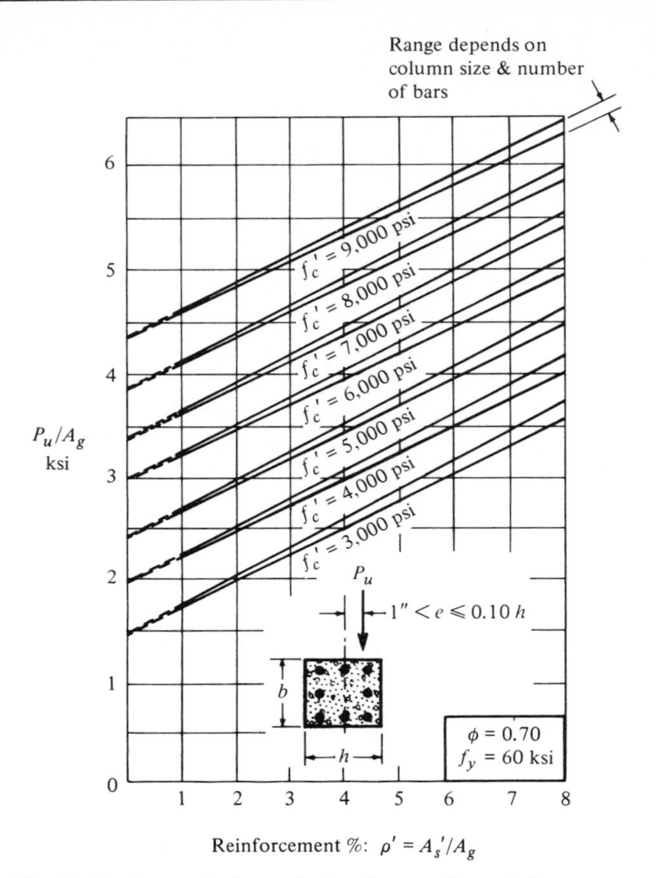

Fig. 10–15 Preliminary design—tied columns for minimum eccentricity.

to total thickness, shape, number of bars, and layout of bars. As a quick check, extend the band for $f_c' = 4{,}000$ psi to $\rho = 0.005$ and read the average $P_u/A_g = 2.1$ ksi. Compare this value to that for the wall-like column in Fig. 10-14 which is an unusual shape. $P_u = 1230$ kips; $A_g = 600$ in.^{2}; $P_u/A_g = 2.05$ ksi.

APPLICATION OF PRELIMINARY DESIGN CHART TO WALLS AND LIGHTLY REINFORCED COLUMNS

The chart in Fig. 10-15 covers the range of reinforcement ratios from $\rho = 0.01$ to $\rho = 0.08$ because these are the limits established in the Code for "compression members" (Section 10.9.1). The user is advised to interpret this requirement as intended for principal columns loaded to full capacity. The ratio $\rho = 0.005$ becomes particularly significant since one of the exceptions provided to the requirement for a minimum $\rho = 0.01$ is that ". . . a reduced effective area A_g not less than one-half of the total area may be used for determining minimum steel area and load capacity" (Section 10.8.4). This provision permits retaining the same size column from footing to roof for most buildings to save formwork changing cost and at the same time reducing the minimum area of vertical bars to a more economical minimum of 0.005 times A_g in the upper stories. For bearing walls designed as columns, the minimum ratio of vertical reinforcement to gross area is 0.0015 for bars larger than #5 (Section 10.16.2). Since walls may be designed without ties (Section 10.16.4) for ratios of vertical reinforcement, $p \leq 0.01$, by the same capacity formulas as columns, it seems reasonable to interpret the intent of the Code that the minimum area of vertical bars in tied columns, $\rho = 0.005$, also permits full utilization of steel and concrete capacities.

LOAD-MOMENT INTERACTION CURVES

From the preceding explanations and examples, it is evident that longhand calculations to apply the basic assumptions and requirements for routine column design are not economically practicable. The working engineer must have access to design aids for short-column capacities. Offices with in-house computer facilities will find subroutine programs readily available, mostly based upon the 1963 ACI Code, but nevertheless applicable for the entire range of $P_u \geq 0.10\ f_c'\ A_g$ under the 1971 Code. The 1971 revisions in the range where $P_u \leq 0.10\ f_c'\ A_g$ are quite simple, and easily adopted into a computer program (or performed manually) (Sections 9.2.1.1-c and -*d*). The ACI Design Handbooks present short-column design aids in the form of interaction curves; also some of the computer output is in this form. It behooves the working engineer to become familiar with this form of design data.

Key Points on Interaction Curves

See Fig. 10-16 for a typical load-moment interaction curve for a "short column" with the key points for construction of same identified. Note that it was not necessary for Code purposes to provide standard nomenclature for all of these key points. The nomenclature in Fig. 10-16, followed elsewhere in this Guide, has been used informally by many writers and should either be familiar to the user or self-explanatory. Briefly, the key points, in order from the top down, are identified and defined as follows:

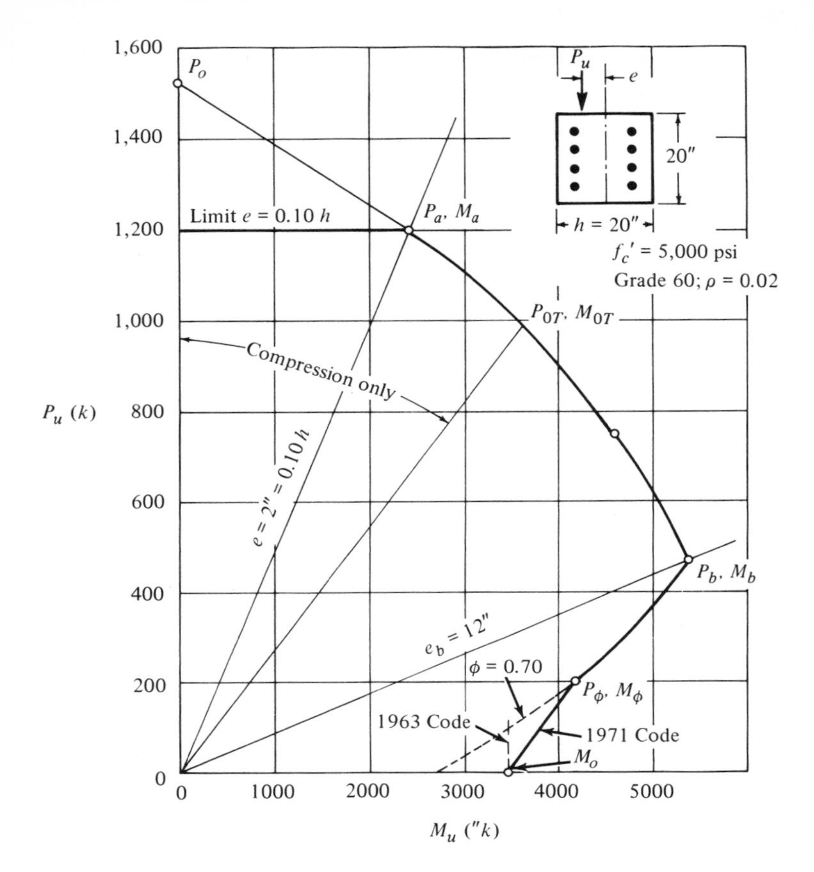

Fig. 10–16 Typical interaction curve.

1. P_o—axial capacity at $e = 0$ (not permitted for use in design; useful only for constructing the curve).

2. P_a—load at minimum prescribed eccentricity, $e = 0.10\ h$ or 1 in. whichever is larger; maximum usable axial capacity allowed on short columns. M_a is the accompanying moment.

3. P_{ot}—load at the point of zero tension in the steel nearest to the "tension face." Compression lap splices used for lesser moments will satisfy the Code requirements for minimum tensile capacity (Sections 7.10.3 and 7.10.5). End bearing splices with a minimum stagger of 50 percent A_g in each face will satisfy the Code for somewhat greater moments.

4. P_b—load at the point of balanced conditions; the compression strain in the concrete reaches 0.003 and the stress in the tensile reinforcement

reaches f_y simultaneously; the point of transition between a failure by compression in concrete or yielding of the steel controlling. M_b is the accompanying moment.

5. P_ϕ = the load at the beginning point for the transition from a ϕ = factor of 0.70 for compression members (columns) to 0.90 for flexural members (beams). M_ϕ is the accompanying moment.

6. M_o—moment at the point of "pure flexure," where $P_u = 0$; $e = \infty$, and $\phi = 0.90$.

Major and Minor Axes—Double Interaction Curves

Several features of the typical load-moment relationship may be observed in Fig. 10-16. The quantitative effect of a moment for eccentricity, $e = 0.10\ h$, in this case reduced the axial load capacity about 22 percent. This percentage is about the average; it may be higher with the steel unfavorably located. The identifying sketch showing axis of bending is necessary for any steel arrangement except the standard 4-bars with one in each corner and those with reinforcement symmetrical about both axes of bending, square or round columns with bars in a circular pattern, and multiples of four bars in all faces of square columns. Except for these doubly symmetrical steel arrangements in square columns, round columns, or special doubly symmetrical column shapes, two load-moment curves are required to define the maximum axial load capacity permitted for use at $e = 0.10\ h$, for bending about the major and the minor axes (see Figs. 10-10 and 10-14). Note the effect of the transition in Fig. 10-16 from $\phi = 0.70$ for the tied column to $\phi = 0.90$ for pure flexure. This transition typically begins at $P_u = 0.10\ f_c' A_g$, but for unusual columns of low f_c', high f_y, high steel ratios, unsymmetrical layout ($A_s' < A_s$), and high ratios of cover to total depth, where $P_b \leq 0.10\ f_s'\ A_g$, the transition begins at P_b (Section 9.2.1.2-d). Note that the 1963 Code transition in the typical example was the more conservative.

Transition ϕ Factors from Column to Beam

The 1963 Code, however, cannot be assumed to provide more conservative capacities in the transition zone of the interaction curves. Use of existing handbooks or computer programs will require a check of values in this range to determine whether they conform to the 1971 Code (see Fig. 10-17). Here is shown the lower portion of the curve for the column in Fig. 10-16 with $\rho = 0.02$ compared to the curves for the same column with $\rho = 0.04$ and $\rho = 0.06$. The curve for the high percentage of steel, $\rho = 0.06$, represents about the extreme difference, since the 1963 Code

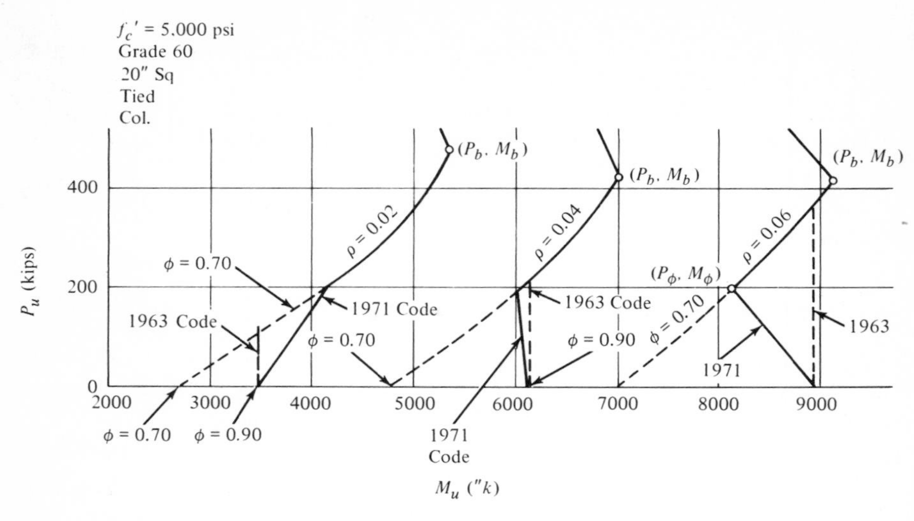

Fig. 10–17 Short columns, 1963 vs. 1971 codes at transition ϕ.

limited the vertical transition to $P_\phi \leqq P_b$. While the problem of providing a smooth transition with no embarrassing, obviously illogical discontinuity was important to writing the Code, it does not represent a condition encountered often in routine practical design.

END BEARING SPLICES

More important features of the Code for practical everyday design are shown in Fig. 10-18. The point of zero tension in the vertical bars on the tension face is easily located by longhand calculations, and is identified in most handbooks in either load tables or interaction curves and in most computer design programs. Under the Code, columns for all load-moment combinations above this point can be designed for either end-bearing splices staggered 50 percent or compression lap splices for 100 percent of the vertical bars (Sections 7.10.3, 7.10.5, and 7.7.2). Since column splice design for multistory columns is a most important problem not yet susceptible to mechanical solutions and practical columns commonly lie above the point (P_{ot}, M_{ot}) this point becomes essential to economical design. Note double interaction curves constructed for values of $M_u > M_{ot}$ showing reduced bending capacity if end bearing splices are used for 50 percent of the bars (stagger 50 percent). The curve for 50 percent end bearing splices (dashed line) must be employed for tensile

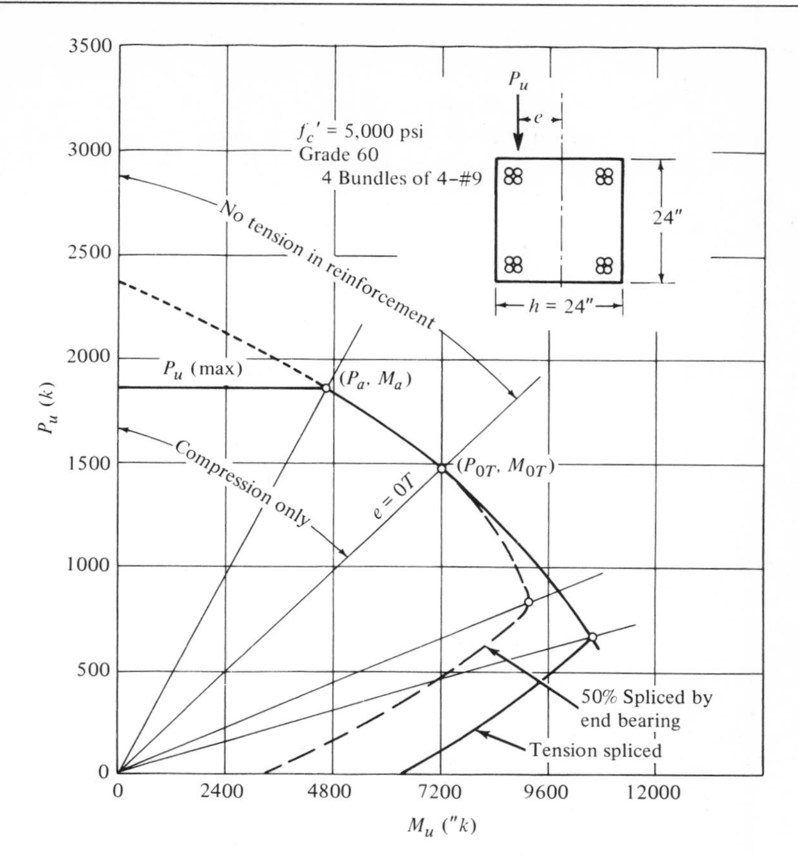

Fig. 10–18 Interaction Curve—End Bearing Splices.

stress larger than 25 percent f_y in the bars of any bundle (Section 7.10.3), and is conservative for all loads causing lesser tension.

BUNDLED VERSUS SEPARATE BARS

For smaller bars, the most economical design is to use compression lap splices of 100 percent of A_s at the splice section, using bars one story in height. For the large bars, #14 and #18, for which lap splices are not allowed (Section 7.5.2), the end bearing splice with 50 percent stagger using bars of two-story height is the economical solution. Tension splices can become quite costly for the high-strength bars in larger sizes. (see Chapter 13 for required lap lengths). In Fig. 10-18, note the reduction in capacity in the zone of high moments with 50 percent A_s spliced by end bearing; few practical columns lie in this zone. Note also the greater efficiency to resist bending in either direction of the four 4-bar bundles in the corners of the column compared to sixteen separate bars in all four faces. The bundled arrangement requires one tie per spacing instead of three per set required for separate vertical bars (Section 7.12.3). See the following section in this chapter on column details.

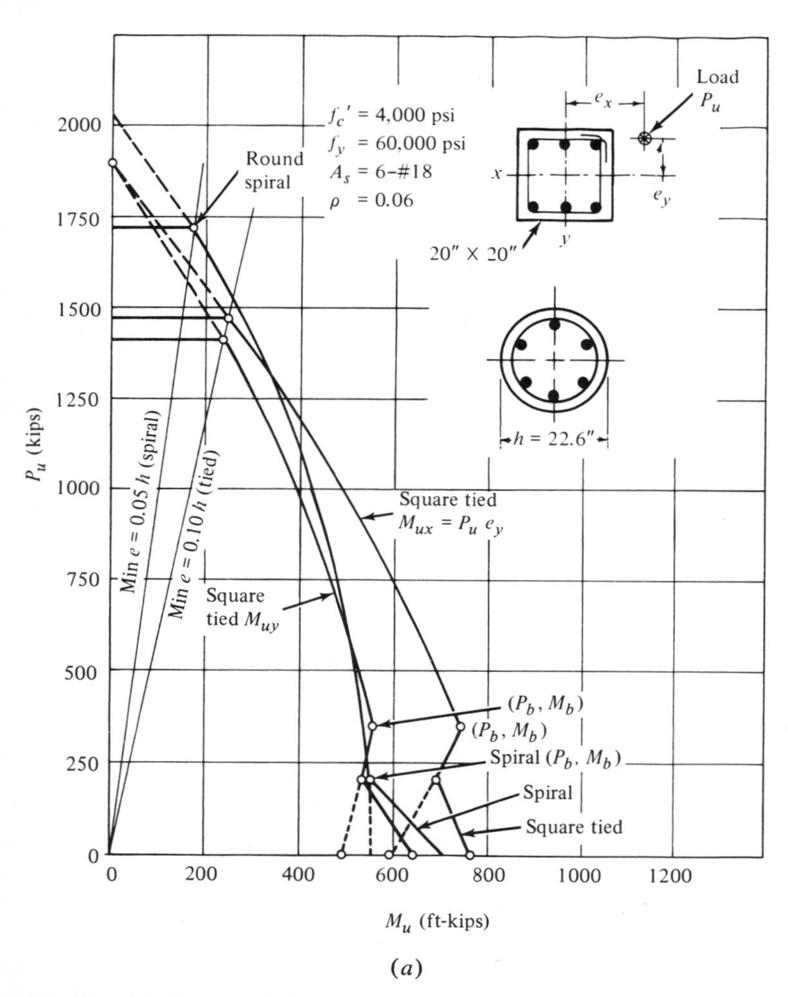

(a)

Fig. 10–19 (a) Square tied vs. round spiral columns; (b) square vs. round tied columns; (c) round columns—spiral vs. ties.

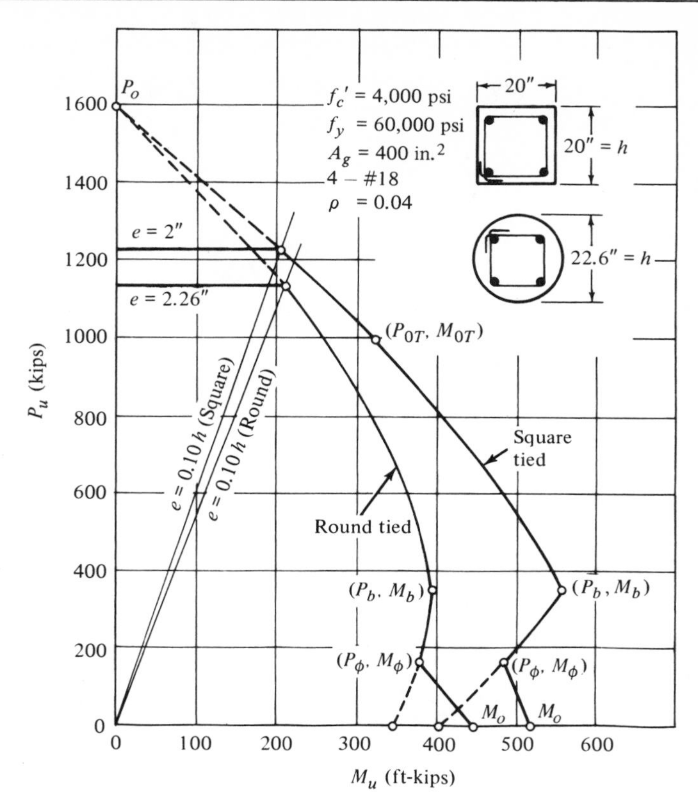

Fig. 10–19 (*b*) *continued*

COMPARISON OF TYPE OF COLUMN—SQUARE, ROUND, TIED, AND SPIRAL

Figure 10-19 contains a number of interaction curves for comparable column designs. The three curves for $\rho = 0.06$ with 6–#18 vertical bars show comparisons for:

1. efficiency of locating the vertical bars in two faces to resist bending in one direction. See Fig. 10-19(*a*) for bending capacity about the major and minor axis.
2. efficiency of square tied columns versus round columns with spirals for moment capacity wherever eccentricity exceeds about 0.10 *h* (see Fig. 10-19(*a*)).

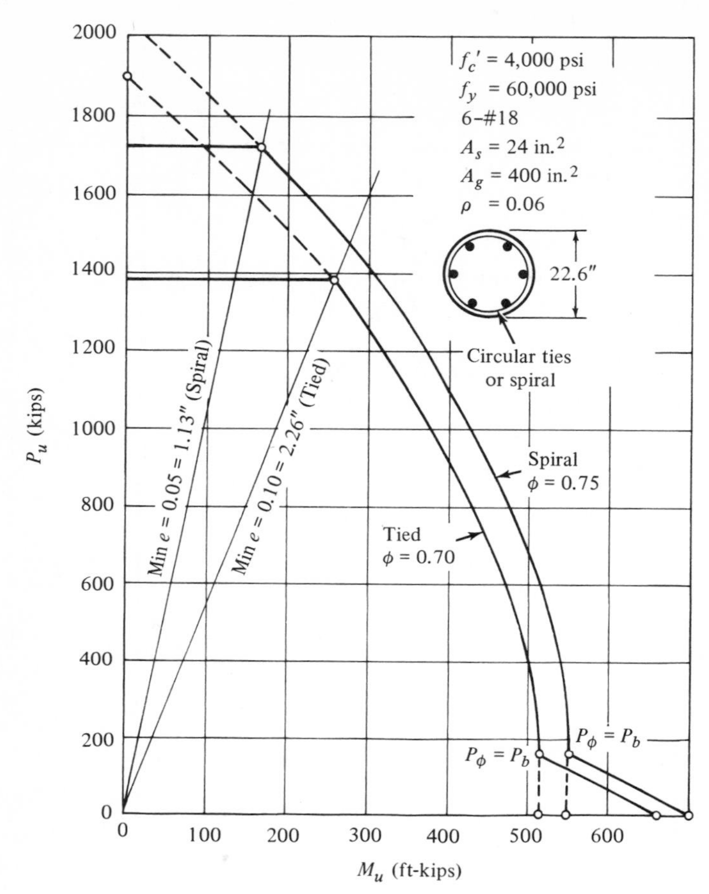

Fig. 10–19 (*c*) *continued*

3. the small gain in capacity for spirals versus circular ties in round columns (see Fig. 10-19(*c*)).

In cases (*a*) and (*c*), the concrete areas and vertical steel are identical. The spiral column requires a 3/8 in. ϕ spiral @ 1-3/4 in. pitch (Sections 10.9.2 and 7.2.2), weighing 13.4 lb per ft, plus two spacers weighing 1.4 lb per ft, plus finishing turns, say total weight equals 15 lb per ft. The tied column requires one #4 tie per set @ 20 in. spacings (Section 7.12.3) weighing 2.4 lb per ft. The unit cost in place of reinforcing steel in light ties and spirals is two to three times that for the heavy vertical bars.

The efficiency per pound in terms of added column capacity is far greater for vertical steel, and thus more capacity at less cost for any loading can be provided by adding up to two more #18 vertical bars to the tied columns of Figs. 10-19 (*a*) and (*c*) than by the use of spirals.

Similar comparisons lead to the conclusion that the cost of spiral reinforcement under the 1971 Code can seldom be justified (in nonseismic areas) unless three conditions exist: (1) the minimum eccentricity may be used, 0.05 h for spirals versus 0.10 h for tied columns, (2) the maximum steel ratio, $A_s'/A_g = 0.08$, is insufficient for the maximum tied column desirable, and (3) the maximum practicable concrete strength, f_c' is specified. When these three conditions exist, the only reinforced concrete design recourse under the Code is to utilize a spiral. Even then for a square column, it will always be more difficult to provide the vertical steel required for a steel ratio of 0.08 in a circular pattern in one circle. ("Square" spirals, sometimes specified, are usually impracticable due to fabricating limitations, and, if attempted, will, of course, increase costs.)

Where any sizable moments exist, the use of spirals even with the larger ϕ-factor, becomes very inefficient due to the inefficient circular pattern required for the vertical bars. In Fig. 10-19 (*b*), the moment capacity is almost 30 percent greater for the square column than for the round with equal ϕ-factors. Note that the four-bar round tied column, is permitted under the Code (Section 10.9.1). Four bars in a circular column can be tied by a standard square tie in a square "pattern" which gives the new code provision especial practical significance.

In Fig. 10-19 (*c*), observe the very slight additional capacity, 0.75/0.70, for most of the load-moment interaction range developed by the spiral versus circular ties. Separate circular ties are specifically permitted and, of course, only one per tie spacing is required (Section 7.12.3). This requirement can also be satisfied by a "continuous circular tie" at a pitch equal to the required tie spacing (Commentary 7.12.3).

SPECIAL PROBLEMS IN COLUMN DESIGN

Some special problems in column design have been created by more sophisticated methods of analysis and design permitting use of lower "safety factors" where the ultimate capacity is more closely assessed and the resulting safety factors are more nearly uniform. When past safety factors varied from (estimated) 0.9 to 9.0, many present requirements such as minimum eccentricity, biaxial bending moment design, distinction between columns "braced" and "unbraced" against sidesway, and reductions for slenderness effects were unjustifiable exercises in precision.

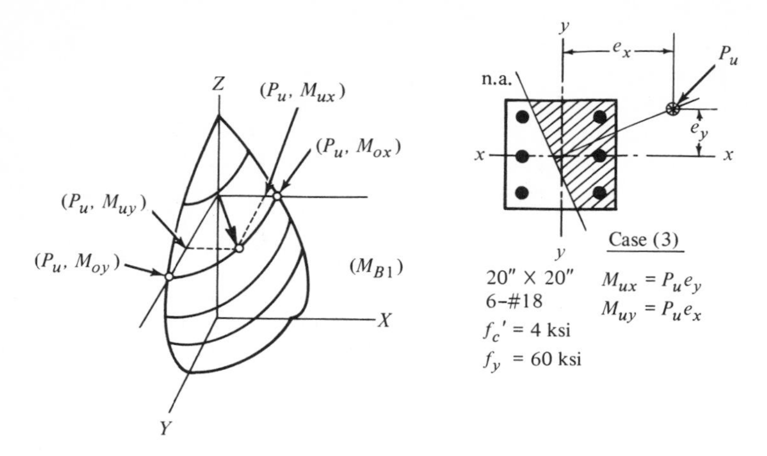

Fig. 10–20 Biaxial load-moment interaction surface.

Biaxial Bending

The problem of minimum eccentricity about *either* principal axis has been discussed. Where bending moments larger than minimum exist about both x and y axes simultaneously with the maximum design load, P_u, the biaxial bending conditions must be considered. This special problem is often encountered in the practical design of corner columns, wall columns with heavy spandrel beams, and columns supporting two-way construction, although possibly for a minority of the columns in a particular structure. To illustrate bi-axial bending, a three-dimensional load-moment curves interaction surface is required (see Fig. 10-20).

The interaction surface for a round column with six or more bars in a circular pattern is formed simply by rotating the planar load-moment interaction curve through 90 deg. All combinations of moments, $P_u e_x$ and $P_u e_y$ at a given design load, P_u, constitute a horizontal slice through the interaction surface (see Fig. 10-21). For the special case of a polar symmetrical (round) column, an exact solution is simple because the curves are circular at each slice or level for the load, P_u.

CASE 1: POLAR SYMMETRY—ROUND COLUMNS WITH VERTICAL BARS IN A CIRCULAR PATTERN. Biaxial load moment capacity is represented by a circular curve with a radius equal to M_u. Uniaxial design moments, $M_{ux} = P_u e_y$ and $M_{uy} = P_u e_x$, occurring simultaneously about both axes are simply combined for the biaxial design moment (see Fig. 10-21).

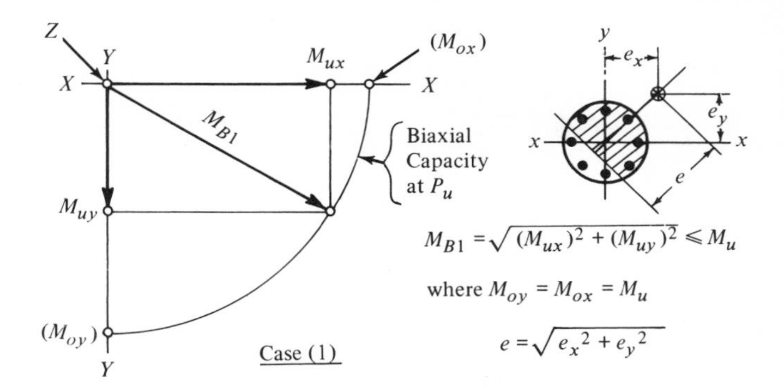

Fig. 10–21 Biaxial bending—polar symmetrical section—case 1.

$$M_{Bi} = \sqrt{(M_{ux})^2 + (M_{uy})^2} \leqq M_u \tag{10-1}$$

where M_u is the uniaxial bending moment, $P_u e$, and $e = \sqrt{(e_x)^2 + (e_y)^2}$. (This analysis is based on the assumption that the vertical bars can be replaced by an equivalent circular tube area. This assumption is sufficiently accurate when six or more bars are present.)

CASE 2: ROUND AND SQUARE COLUMNS WITH EQUAL BENDING CAPACITIES ABOUT BOTH PRINCIPAL AXES

($M_{ox} = M_{oy}$) For round columns with four bars, square columns with four bars (or bundles of bars) one in each corner, or square columns with equal steel in all four faces; the uniaxial bending capacity about the X-axis, M_{ox}, is equal to the uniaxial bending capacity about the Y-axis, M_{oy}. The curve of biaxial capacity is the horizontal slice at any load level, P_u, through the biaxial interaction surface. This curve is nearly circular, even though the biaxial moment capacity is not exactly constant. For a rapid practical solution note that the straight line connecting M_{ox} and M_{oy} always lies *inside* the biaxial capacity curve. When using longhand solutions or a handbook giving the uniaxial bending capacity, M_{ox} ($M_{ox} = M_{oy}$), simply add the design moments, $M_{ux} + M_{uy}$. Use the sum as the design moment, M_u. Select a column with a capacity for uni-axial moment, M_{ox}, equal to or greater than M_u.

$$M_u = M_{ux} + M_{uy} \leqq M_{ox} \tag{10-2}$$

(See Fig. 10-22.)

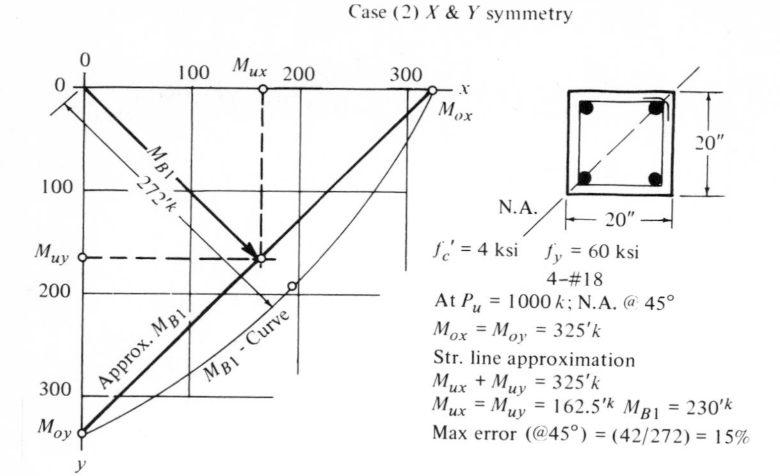

Fig. 10–22 Biaxial capacity approximation for columns with symmetry about X- and Y-axes—case 2.

CASE 3: COLUMNS WITH UNEQUAL BENDING CAPACITIES

(M_{ox} = M_{oy}) Rectangular columns or square columns with unequal amounts of steel about X or Y axes have more complicated biaxial load-moment interaction surfaces. One horizontal slice for an unsymmetrical square column is shown in Fig. 10-23. The simple solution proposed for case 2 becomes the more general form:

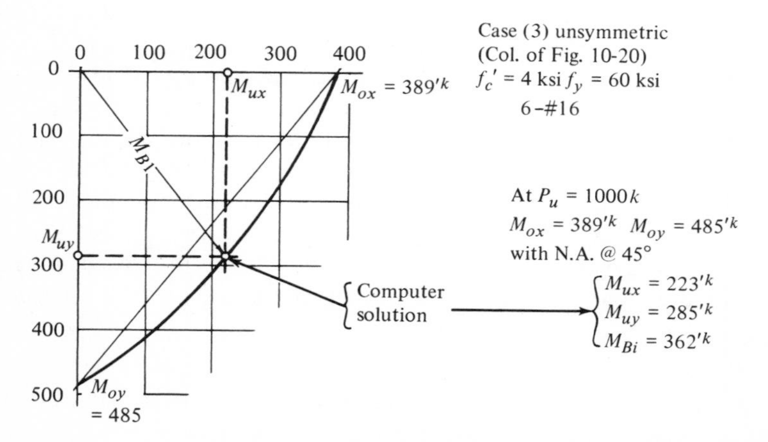

Fig. 10–23 Biaxial capacity—unsymmetric section—case 3.

$$\frac{M_{ux}}{M_{ox}} + \frac{M_{uy}}{M_{oy}} \leq 1 \quad \text{(ACI Code 1963, Section 1407-c, Eq. 14-14)}$$

where M_{ux} and M_{uy} are the design moments occurring simultaneously, and M_{ox} and M_{oy} are the uniaxial moment capacities. Since M_{ox} is not equal to M_{oy}, the equation cannot be reduced to the simple form of Eq. 10-2. In the 1963 Code, Eq. 14-14 was limited to columns controlled by tension (where axial loads are low) and for working stress design only, probably because it is a better though still conservative approximation where axial load is low. To illustrate the application of Eq. 14-14, consider the column of Fig. 10-20. The computer solution at a load, P_u, of 200 kips gives moment capacities, M_{ox} = 534 ft-kips and M_{oy} = 690 ft-kips. Assume design moments with the 200 kip loading are: M_{ux} = 300 ft-kips; M_{uy} = 300 ft-kips.

$$\frac{M_{ux}}{M_{ox}} + \frac{M_{uy}}{M_{oy}} = \frac{300}{534} + \frac{300}{690} = 0.562 + 0.435 = 0.997 < 1.00$$

It will be observed that this solution lies just inside the straight line of a graphic solution similar to that in Fig. 10-23.

To determine the maximum error conservatively underestimating the biaxial capacity of a column by the use of Eq. 14-14, see Fig. 10-23. The load, P_u = 1,000 kips, is well within the zone where compression controls. The curve shown is drawn from a computer solution. The biaxial capacity, M_{B1}, is the combination of moments which causes the column to bend about a neutral axis at 45° to the principal axis, and should represent about the maximum difference between the exact solution curve and the straight line approximation. M_{B1} = 362 ft-kips; the straight line (scaled) is about 312 ft-kips. The conservative error = 50/362 = 14 percent.

A more accurate solution still suitable for longhand calculations is

$$\frac{1}{P_u} = \frac{1}{P_{ux}} + \frac{1}{P_{uy}} - \frac{1}{P_0} \quad \text{(1963 Code Commentary, Section 1905)}$$

P_u is the load capacity under biaxial bending; P_{ux} and P_{uy} are the uniaxial capacities with e_y and e_x, respectively, and P_o is the axial load capacity at $e = 0$ (see Fig. 10-24). The solution in Fig. 10-24 using this equation was developed for a single value, P_u = 1,000 kips, with various uniaxial load capacities from the *CRSI Handbook,* 1972, reversing the normal design procedure which would solve for the safe capacity, P_u. The solution for all combinations at the single load level is presented so that the reader may compare the accuracy of this equation to other solutions. This equation

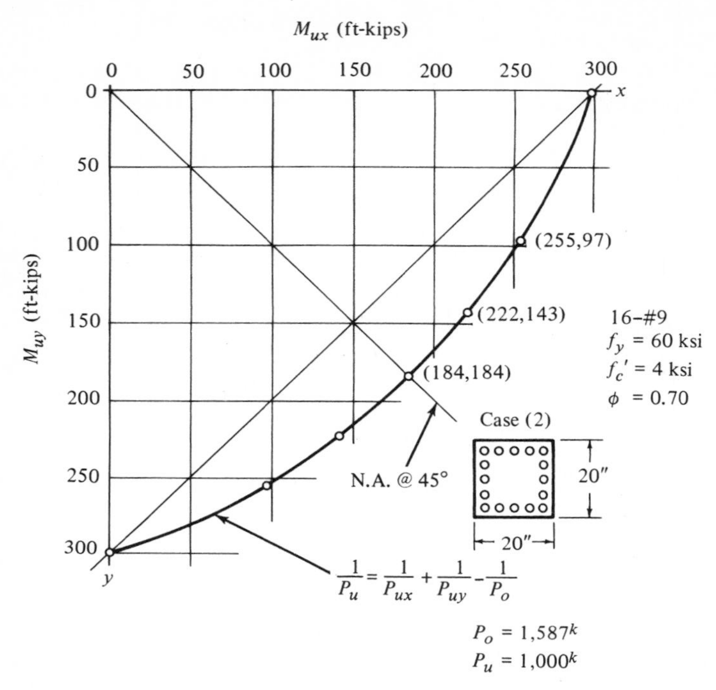

Fig. 10–24 Biaxial capacity—manual solution.

may also be used where more exact solutions are desired for the simpler cases (1 and 2). It is applicable wherever values of $P_u \geq 0.10\ P_o$. For lesser values, use Eq. 10-2 or the general form of Eq. 10-2 (ACI Eq. 14-14).

EXAMPLE. Consider the column in Fig. 10-20. The straight line approximation will be compared to the more accurate longhand solution for the design load, $P_u = 200$ kips, and the design moments, $M_{ux} = M_{uy} = 300$ ft-kips. The following data is typical of that presented in the 1972 *CRSI Handbook* or may be interpolated from the interaction curves presented in other handbooks: $P_o = 1900$ kips; at an eccentricity, $e_y = 18$ in., $P_{ux} = 366$ kips; at $e_x = 18$ in., $P_{uy} = 471$ kips. Check $0.10\ P_o = (0.10)(1900) = 190\ k < P_u = 200$ kips. Determine the safe axial load, P_u.

$$\frac{1}{P_u} = \frac{1}{366} + \frac{1}{471} - \frac{1}{1900}; \quad P_u = 231 \text{ kips.}$$

Again, the straight line approximation is conservative. For the load of 200 kips, the uniaxial capacities for moment are: $M_{ux} = 534$ ft-kips and $M_{uy} = 690$ ft-kips. If these points are plotted, the straight line connecting them passes through $M_{ux} = M_{ux} = 300$ ft-kips. The straight line approximation is $\frac{231 - 200}{231} = 13.4$ percent (conservative).

Compare the manual solution to the computer solution (Fig. 10-23), for the same column at a load level, $P_u = 1{,}000$ kips. The eccentricity $e_x = 285/1{,}000 = 0.285$ ft; similarly, $e_y = 0.223$ ft. From uniaxial solutions, loads interpolated for these eccentricities are: $P_{uy} = 1269$ kips at $e_x = 0.285$ ft, and $P_{ux} = 1291$ kips at $e_y = 0.223$ ft. $P_o = 1900$ kips.

$$\frac{1}{P_u} = \frac{1}{1269} + \frac{1}{1291} - \frac{1}{1900} = 0.001037. \quad P_u = 965 \text{ kips.}$$

The error, conservative, between the longhand solution and exact computer solutions is $(1{,}000 - 965)/1000 = 3.5$ percent.

It will be noted that the last equation, although simple enough for longhand use, requires the complete load-moment interaction curve data to solve. For sections with unequal capacities in bending about the principal axes—the general case—both interaction curves or complete tabulations of capacities about both axes are required. Thus, the exact solution is convenient only if the user has a computer-prepared handbook available. The straight line approximation is the only solution convenient for use requiring only the data shown for Figs. 10-21 through 10-23.

The designer will find curves for the "exact" solution of biaxial bending load capacities most helpful for further study, if desired. Not all will give you exactly the same answers for biaxial capacity depending upon which form of stress block (rectangular or parabolic) was used; whether the bar areas were deducted from the concrete in the stress block; whether the exact depth for each bar size with minimum cover and ties were used or an average value; etc.

Slender Columns

The Code requires column capacity reductions for slenderness effects. Slenderness effects are considered to include buckling and elastic shortening and the secondary moment due to lateral deflection. Two procedures for reducing design capacity to allow for slenderness effects are permitted. The preferred method is a comprehensive analysis of the structure, and ap-

[1] *CRSI Handbook, 1972;* Section 2, "Design of Columns."

plication of the appropriate increases in design moment resulting from lateral deflections in a secondary stress analysis (Section 10.10.1). The approximate method is a column-by-column correction based upon stiffnesses of the columns and beams, applied primary design end moments, and consideration of whether the entire structure is "laterally braced" against sidesway by definition (Sections 10.10.2 and 10.11).

For purposes of applying Code requirements to design columns for slenderness, one might consider structures classified as follows: (1) very tall, where the slenderness provisions determine the need for shear walls or other special lateral bracing, (2) tall enough so that slenderness effects are a major factor but special lateral bracing is not required, (3) short, where lateral load effects are minor, and (4) special, such as two-column multistory frames. The vast majority of buildings fall into class 3 where the approximate evaluation of slenderness effects (Section 10.11) is appropriate. Since most tall structures involve a number of equal size, equal height columns and some lateral bracing, the only practical approach for classes 1 and 2, as well as special structures, is a complete analysis, the Code-preferred method (Section 10.10.1). Available computer programs* incorporate portions of this approach, such as to compute lateral deflections, and to add secondary moments resulting therefrom to the primary load moments or to include effects of axial shortening. The authors recommend that the effects of both axial shortening and the secondary moments due to elastic lateral deflections should be included for a comprehensive analysis.

For the common structures, with height-to-base width ratios less than two, the approximate evaluation of slenderness by longhand calculation can be particularly rapid. In such structures, large bending moments in columns are unusual. With ordinary story heights, column sizes, and spans, each lateral force resisting frame will usually have a sufficient number of columns so that the wind moment is small, and only in edge columns with no cantilever will gravity load moment become an important factor. Most columns in such frames can be designed at a low steel ratio for the minimum eccentricity, $e = 0.10\ h$. See preliminary selection chart, Fig. 10-15.

An approximate evaluation of slenderness effects and design to resist same then consists essentially of four steps:

1. Determine if frame is "braced" against sidesway.
2. Determine effective length factor, k.

* ICES-STRUDL program for IBM 360 by CRSI-PCA-MIT; IBM STRESS for IBM 1130; and PCA "Lateral Load Analysis of Multi-Story Frames with Shear Walls."

3. Determine design moment magnification factor, δ_s, about both axes separately.
4. Increase steel ratio as required for same vertical design load, P_u, and magnified design moment, $\delta_s M_u = M_c$.

STEP 1. Compare stiffness of any walls or other bracing to that of parallel frames. If walls or lateral bracing present have a stiffness greater than six times (Commentary 10.11.3) that of the parallel frames, the frames can be considered to be braced.

STEP 2. For braced frames, $k = 1.0$ (Section 10.11.3). For unbraced frames, k must be greater than 1.0 (Section 10.11.3) (see chart, Fig. 10-25, and "determination of the effective height," unbraced frames, page 298).

STEP 3. Compute column stiffness, with consideration of longtime load effects and steel ratio (Section 10.11.5). Two formulas are prescribed:

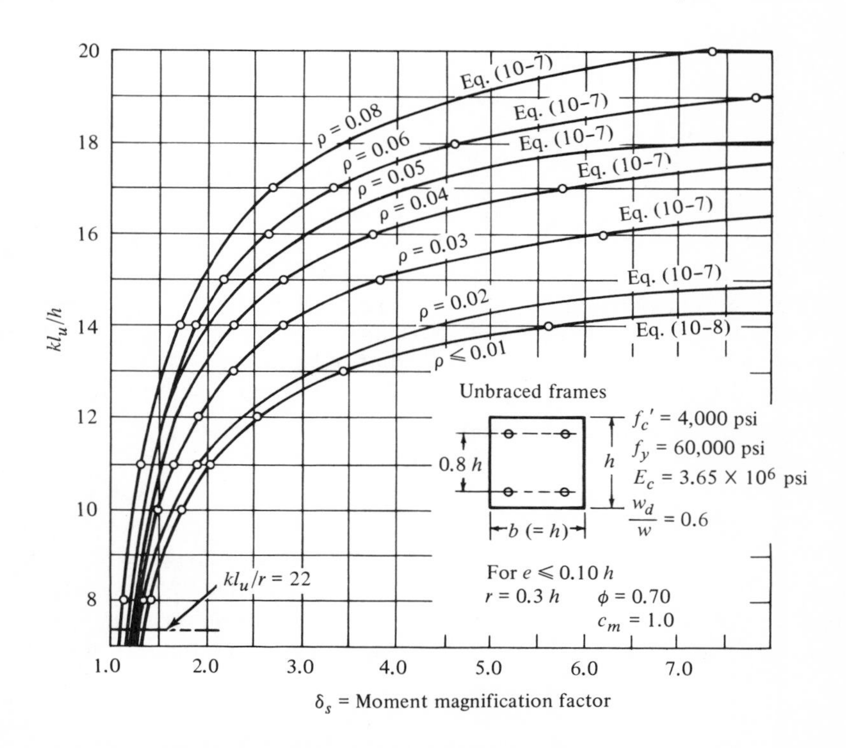

Fig. 10–25 Slender columns—Unbraced.

$$EI = \frac{E_c I_g/5 + E_s I_s}{1 + \beta_d} \quad \text{Code Eq. (10-7)}$$

and

$$EI = \frac{E_c I_g/2.5}{1 + \beta_d} \quad \text{Code Eq. (10-8)}$$

where subscripts c = concrete; s = steel; g = gross; and β_d = ratio of the maximum design dead to maximum design total load, always positive. Equation (10-8) is applicable where $\rho \leqq 0.01$ and Eq. (10-7) for larger values of the steel ratio (Commentary Section 10.11.5).

Using the value of EI so computed, the critical (buckling load), P_c is computed as

$$P_c = \frac{\pi^2 EI}{(kl_u)^2} \quad \text{Code Eq. (10-6)}$$

The moment magnification factor, δ_s, is defined as

$$\delta_s = \frac{C_m}{1 - P_u/\phi P_c} \geqq 1.0 \quad \text{Code Eq. (10-5)}$$

in which $C_m = 0.6 + 0.4\,(M_1/M_2) \geqq 0.4$ Code Eq. (10-9) where M_2 is the larger end moment (+); and M_1 is the smaller, (+) for single curvature and (−) for double curvature.

For all other cases $C_m = 1.0$.

When design is based on minimum specified eccentricities, M_2 of Code Eq. (10-4) must be based on the specified minimum eccentricity. Where computations show no eccentricity at both ends of the column, conditions of curvature must be based on a ratio of $M_1/M_2 = 1.0$. Where the eccentricities were calculated and are less than the minimum, $e \leq 0.10\,h$, for design, computed moments shall be used to compute C_m (Section 10.11.6).

Short cut solutions to determine δ_s (Step 3) for conditions of $e \leq 0.10\,h$, steel concentrated in the two faces parallel to the neutral axis or neglected in other faces, for given grades of steel and f_c' are easily prepared. The charts in Fig. 10-26 present direct solutions for δ_s, using normal weight concrete, $f_c' = 4{,}000$ psi, and $f_y = 60{,}000$ psi and $\beta_d = \dfrac{1.4D}{1.4D + 1.7L} = 0.6$ for square sections.

Step three, determination δ_s, becomes very simple with charts such as these. For frames not braced against sidesway, enter Chart A with steel ratio, ρ, and $\frac{kl_u}{h}$ and read δ_s directly. For braced frames, enter appropriate Chart B, C, D, E, or S_s with ρ, $\frac{kl_u,}{h}$ and C_m and read δ_s directly.

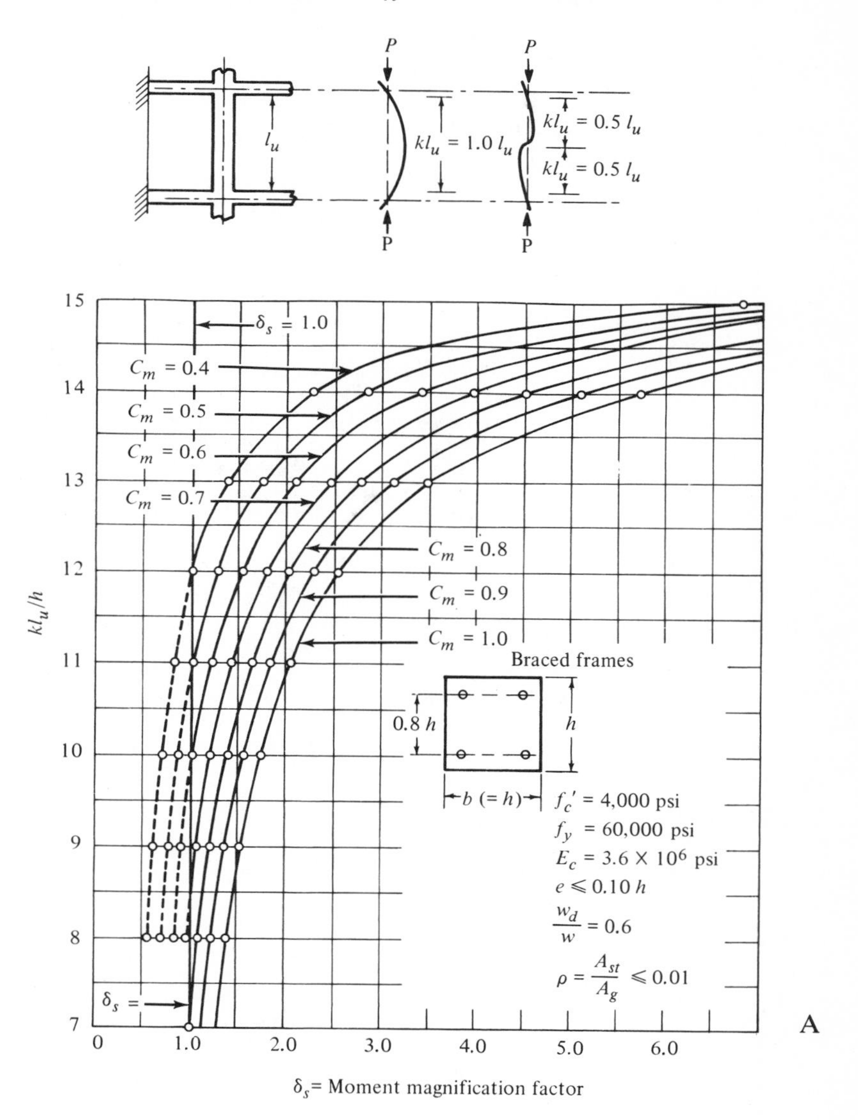

Fig. 10–26 Effective heights limits—braced frame and slender columns, charts A to E.

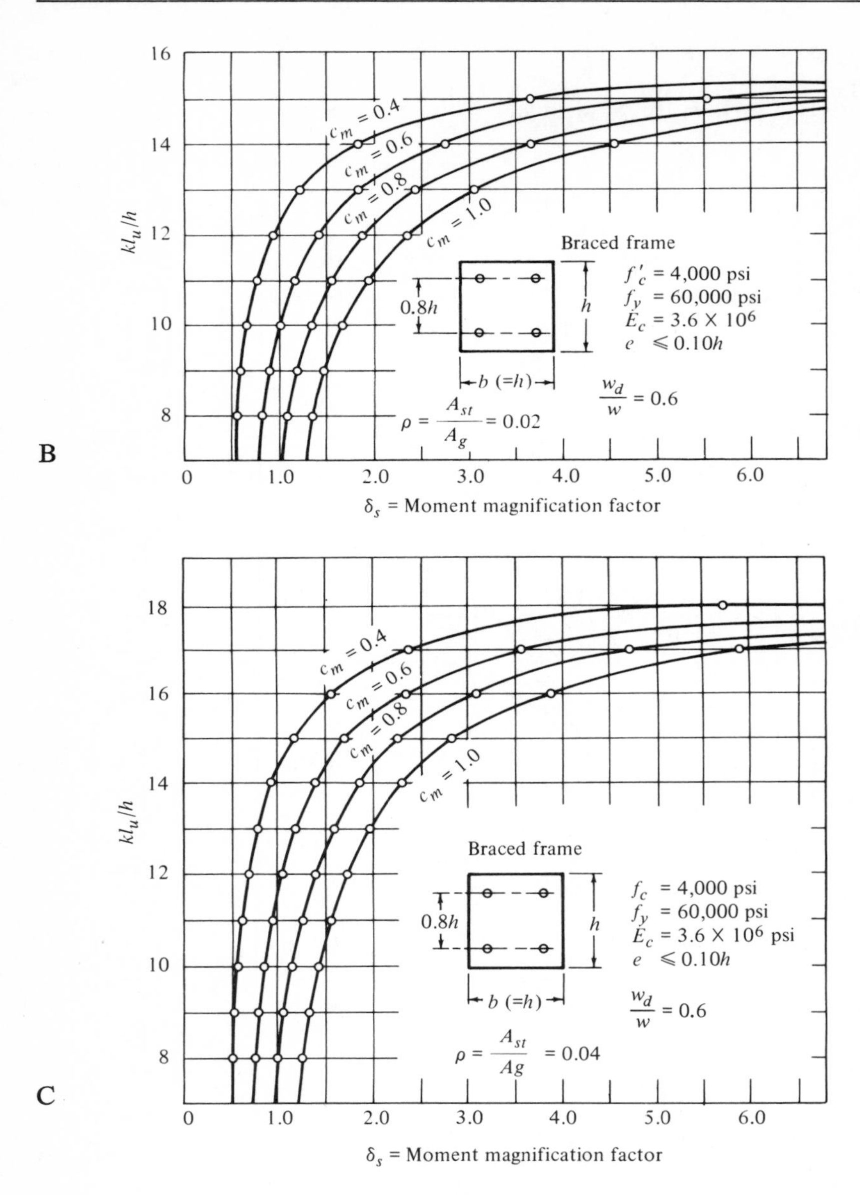

Fig. 10–26 *continued*

STEP 4. This consists of selecting a higher steel ratio to satisfy load-moment conditions of: P_u, $(\delta_s M_u)$.

Example: See page 300.

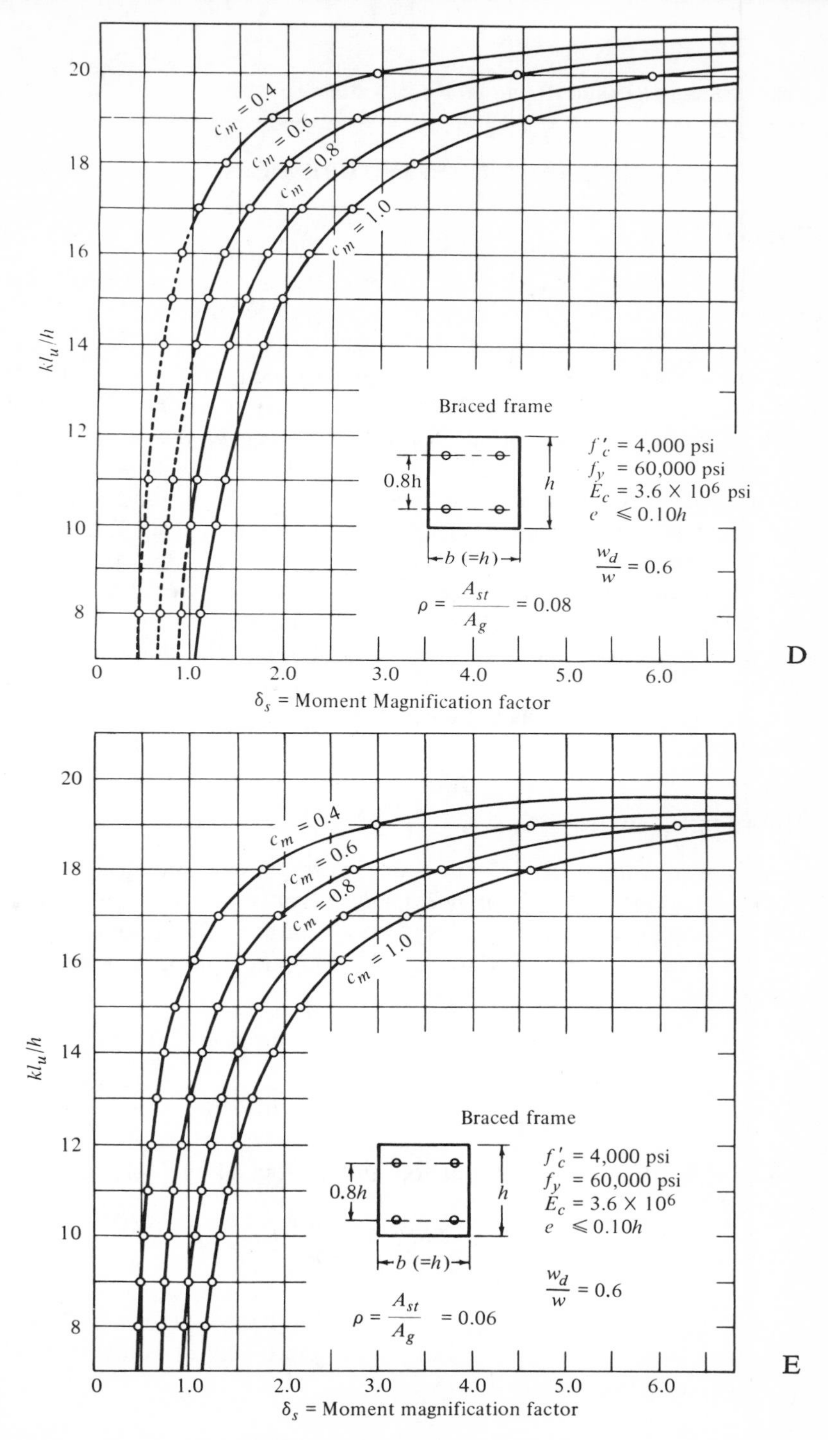

Fig. 10–26 *continued*

DETERMINATION OF THE EFFECTIVE HEIGHT

Braced frames

In braced frames, neglecting axial shortening of members, the intersections of the column and beam centerlines are considered as fixed points by definition. The effective length of a member in a buckling curve about its centerline, defined as l_u, is the length between points of contraflexure, reversals of curvature (hinges). The column can bend only in single curvature between the fixed points for $kl_u \leqq l_u$, or in double curvature for $kl_u \geq 0.5\ l_u$. (see Fig. 10-26).

In reinforced concrete frames consisting of beams and columns with average ratios of column to beam stiffness varying from 0.2 to 4.0, the effective height, kl_u, in braced frames can vary from about 0.6 l_u to 0.9 l_u. The same column under different patterns of live load can have different effective height as it bends from single to partial or full double curvature. As a simple, quick, and safe design procedure, the authors recommend using $kl_u = 1.0$ for all columns in braced frames. Note that the Code required l_u to be taken as the *clear* distance between lateral supports (Section 10.11.1).

Unbraced frames

In unbraced frames, the column-beam intersections can deflect laterally and buckle as a unit so that kl_u may extend several stories in height, particularly in frames with flexible beams and stiff columns. The following formulas provide sufficient accuracy for design and are appropriate for practical use:*

$$k = \frac{20 - r'}{20} \sqrt{1 + r'} \qquad \text{for } r' < 2.0$$

$$k = 0.9\sqrt{1 + r'} \qquad \text{for } r' \geq 2.0$$

where r' is the average ratio at top and bottom of the column,

$$r' = \frac{\Sigma K_{\text{col}}}{\Sigma K_{\text{beam}}} = \frac{\Sigma E_{cc}\ I_c/l_c}{\Sigma E_{cb}\ I_b/l_1}$$

Using $E_c I_c$ computed per Eq. 10-7 or 10-8 for the columns, and Code Eq. 9-4 for beams (see page 65). For frames consisting of equivalent columns and flat slabs, use $r' = \alpha_{ec}$.

* Richard W. Furlong, "Column Slenderness and Charts for Design," *ACI Journal* (Jan. 1971), 9-18.

DERIVATIONS AND BASIS OF CHARTS

As an aid to readers who may wish to extend these charts for other concrete strengths, D/L ratios, round columns, $\phi = 0.75$ for spirally reinforced columns, or lightweight aggregate concrete (lower E_c values), the derivation is as follows:

1. For $\rho \leqq 0.01$, use Code Eq. (10-8); $EI = E_c I_g/4$.
2. For $\rho \geqq 0.02$, use Code Eq. (10-7); assume an average value for the ratio of the center-to-center of bars to overall column dimension. Take $gh = 0.8\ h$, which is closely applicable to a wide variety of column sizes, cover depths, and bar sizes.
3. For P_u at $e = 0.10\ h$, take the value of P_u/A_g for $f_c' = 4{,}000$ psi from the preliminary design chart (Fig. 10.15). For a square tied column, $P_u = (P_u/A_g)(h)^2$. (*Note:* Fig. 10-15 is based on $\phi = 0.70$.)
4. Consider all steel concentrated in the corners of a square column.
5. Compute coefficients for $E_c = 3.6 \times 10^6$ psi; $\phi = 0.70$; and $\dfrac{1.4D}{1.4D + 1.7L} = 0.6$.

It will be noted that other values for E_c can be substituted directly into the tabulated coefficient for EI only for $\rho \leqq 0.01$, which utilizes Code Eq. 10-8. The coefficient itself will change when based on Code Eq. 10-7. Coefficients for P_c which include the term EI will require revision proportional to changes in the coefficient for EI.

6. Points in the charts of Figs. 10-25 and 10-26 are located using the tabulated coefficients very simply, and will form a family of similar curves in each case. As a sample, consider a 12 in. square column, $f_c' = 4$ ksi; $\rho = 0.08$; braced so that $k = 1.0$; with unsupported height $= 17$ ft; where $C_m = 1.0$ (conservative for small values of eccentricity at each end).

$$\left(\frac{h}{kl_u}\right)^2 = \left(\frac{1}{17}\right)^2$$

From Table 10-1 coefficients, $P_c = 2700 \left(\dfrac{h}{kl_u}\right)^2$; $P_u/\phi = 5.86$.

$$\delta_s = \frac{C_m}{1 - \dfrac{P_u/\phi}{P_c}} = \frac{1.0}{1 - \dfrac{5.86}{2700}(17)^2}$$

$$\delta_s = \frac{1}{1 - 0.628} = 2.69$$

See Chart E, Fig. 10-26. For $(kl_u)/h = 17$ and $C_m = 1.0$, read $\delta_s = 2.7$.

TABLE 10-1 Square Column Slenderness Coefficients
(f'_c = 4 ksi; Grade 60 Bars; $E = 3.6 \times 10^3$ ksi; $b = h$; $e \leqq 0.10\ h$)

$p = A_{st}/A_g$	EI[a]	P_c (ksi)	P_u/ϕ (ksi)
≤ 0.01	$0.0209\ E_c h^4$	$750\ (h/k\ell_u)^2$	3.15
0.02	$0.0264\ E_c h^4$	$900\ (h/k\ell_u)^2$	3.57
0.03	$0.0344\ E_c h^4$	$1{,}200\ (h/k\ell_u)^2$	3.93
0.04	$0.0425\ E_c h^4$	$1{,}500\ (h/k\ell_u)^2$	4.29
0.05	$0.0510\ E_c h^4$	$1{,}800\ (h/k\ell_u)^2$	4.72
0.06	$0.0585\ E_c h^4$	$2{,}100\ (h/k\ell_u)^2$	5.07
0.07	$0.0665\ E_c h^4$	$2{,}400\ (h/k\ell_u)^2$	5.50
0.08	$0.0745\ E_c h^4$	$2{,}700\ (h/k\ell_u)^2$	5.86

[a] For $\rho \leqq 0.01$, EI per Code Eq.]0–8; for $\rho \geqq 0.01$, EI per Code Eq.]0–7.

For a square column, radius of gyration, r, equals 0.3 h (Section 10.11.2). Slenderness effects may be neglected for columns with ratios of $kl_u/r < 22$ in all frames (Section 10.11.4). Note charts begin with ratios of $kl_u/h = 7$.

EXAMPLE 1: SLENDERNESS EFFECTS—UNBRACED FRAME

Consider an interior column in a low building or the upper stories of a high-rise building supporting a flat plate floor. The flat plate design requires a column 12 × 12 in. minimum size square column. For a 12 × 12 square column, the ratio of stiffnesses, α_{ec}, column to floor system at the ground floor and second floor is 0.45. The columns are not braced against sidesway. The total design load after live load reductions is 372 kips; unbraced first floor height is 9′–4″. There are a sufficient number of columns per frame in both directions so that the eccentricities due to wind moments are less than 0.10 h and interior columns will be designed for minimum eccentricities about both axes. Grade 60 bars and, if feasible, concrete $f_c' = 3750$ psi, will be used. Note the variation in f_c' from that used in charts; for this small difference, correction is not significant.

From any handbook, select a short column design at lowest steel ratio: 12 × 12 in. square tied column, four # 9; P_u = 374 k.* For the slenderness effect compute effective height ratio, k, as follows:

$$k = \frac{20 - r'}{20}\sqrt{1 + r'}; \qquad \text{take } r' = \frac{K_{\text{col}}}{K_{\text{floor}}} = 0.45$$

$$k = 1.174; \qquad kl_u = (1.174)(9.33) = 11.0$$

(See Fig. 10-25, Unbraced.) Enter with kl_u/h = 11.0, and ρ = 0.0278.

* Selected from the *CRSI Handbook, Ultimate Strength Design,* 1970.

Read $\delta_s = 1.66$. Compute magnified eccentricity, $(e) = 1.66 \times 1.2$ in. = 2.0 in. Return to the handbook and select a short column design for P_u = 372 k and e = 2.0 in.

$$12 \text{ in.} \times 12 \text{ in., four \#11 bars, capacity } P_u = 374\ k$$

Since a square column with four vertical bars has equal bending capacity about both principal axes, separate slenderness reduction for each is unnecessary in this simple example.

EXAMPLE 2: SLENDERNESS EFFECTS—BRACED FRAME

Consider an edge column in a low building with end walls and stairwell walls (shear walls) which provide lateral bracing in the E-W direction (Fig. 10-27). Vertical load is P_u = 390 kips for the first floor columns which have a clear height, h, of 14 ft 6 in. The bending moment in the E-W direction computed for a preliminary column size of 10 in. × 14 in. was 98 ft kips at the top of the first floor edge column; individual column footings were considered hinged, moment at the bottom equal zero. It is desired to maintain the 10 in. width of these columns in the N-S direction. Use Grade 60 bars; f_c' = 5,000 psi for columns. Short column design requirements:

$$\begin{aligned} P_u &= 390\ k \\ M_{ux} &= 98 \text{ ft-kips} \\ e_x &= 3.03 \text{ in.} \\ M_{uy} &= (390)(0.10)\left(\frac{10 \text{ in.}}{12}\right) \\ &= 32.5 \text{ ft-kips} \end{aligned}$$

Selection a short column from a design handbook, 10 × 14, 4–#11, $\rho = 0.0445$, capacity, P_u = 393 kips at e = 3 in. Compute ratio, M_1/M_2, smaller-to-larger end moment. $M_1 = 0$, M_2 = 98 ft-kips; $M_1/M_2 = 0$. Take $k = 1.0$. $kl_u/h = \dfrac{14.5 \times 12}{14} = 12.4$. $kl_u/r = \dfrac{(1.0)(12.4)}{0.3} = 41.4$ $> 34 - 12\ (M_1/M_2)$, and so slenderness effects must be considered. Compute the effect of end restraints, $C_m = 0.6 + 0.4\ (M_1/M_2) = 0.6$. In Fig. 10-26, select Chart C for braced frames with $\rho = 0.04$, enter with $C_m = 0.6$ and $kl_u/h = 12.4$ and read $\delta_s = 1.2$; repeat with chart for ρ = 0.06, and read $\delta_s = 1.0$. Use $\delta_s = 1.1$. $(\delta_s)\ (e) = 1.1 \times 3.03$ in. = 3.3 in.

Select a 10 × 14 column, 6–#11 vertical bars, ρ = 0.0668 with a

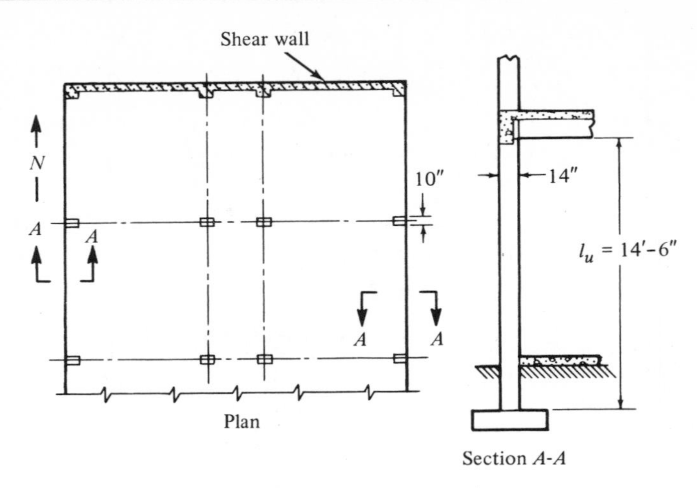

Fig. 10–27 Column location.

design capacity as a short column of $P_u = 390$ k at $e = 3.3$ in. Note that use of slenderness coefficients from charts for $f_c' = 4{,}000$ psi is conservative for designs using $f_c' = 5{,}000$ psi since δ_s is reduced slightly with increased E_c.

Again as in the preceding example, by inspection it is unnecessary to compute biaxial slenderness reductions in capacity about each axis and biaxial bending capacity. The N-S moments are small ($e = 0.10\ h = 1.0$ in.), the column capacity for bending about the minor axis is $P_u = 586$ kips at $e = 1.0$ in., and by inspection, the full load can occur only when panels on both sides of the E-W (minor) axis are fully loaded, and full vertical load and biaxial bending cannot occur simultaneously.

Variation from eccentricity, e ≧ 0.10 h

The question of error involved in applying the charts of Fig. 10-26 to conditions where $e \geqq 0.10\ h$ as in Example 2 can be readily determined, and corrections applied. The moment magnification factor, δ_s, for slenderness effects increases with larger values of column capacity, P_u.

$$\delta_s = \frac{C_m}{1 - \dfrac{P_u/\phi}{P_c}} > 1.0 \quad \text{Code Eq. (10-5)}$$

Since the capacity, P_u, associated with the values of $e \geqq 0.10\ h$ is smaller than P_u with $e = 0.10\ h$, it is conservative to use the charts for δ_s with $e \geqq 0.10\ h$, but, as the eccentricity becomes much larger than $0.10\ h$, it obviously may become overconservative and wasteful.

To check the case ($\delta_s = 1.10$) of Example 2, from a handbook determine that for $e = 0.10\ h$, 10×14 column, four #11 bars, $P_u = 515$ kips, and for $e = 3.0$ in., $P_u = 393$ kips:

$$\delta_s = 1.10 = \frac{1}{1 - 515/\phi P_c}$$

$$(1 - 515/\phi P_c) = 1/1.10 = 0.91$$

$$(515/\phi P_c) = 0.09 \text{ compared to } (393/\phi P_c) = 0.0764$$

$$(1 - 393/\phi P_c) = 0.9236$$

Corrected for P_u at $e = 3.0$, $\delta_s = 1/0.9236 = 1.08$ vs. 1.10.

Obviously, corrections for small increases in eccentricity greater than $0.10\ h$, when using the charts of Fig. 10-26 are not significant. Such refinements in the elaborate analysis for slenderness effects, in view of the much larger uncertainties involved in the assumptions and tests upon which the theory is based, are not warranted. The expenditure of design time for more definite and more significant refinements arising from consideration of reductions in the effective depth, d, at splices, routinely neglected by many designers, is much more easily justified. See the following section.

Column Details

Under the 1971 ACI Code, the design of columns has been rationalized to reconcile theory and test results and refined to result in nearly uniform safety against overloads. Computer programs and computer-prepared handbook designs, not only available but necessary for practical design, should not, however, induce such confidence in the apparently authoritative precision of calculations, that the art of design is neglected. None of the design aids to the best of the authors' knowledge consider the effect of practical details such as splices and construction joints, always present in real building columns. With the more precise estimation of strength, consideration of these practical details and appropriate corrections to design capacity, often considered negligible under past codes, becomes more important. Most of the Code requirements for column details have been consolidated into six sections of the Code (Sections 7.6, 7.7, 7.10, 7.11, 7.12, and 7.14). Some of these common requirements are shown in Fig. 10-28. Note the reduction in the effective depth, minimum one bar diameter to the tension steel at the base of the upper story column. Even

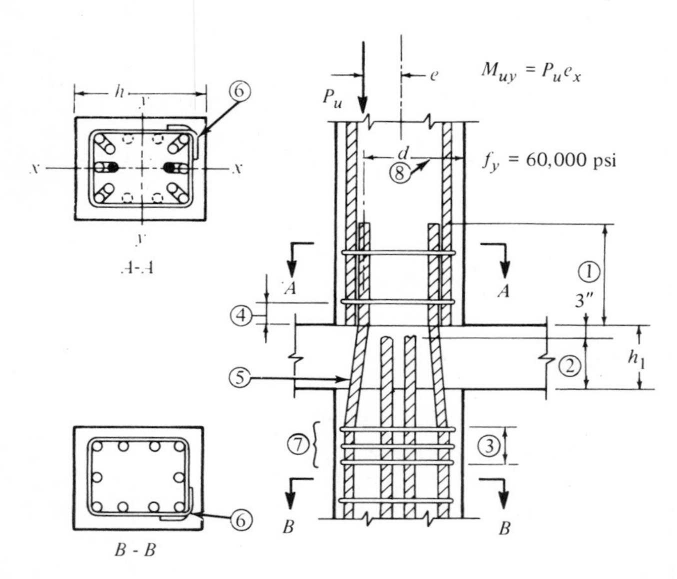

Fig. 10–28 Compression lap splices—100% A_s—with offset bent vertical bars.

Notes to Fig. 10-28:

(1) Compression lap lengths = 30 bar diam (see Table 13-2). Within the minimum tie area = 0.0015 *bs*, lap length = 25 bar diam (see Table 13-3). For #11 or smaller bars only (Section 7.5.2.).

(2) Standard practice detail length is (h − 3 in.) unless designer specifies differently (ACI 315). This dimension must be 1.0 $l_d \geqq 8$ in. for compression if cutoff bar is required for compression just below the floor (Section 12.6). For compressive l_d lengths, see Table 13-1.

(3) Extra ties must be located within 6 in. of the lower bend point of off-set verticals. (Selection 7.10.1).

(4) The first tie must be located no more than one-half tie spacing above floor (Section 7.12.3).

(5) The maximum slope from the vertical is 1:6 (Standard detail ACI 315 and Section 7.10.1).

(6) Alternate hook position (ACI 315 recommended note).

(7) These ties designed to resist 1.5 times the horizontal component of the compressive force in the offset bent bars = (1.5/6) $f_y A_s'$ (Section 7.10.1).

(8) The reduced depth, d, at the splice equals the column thickness minus the concrete cover minus the tie dia. and minus one and one-half bar dia. (Section 7.14.1).

where the column size is unchanged it is standard practice to offset bend the vertical bars from below, but not necessary except in some round columns where spacing limitations permit two sets of verticals in one ring. Since provision of a minimum tensile capacity is required even for columns designed to carry only compression (Section 7.10), this correction should not be overlooked.

The art of design is involved in maintaining the design capacity at full value through all critical sections of a column as it is a member constructed in pieces with the joints and splices customarily located at points of maximum moment. One of the problems arises with the use of higher strength concrete in the column than in the floor system or footing for overall economy. Where the column concrete, f_c', is not more than 40 percent greater than f_c' for the floor system, the problem may be neglected (Section 10.13). If it is desired to use a column concrete strength in excess of this moment, the designer has three options:

1. Specify that the column strength concrete be placed in the floor system over the column (area A_g) and for an area around it equal four times column area (4 A_g). This operation involves the placing of two different mixes of concrete simultaneously and requires continuous dependable field inspection better than will usually be supplied. It has the advantage of simplicity and provides a small bonus shear capacity in the surrounding floor, but the authors recommend it for use only where field inspection is provided (see Fig. 10-29).

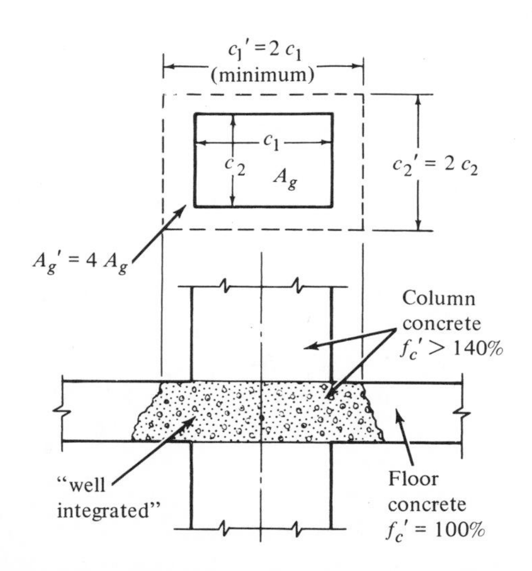

Fig. 10–29 Two-strength floor (not recommended).

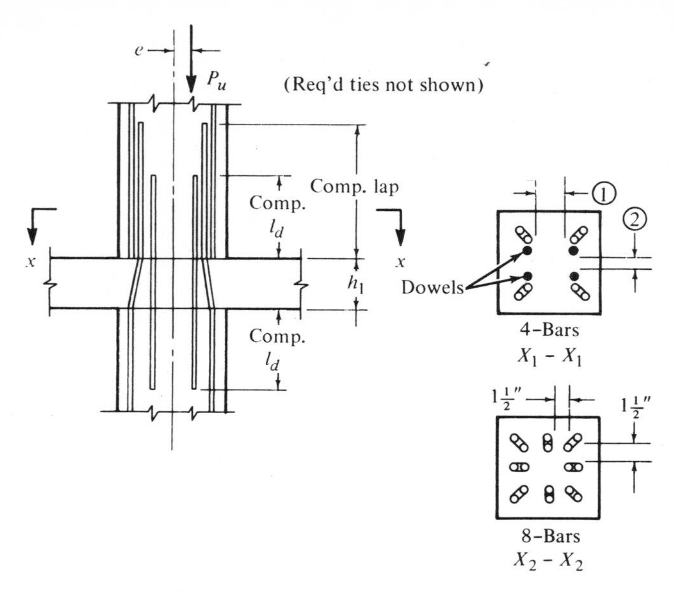

Fig. 10–30 Offset verticals plus added dowels (not recommended).

Notes to Fig. 10-30:
(1) Dimensions (1) and (2) must be at least one and one-half bar diam $\geqq 1\frac{1}{2}$ in. clear between any group of dowels or bottom bar and top bar. If the splices were at minimum spacing, *Section* $X_2 - X_2$, dowels are impractical (Section 7.4.4).
(2) If spacing of vertical bars permits the addition of dowels, required dowel length is equal to 2 times l_d plus the floor depth, h (see Table 13-1).

2. The second option is to add dowels or spirals through the floor system to increase the capacity by the required amount. Spirals are seldom practicable and generally uneconomical for nonseismic design. If the additional capacity is to be provided by dowels, these dowels must be developed for compression on either side of the weaker floor system depth (see Fig. 10-30). For the usual column designed for $e = 0.10\ h$ about both principal axes and in either direction, the added dowel area may be located symmetric about both axes. Where a column is required to resist a heavy bending moment about one axis, the added dowels should be located within the plan of the column near the centroid of the assumed stress block.

The added dowels are easily overlooked, especially because they will usually be added into the fresh concrete at the conclusion of casting the lower column. If they are used to supplement offset verticals from below,

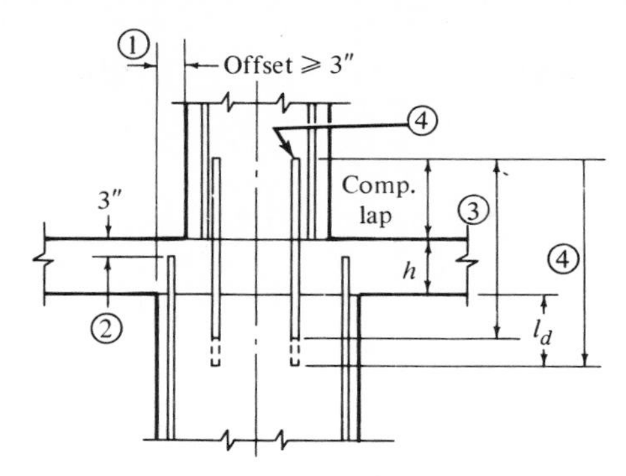

Fig. 10–31 Separate dowels at offset faces.

Notes to Fig. 10-31:

(1) Where face of column above is offset 3 in. or more from the face of the column below, offset bars are not permitted; the splice must be made by separate dowels (Section 7.10.2).

(2) Cutoff column verticals stop 3 in. below finished floor unless otherwise noted (ACI 315 standard practice).

(3) Length of dowel equals 2 times the required lap for tension or compression plus 3 in.

(4) If f_c' for the column concrete is more than 40 percent greater than for floor concrete, added compression capacity required may be provided by increasing the size of the separate straight dowels. For this purpose, the extension of the dowel below the floor must 1.0 l_d (Section 10.13.1-b). Dowel length will be as determined above or one compression lap length plus the floor thickness plus 1.0 l_d, whichever is larger. Note here also that it will be coincidental if midpoint of the dowel length is located at the finish floor line; designer must indicate embedment required; otherwise placing practice is to extend half the length above finished floor.

the authors recommend provision of complete details showing location both in plan and elevation in the column schedule as well as field inspection before casting the floor. Provision of added capacity under this option is simpler where separate dowels are to be used in any event (see Fig. 10-31). Where separate dowels are provided for the splice, the added capacity can be supplied simply by requiring the use of larger size dowels.

3. The third option is to compute a reduced column capacity based on a composite value of f_c' equal to 75 percent of the column concrete and 35 percent of the floor concrete f_c'. This option is self-defeating for economy, except for columns controlled by tension.

In general for practical design the authors recommend selection of concrete strengths within the limit of 40 percent so that this problem can be neglected since all three options require special conditions to be effective.

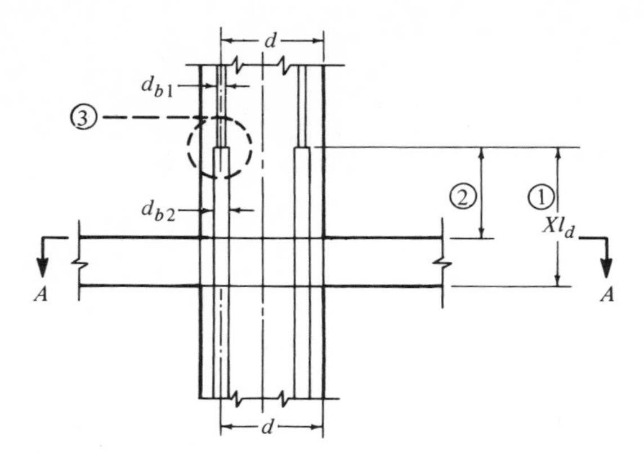

Fig. 10–32 Butt splices of larger to smaller bars.

Splices in column verticals

The difficult design requirements for splices of column verticals involve tensile capacity (Sections 7.10.3, 7.10.4, and 7.10.5). In columns controlled by tension, where all splices are located near points of maximum moment, "full" tensile splices are required. Full tensile splices may consist of: (1) Class C tension laps; (2) full welding, butt welds developing 125 percent of f_y; (3) full positive connections, mechanical sleeve devices developing 125 per cent of f_y (Sections 7.6.3, 7.5.5.1, and 7.5.5.2). Whenever the calculated tension in a bar exceeds one-half f_y, a full-tension splice is required (Section 7.10.4) regardless of the percentage spliced. Whenever the calculated tension in a bar is less than one-half f_y any type of splice may be used, but the tensile capacity of the splice and any unspliced bars

Notes to Fig. 10-32:
(1) *Compression controls.* If the larger bar is required to carry the additional load of the floor slab compression at the top of the lower column, the distance from the bottom of the slab to the splice must be sufficient to develop this additional load. The fraction, $X = [1 - (d_{b1}/d_{b2})^2]$, or l_d for full compression development should be adequate.

Tension controls. Similarly, if the larger bar is required to develop full tensile capacity at the top of the lower column, the extension above the bottom of the floor, must be Xl_d for full tension development (Section 12.1.1). See Table 13-4 for l_d.
(2) If column $f_c' > 140$ percent floor f_c', the difference between the areas of the larger and smaller bars can be considered effective as added dowel area through the floor provided the extension above the floor is Xl_d for compression (Sections 12.1.1, 12.6.1, and 10.13.1-b). See Table 13-2 for compression l_d. This procedure is recommended by the authors in lieu of adding separate dowels when conditions permit its use.
(3) Note that there will be loss of available effective depth, d, for the smaller bar at a concentric butt splice of larger to smaller bars where cover is established by the larger bar. The loss $= 1/2\ (d_{b2} - d_{b1})$. A correction should be considered to computed capacity if tension controls *and* if the splice is located near a point of maximum moment. Location of splices in the center third of the column story height will avoid the point of maximum moment for the usual condition of double curvature. Otherwise, the indicated correction should be considered.

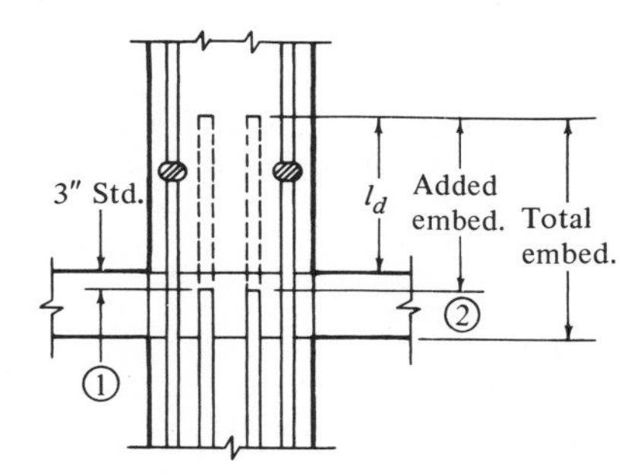

Fig. 10–33 Butt splice same size bars—reduced quantity.

Notes to Fig. 10-33:
(1) Standard detailing practice is to cut off bars as shown when discontinued in the next story unless otherwise dimensioned by the engineer (ACI 315).
(2) If the cutoff bars were required at the top of the lower column for stress added by the floor system, the total extension above the bottom of the floor beam or slab should be 1.0 l_d for tension. See Table 13-1 for compression; Table 13-4 for tension.
If the column concrete $f_c' > 140$ percent floor concrete f_c', the cutoff bars can be extended to avoid use of separate dowels. To be considered fully effective at the top of the floor slab, the extension required is 3 in. plus l_d for compression (Sections 10.13.1-b, and 12.6.1). This device is recommended by the authors where conditions permit instead of using separate dowels.

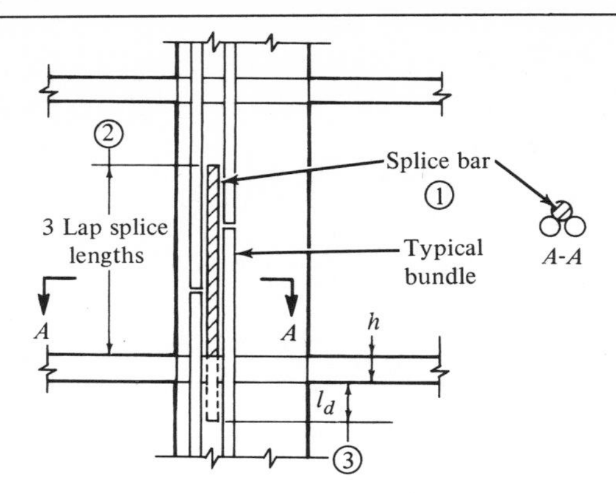

Fig. 10–34 Bundled bars—lap spliced 2-bar bundle.

Notes to Fig. 10-34:
(1) The splice bar must be the same size as the bars in the 2-bar bundle if the lap splice is for full development of the bundle, whether for tension or compression. This method of making a splice technique has been used with sheared ends; or, for partial tension capacity only, with endbearing splices using a smaller size splice bar.
(2) The first splice (left bar of bundle in Figure) is most conveniently located one required lap length above the floor. The splice bar is a minimum of 3 times the lap length required for a bar in a 3-bar bundle which equals 3.6 times the individual bar lap length required (Section 12.7.1).
(3) If column concrete $f_c' >$ 140 percent floor concrete f_c', extend the splice bar a distance l_d below the floor for compression and it will also serve in place of an added separate dowel. (Section 10.13.b).

Notes to Fig. 10-35:
(1) For complete detail examples and explanations of accepted industry practice, see "Proposed Revision of the Manual of Standard Practice for Detailing Reinforced Concrete Structures," 1970, by ACI Committee 315.
(2) In any rectangular column the four corner bars are preferably enclosed by a single, one-piece tie to aid in holding the bars in position during erection. For extremely long wall-like columns, a two-piece single tie or overlapping separate rectangular ties may be more convenient to hold the shape of the outer bars in plan. Each bar in the corner of any column must be held in the corner of a tie with an included angle not more than 135° (Section 7.12.3).
(3) The 4-bar column detail shows the preferred standard tie with overlapping 90° hooks forming the splice (Section 7.11 and ACI 315). The position of this splice should be staggered at each tie spacing by rotating the tie 90° or 180°. A note to this effect should be added to the detail on the structural drawings and should be repeated by detailers on the placing drawings. (For the confinement area in seismic "ductile frames," overlapping longer 135° hooks are usually required.)
(4) The 6-bar column detail shows the usual standard tie required. The intermediate bars do not require the support of a corner of a tie if the clear spacing (X) on both sides is not more than 6 in. to a bar in the corner of a tie. If $X >$ 6 in., the two bars are most easily held by a "candy-stick" tie (the 2-bar tie shown in the

"wall-like column" detail). Again, it is recommended that the designer add the note to alternate the position of the 90° and 180° hooks at each tie spacing.
(5) The 2-piece tie and the candy-stick tie are recommended for all interior ties in columns using large, butt-spliced vertical bars, particularly with staggered splices, where in-place assembly on free-standing verticals is required (Section 7.12.7 and ACI 315). The required lap for 2-piece ties is a Class A tension lap (Section 7.12.7). See tension lap splices, Table 13-8, top bars.

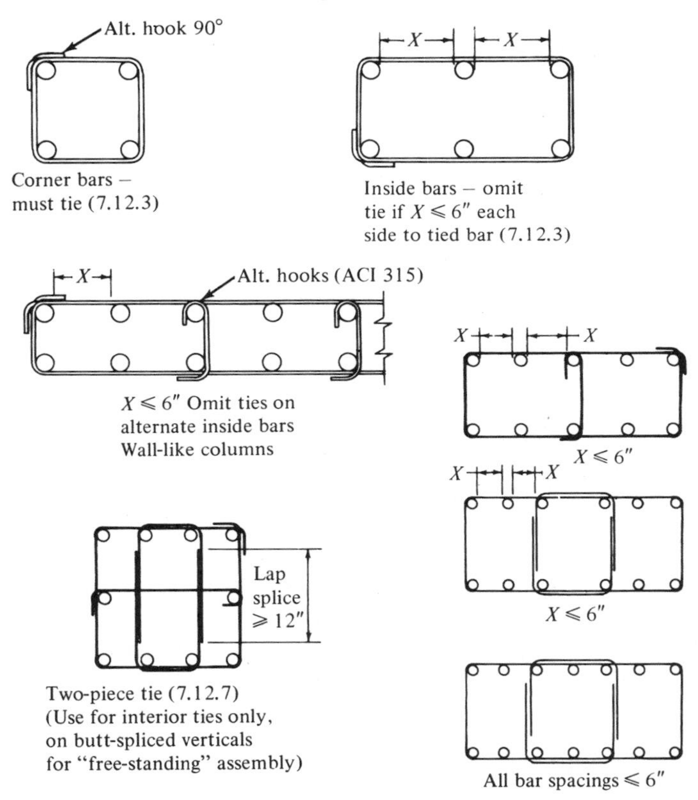

Fig. 10–35 Column tie detail requirements (Section 7.12.3).

in that face of the column must be twice the calculated tension in that face of the column (Section 7.10.3), and in no case less than one-fourth $f_y A_s$ (Section 7.10.5).

Columns with eccentricity inside the kern ($e \approx 0.17\ h$ for square and $0.13\ h$ for round) have no tension, and if the bars are #11 or smaller, compression lap splices for all bars at one section will provide 25 percent tensile capacity or more, satisfying the minimum requirement (Section

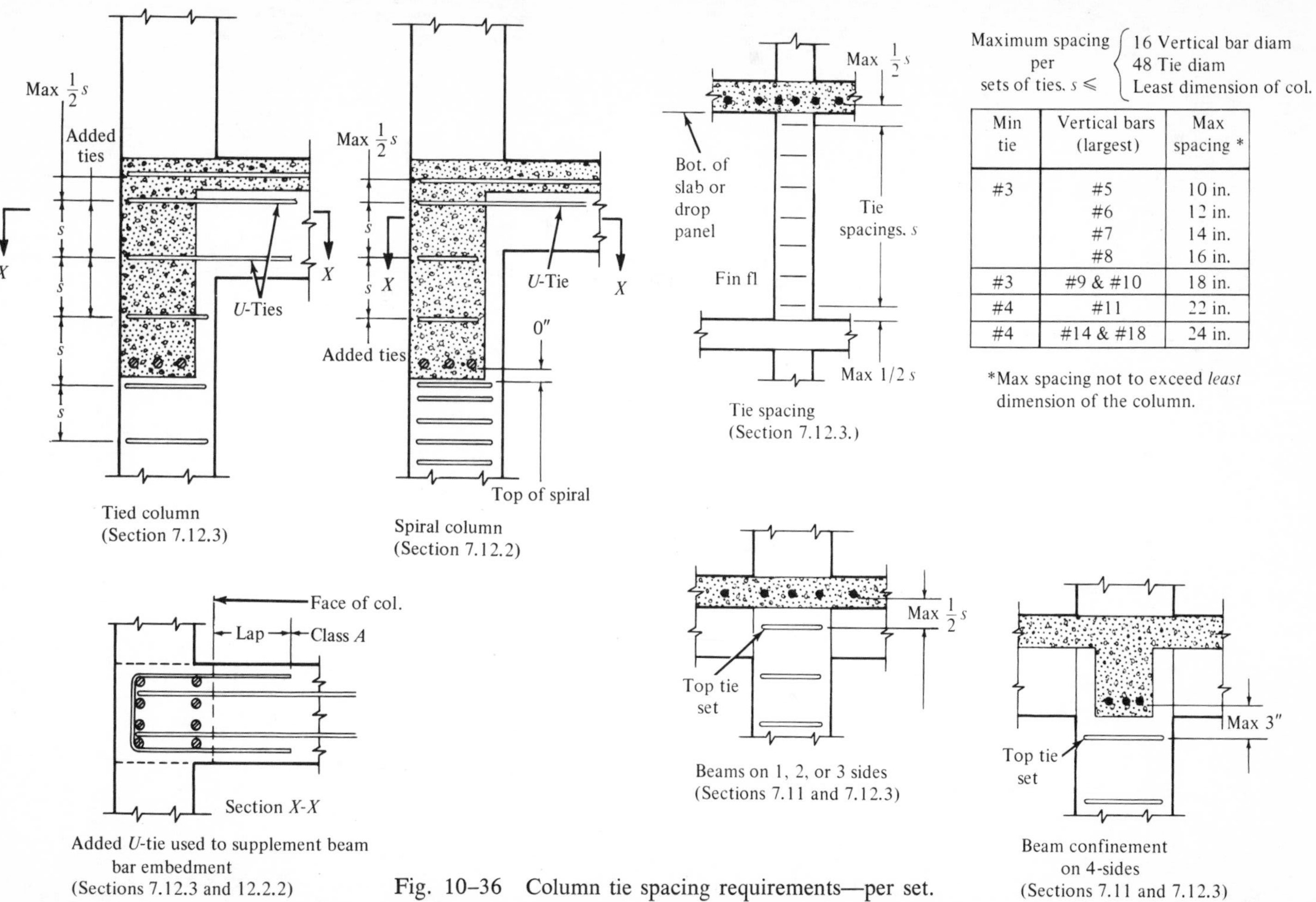

Min tie	Vertical bars (largest)	Max spacing *
#3	#5 #6 #7 #8	10 in. 12 in. 14 in. 16 in.
#3	#9 & #10	18 in.
#4	#11	22 in.
#4	#14 & #18	24 in.

*Max spacing not to exceed *least* dimension of the column.

Fig. 10–36 Column tie spacing requirements—per set.

7.10.5). If the bars are #14 or #18 in size, end bearing butt spices can be used, but must be staggered so as not to splice more than 75 percent of the bar area in any face of the column. For larger eccentricities, up to those causing tension of one-half f_y in the steel, various combinations of staggered end bearing splices, compression lap splices, tension lap splices, separate dowels, welded splices, or positive connections with less than 125 percent f_y capacity may be used.

Notes to Fig. 10-36:
(1) Maximum tie spacings are governed by three criteria, whichever is least. In no case shall ties or set of ties be spaced further apart than the least dimension of the column. The other two criteria are 16 vertical bar diameters or 48 tie diameters (Section 7.12.3). The latter two criteria are easily tabulated for required minimum tie sizes and all practicable sizes of verticals. See tabulation.
(2) The lowest set of ties and the upper set are required to be within one-half tie spacing of the finished floor below and the lowest horizontal reinforcement in the slab or drop panel above, respectively. This requirement is not contingent upon the slab or drop panel providing concrete confinement on all four sides of the column and so it will apply to edge as well as interior columns.
(3) With beams or brackets to provide concrete confinement of the joint area on *all* (four) sides at the top, the top tie set must be within 3 in. of the lowest horizontal reinforcement in the beams or brackets. Note that when one or more sides are not confined by concrete, the ties must be within one-half tie spacing of the lowest steel in the slab.
(4) When only partial confinement by concrete beams is provided, added ties are required up to a maximum of one-half tie spacing, *s*, from the lowest *slab* steel (Sections 7.11 and 7.12.3). Above the normal top of the spiral in spiral columns, the same requirement applies (Section 7.12.2). Although the Code requires that the spiral extend *to* the level of the level of the lowest horizontal reinforcement in the deepest beam present (dimension 0″), in practice it is usually ended at the construction joint between the lower column and the floor system as the drawing shows. *IMPORTANT*: it is *not* accepted practice to supply ties designated as "added"; if desired, complete details for added ties *must* be prepared by the engineer and typical details shown on design drawings (ACI 315).
(5) A device is illustrated here to satisfy two code purposes with the same added tie. Where the added ties are required only at one unconfined face as shown, such ties need not be closed but may be U-shapes extended into the beam. Properly lapped past the face of the column with the beam bars, they may then be also considered effective additions to either the top or bottom beam bars, depending upon depth position. This device is recommended where depth of the column and/or spandrel is too shallow or congested to permit placing sufficiently long hooked ends on the beam bars required anchorage (Sections 7.11 and 12.2.2).

11 walls

GENERAL

Reinforced concrete walls are defined (Section 2.1) as vertical elements used primarily to enclose or separate space. The design of such walls must be based on lateral forces or any other loads to which they may be subjected (Section 14.1.1). A rational method of design for walls subject to flexure or to both flexure and axial compression is provided (Section 10.16). An empirical method of design for bearing walls with small moment (resultant compressive force within the middle-third of the wall thickness) is permitted (Section 14.1.2). The empirical design method is explained in detail (Section 14.2). Nonload bearing walls whether precast or cast-in-place, ordinary reinforced or prestressed, must be designed by the rational method, taking into account all loading conditions (Sections 10.16, 16.3, and 18.1.3).

SPECIAL WALLS

Walls which principally resist horizontal shear forces (shear walls in low-rise buildings) in the plane of the wall and parallel to the length of the wall must conform to the requirements for shear (Section 11.16). Shear walls which resist forces from seismic accelerations must conform to the requirements for special shear walls (Section A.8). Special requirements for walls designed as grade beams are provided (Section 14.3). In addition to the general provisions (Section 10.16), precast concrete wall panels must conform to requirements for precast concrete wall panels (Section 16.3).

COVER

Minimum concrete cover required for the protection of reinforcement (Sections 7.14.1, 7.14.1.1, and 7.14.1.2) for purposes other than fire resistance is shown in Fig. 11-1.

DESIGN LOAD COMBINATIONS

Walls must be proportioned for adequate strength using either the load factors (Section 9.3) and the capacity reduction factors (Section 9.2.1),

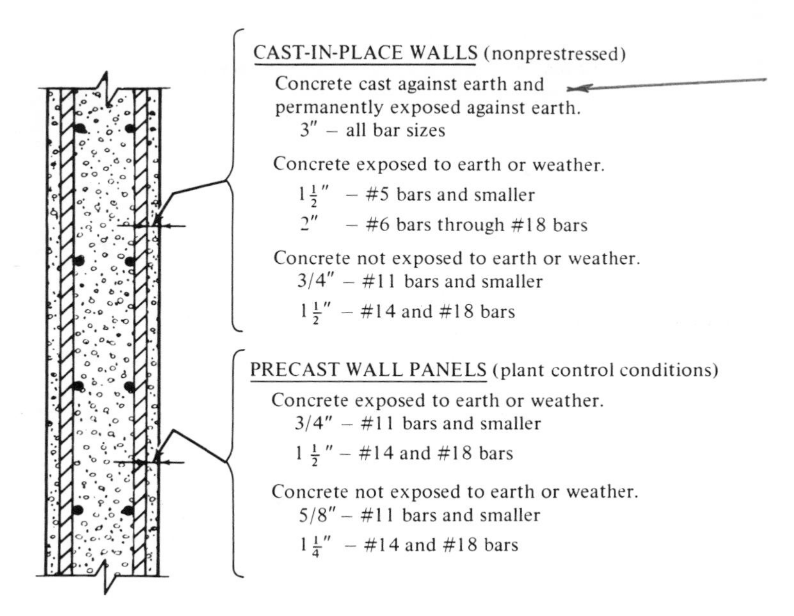

Fig. 11–1 Concrete Protection for reinforcement.

or the alternate method (Section 8.1.2). When subjected to lateral earth pressure, H, dead load, D, and/or live load, L, the design load, U, must not be less than $1.4\ D + 1.7\ L + 1.7\ H$ (Section 9.3.4). When subjected to lateral pressure, F, from liquids, the design load, U, must not be less than $1.4\ D + 1.7\ L + 1.4\ F$. The vertical pressure from liquids is considered as dead load (Section 9.3.5).

When D or L reduce the effect of H or F, such as in the design of the heel or toe of a retaining wall, the load factors for strength design must be reduced from 1.4 to 0.9 for D and from 1.7 to zero for L (Sections 9.3.4 and 9.3.5).

NONLOAD BEARING WALLS

Nonload bearing walls can be defined as vertical elements which support no vertical load other than their own weight. These include basement walls, cantilever retaining walls, area enclosure walls, exterior precast panel walls, tilt-up walls and cast-in-place walls for hydraulic structures. Such walls are subjected primarily to flexure or to flexure and relatively small axial compression loads with the resultant compressive forces outside the kern ($e > h/6$) of the section and must be designed by the rational method (Section 10.16). Design for shear forces perpendicular to the face of such walls must conform to the requirements for slabs (Section 11.10).

Minimum vertical reinforcement (Sections 10.16.2, 10.16.3, and 10.16.4) and horizontal reinforcement (Section 10.16.5) for nonload bearing walls with Grade 60 reinforcement of #5 bars or smaller are shown in Fig. 11-2. When bars larger than #5 are used, the minimum vertical and horizontal reinforcement must be increased 25 per cent (Sections 10.16.2 and 10.16.5-b). The Code does not require that minimum reinforcement be placed in the two faces of a wall (Section 10.16).

Thickness h (inches)	IN SINGLE LAYER ONLY[a]		IN BOTH FACES[b]	
	Horiz. $\rho = 0.0020$	Vert. $\rho = 0.0012$	Horiz. $\rho = 0.0010$	Vert. $\rho = 0.0012$
6	#3 @ 9	#3 @ 18	—	—
8	#4 @ 12	#3 @ 11	—	—
10	#5 @ 15	#4 @ 16	#4 @ 18	#3 @ 18
12	#5 @ 12	#4 @ 13	#4 @ 16	#3 @ 15
14	#5 @ 11	#5 @ 18	#4 @ 14	#3 @ 13
16	#5 @ 10	#5 @ 16	#4 @ 12	#3 @ 11
18	#5 @ 8½	#5 @ 14	#5 @ 17	#4 @ 18

[a] Bars arranged in two-way mat, usually centered in wall.
[b] Bars arranged in identical two-way mats, one near each face of wall.

Note: When steel is placed in two layers, maximum spacing limitations apply separately, but minimum steel requirements (ρ) apply to the sum of steel areas in both faces.

Although a literal application of the requirements for minimum reinforcement (Sections 10.16 and 18.1.3) would apply to prestressed nonload bearing panel walls, the authors would suggest that nonprestressed reinforcement be added to such wall panels *only* if required for excess tension under a design loading, including temperature and shrinkage effects. One additional exception for nonload bearing walls is the thin "serpentine" cantilever enclosure wall which may be considered to be a simple thin shell structure (Section 19.1.3). Again a strict application of Code requirements would require minimum steel as in a plane wall (Sections 19.1.2 and 10.16). The authors suggest that vertical steel be provided *only* in such amounts and at such locations as required by structural analysis, considering effects of all design forces without regard to minimum steel areas or spacings prescribed for plane walls.

LOAD BEARING WALLS, EMPIRICAL DESIGN

Load bearing walls are used primarily as compression members. When compression loads are eccentric the wall must be capable of resisting the

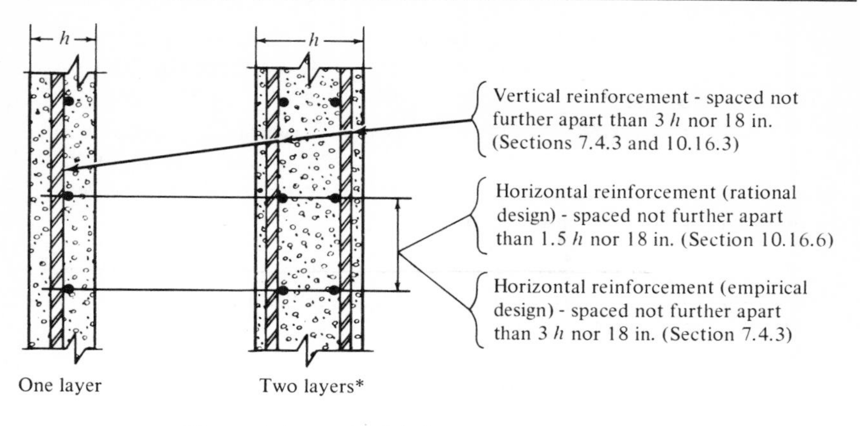

Fig. 11–2 Minimum wall reinforcing and maximum spacing.

compression load and the flexural moment resulting from the eccentricity. Load bearing walls can all be designed by the rational method (Section 10.16), or by the empirical method (Section 14.2) only if the eccentricity of the load is equal to or less than one-sixth of the wall thickness ($e < h/6$). When designed by the empirical method the ratio of the unsupported length to thickness l_c/h cannot be more be more than 25 (Section 14.2-d), or 30 for panel and enclosure walls (Section 14.2-j), unless it can be shown by a structural analysis that adequate strength and stability will exist at greater ratios (Section 14.1.2). A "panel or enclosure" bearing wall may be defined as a wall provided with lateral bracing in a vertical plane (in addition to lateral bracing of floor or roof construction). Ordinarily, such bracing consists of full height cross-walls or pilasters. The minimum thickness of such walls can be less than that for bearing walls braced only by floor and roof members, since two-way slab behavior is possible. Under the empirical method of design, such walls must be anchored at all floors and pilasters (Section 14.2-i), thus ensuring some two-way effectiveness of reinforcement.

Minimum vertical and horizontal Grade 60 reinforcement of #5 bars and smaller for load bearing walls designed by the empirical method (Section 14.2-f) is the same as for load bearing walls designed by the rational method, except that for walls thicker than 10 in. the reinforcement must be placed in two layers (Section 14.2-g). It is not required that one-half of the reinforcement be placed in each layer. If cover requirements are met, a minimum of one-third and a maximum of two-thirds of the reinforcement can be placed in one layer. The minimum quantity of reinforcement for load bearing walls designed by the empirical method may be waived where structural analysis shows adequate strength (Section 14.1.2).

Wall capacity by the empirical method (Section 14.2-b, Eq. 14-1) is based on the average stress, P_u/A_g, as shown in Fig. 11-3 and not on the peak stress that results from eccentricity of the load. In Fig. 11-3 the compression load capacity is given for various l_c/h ratios in terms of the average stress and concrete strength.

$$\frac{P_u}{A_g f_c'} = 0.55\phi \left[1 - \left(\frac{l_c}{40h}\right)^2\right] \quad \text{(Section 14.2)}$$

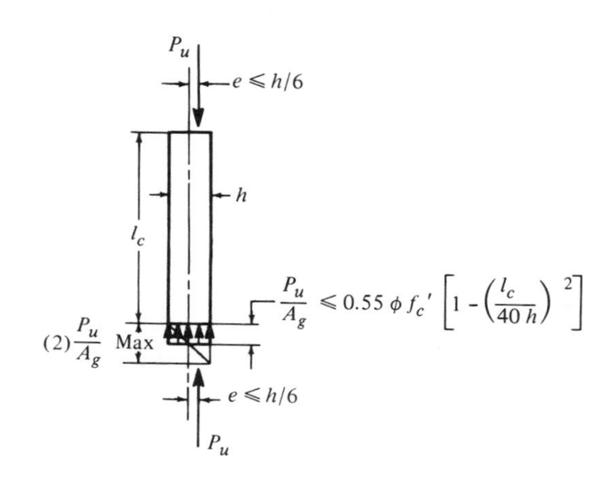

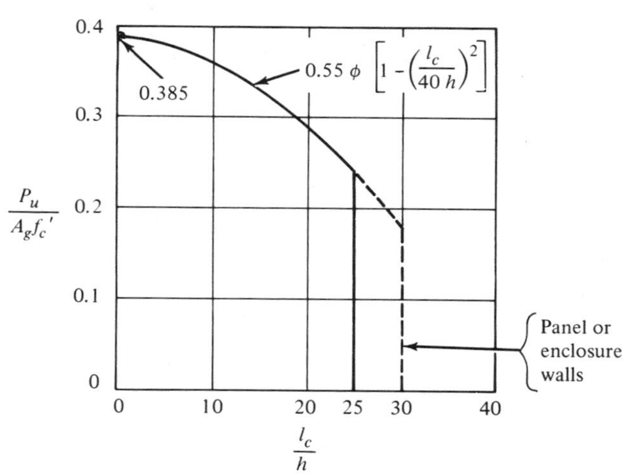

Fig. 11–3 Walls by empirical design.

LOAD BEARING WALLS, RATIONAL DESIGN

Wall capacity by the rational method (Section 10.16) uses the general principles and requirements for column design (Section 10.3) to determine the design strength and takes account of the slenderness effects (Section 10.10 or 10.11). Minimum vertical and horizontal reinforcement for load bearing walls designed by the rational method (Section 10.16) is the same as for nonload bearing walls, as shown in Fig. 11-2. Lateral ties are not required when the area of the vertical reinforcement is equal to or less than 1 percent of the concrete area, *or* where it is not required as compression reinforcement (Section 10.16.4) (see also Chapter 10, Figs. 10-1, -2, -3 and -4, and Fig. 10-14).

COMPARISON OF WALL DESIGN METHODS

A comparison of the rational design strength and the empirical design strength of a 12 in. thick reinforced concrete load bearing wall for various

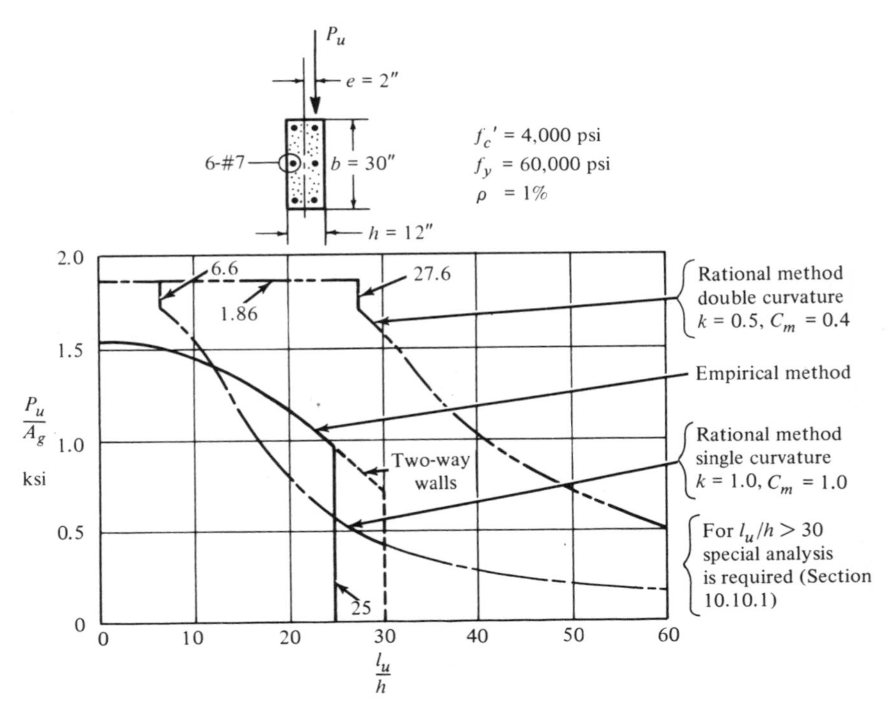

Fig. 11–4 Wall with flexure and axial compression.

l_u/h ratios is shown in Fig. 11-4.* The wall of this example is braced against sidesway. It is reinforced with 1 percent vertical steel and is subjected to a compression load with an eccentricity of 2 in. This eccentricity is the maximum permitted under the empirical design method. Figure 11-4 shows that for single curvature, $k = 1.0$ and $C_m = 1.0$ (Sections 10.11.3 and 10.11.5) the rational method of design provides greater strength for l_u/h ratios less than 12. It also shows that the rational design method cannot be used for l_u/h ratios greater than 30 for single curvature and 60 for double curvature with equal end moments unless a special analysis is made. For double curvature, $k = 0.50$ and $C_m = 0.40$, the rational method of design provides greater strength for all l_u/h ratios.

The rational design strength of the wall shown in Fig. 11-4 was determined from a 12 in. by 30 in. section of wall with $f_c' = 4{,}000$ psi and reinforced with six #7 vertical bars as shown in Fig. 11-4. The radius of gyration was assumed equal to 0.3 h (Section 10.11.2) and the modulus of elasticity of the normal weight concrete was taken equal to 57,000 $\sqrt{f_c'}$ (Section 8.3.1). A load-moment interaction diagram was generated for this section of wall as a short column about the weak axis neglecting slenderness effects. This envelope is shown in Fig. 11-5.

SLENDERNESS EFFECTS

An approximate evaluation of slenderness effects was made (Section 10.11) for the wall examples designed by the rational method, but kl_u/r values greater than 100 were not considered to avoid making a special analysis (Section 10.10.1). Because the wall of the example is braced against sidesway in the weak direction, slenderness effects can be neglected in this direction for kl_u/r values less than 34 − 12 $M_{1/2}$ (Section 10.11.4). For single curvature and $M_1 = M_2$, $k = 1.0$ and $C_m = 1.0$, slenderness effects in the weak direction of the wall can be neglected for kl_u/r values less than 22 (l_u/h values less than 6.6). For double curvature and $M_1 = M_2$, $k = 0.5$, and $C_m = 0.4$, slenderness effects in the weak direction can be neglected for kl_u/r values less than 46 (l_u/h values less than 27.6).

The load-moment interaction curve for the example wall section as a short column is shown in Fig. 11-5 and is applicable for l_u/h ratios less than 6.6 for single curvature and 27.6 for double curvature. As shown in Fig. 11-5, for l_u/h ratios greater than 6.6 for single curvature and 27.6 for double curvature, design moments, $P_u e$, must be increased by the magnification factor, δ_s (Section 10.11.5).

---

* See also Chapter 10 for other comparisons. In Chapter 14, $l_u = l_c$.

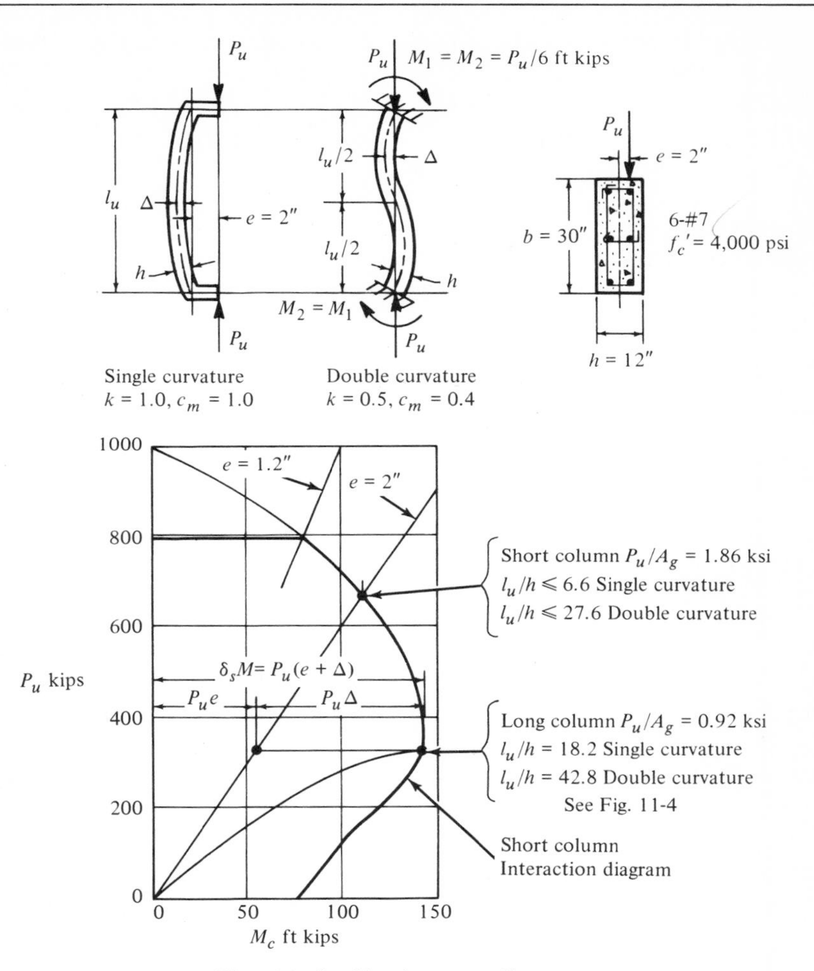

Fig. 11–5 Slenderness effects.

$$\delta_s = \frac{C_m}{1 - \dfrac{P_u/\phi}{P_c}} \quad \text{(Code Eq. 10-5)}$$

The critical load P_c, for Eq. 10-5 is determined from the Euler formula,

$$P_c = (\pi)^2 EI/(kl_u)^2 \quad \text{(Code Eq. 10-6)}$$

EI for use in Eq. 10-6 will be taken in accordance with the preferred equation for columns with low percentages of steel (Code Commentary Fig. 10-4) as

$$EI = \frac{E_c I_g}{(2.5)(1 + \beta_d)} \quad \text{(Code Eq. 10-8)}$$

If dead load, D, is equal to 2.43 L, $\beta_d = 1.4\ D/(1.4\ D + 1.7\ L)$. $\beta_d =$ 0.667 and EI for the example wall becomes

$$EI = \frac{(57{,}000)(\sqrt{4{,}000})(1/12)(30)(12)^3}{(2.5)(1 + 0.667)}$$

$$= 3{,}740{,}000{,}000 \text{ lb in.}^2$$

with EI known, kl_u can be solved for in terms of P_c

$$(kl_u)^2 = 3{,}740{,}000{,}000\ (\pi^2/P_c)$$

The available capacity for moment magnification at any particular eccentricity can be obtained directly from the load-moment interaction curve. Determine the ratio of the maximum moment capacity for any given load to the moment of that load at the particular eccentricity. Figure 11-5 shows that the available capacity for magnification of moment at eccentricity of 2 in. for balanced load, $P_b = 333$ kips (Section 10.3.3) is equal to $145/55 = 2.64$. The maximum moment magnification factor at load $P_b = 536\ k$ for $e = 2$ in. is $134/82 = 1.63$, etc. Critical loads, P_c, and corresponding l_u/h ratios and moment magnification factors for both single and double curvature conditions were calculated for specific design loads, P_u. Compressive stress ratios, P_u/A_g were also calculated for each specific design load and compared in Fig. 11-4 with stress ratios for walls designed by the empirical method. Two such sample calculations for single curvature follow.

Single Curvature Calculations

1. ($P_u = 333$ kips, $\delta_s = 2.60$, $k = 1.0$, and $C_m = 1.0$)

Calculate P_c

$$\delta_s = \frac{C_m}{1 - \dfrac{P_u/\phi}{P_c}} = \frac{1}{1 - \dfrac{333}{0.70\ P_c}}$$

$$P_c = 773 \text{ kips}$$

Calculate l_u

$$(kl_u)^2 = \pi^2 EI/P_c$$

$$l_u^2 = 3{,}740{,}000{,}000\,(\pi)^2/773{,}000$$

$$l_u = 218 \text{ in.}$$

Calculate l_u/h

$$l_u/h = 218/12 = 18.2$$

Calculate P_u/A_g

$$P_u/A_g = 333/(12)(30) = 0.92 \text{ ksi}$$

2. ($P_u = 536$ kips, $\delta_s = 1.505$, $k = 1.0$, and $C_m = 1.0$)

Calculate P_c

$$1.505 = \frac{1}{1 - \dfrac{536}{0.70\,P_c}}$$

$$P_c = 2{,}280 \text{ kips}$$

Calculate l_u

$$l_u^2 = 3{,}740{,}000{,}000\,(\pi)^2/2{,}280{,}000$$

$$l_u = 127 \text{ in.}$$

Calculate l_u/h

$$l_u/h = 127/12 = 10.6$$

Calculate P_u/A_g

$$P_u/A_g = 536/(12)(30) = 1.49 \text{ ksi}$$

Note that the limiting loads for walls designed by the empirical design method for the usual l_u/h range lie between the limits for walls designed by the rational method. The empirical design method is applicable to both single and double curvature conditions, but the limiting moment at $e = h/6$ effectively precludes large sidesway or large l_u/h ratios, except for nearly concentric loads. The limiting l_u/h ratio for use with the empirical method is 25 (Section 14.2-d). For panel walls due to two-way action, the limiting l_u/h ratio for empirical design is increased to 30 (Section 14.2-j), which is also the limit for rational design of walls in single curvature without a special analysis.

SPECIAL WALLS

Plain Concrete

Plain concrete (outside the scope of ACI 318-71) is frequently used for walls of all types. If closely spaced "control joints" or other provisions for unrestrained shrinkage are not provided or if shrinkage cracking is objectionable, horizontal temperature steel at least equal to the minimum required should be used (Section 7.13). Plain concrete will usually suffice for walls with l_c/h ratios equal to, or slightly larger, than those permitted for unreinforced masonry under local codes. For strength design, the prescribed capacity reduction factor, ϕ, equals 0.65 (Section 9.2.1.5). The flexural tensile and shear stress permitted for plain concrete footings should not be exceeded (Section 15.7.2).* Such designs have been accepted for particular applications when justified (Section 14.1.2) and submitted for approval under procedures established by the Code (Section 1.4).

Precast Concrete Panels or Precast Crib Retaining Walls

A variety of precast concrete walls are employed, the design of which is outside the scope of the specific Code provisions for walls (Chapter 14 and Section 10.16). Again, such designs can be approved when the design is submitted under the procedures of the Code (Section 1.4).

Deep Beam Walls

A number of concrete wall applications behave as deep beams and must be designed accordingly (Section 10.7). One common application in this category is deep beam grade walls spanning as simple beams between end supports (footings) and carrying their self-weight plus other wall or column loads above. Another is deep spandrel walls supported full depth at each end and resisting small lateral wind forces, as well as carrying their self-weight. The Code provisions for deep beams ensure that reinforcement

* For further information, see "Proposed Recommended Stresses for Unreinforced Concrete," *ACI Journal*, Nov. 1942; "Structural Plain Concrete," *ACI Journal*, April 1967; and (proposed) "Building Code Requirements for Structural Plain Concrete," ACI Committee 322, *ACI Journal*, May 1971.

for wall-like deep beams will be at least equal to that required for walls (Section 10.7).

Two-Way Reinforced Walls

Walls for underground enclosures to resist lateral forces due to earth or liquid pressure frequently are supported on three or four sides (basements, sewage and water structures, etc.). A realistic two-way elastic analysis will usually reduce overall reinforcement requirements and cracking since two-way reinforcement will conform closely to the elastic analysis of bending. The specific requirements for two-way slab systems based on supports such as columns, or beams between columns (Chapter 13), are difficult to extend to walls (See Chapter 8 of this Guide). Within the general Code requirements (Section 10.16 and the remainder of Chapter 10), two-way reinforcement proportioned by any of the methods specified in ACI 318-63, should be acceptable under ACI 318-71 as a "rational" method.

12 footings

GENERAL

Ultimate soil pressure or ultimate pile capacity as shown by tests is normally not used in design, except to establish an allowable soil pressure, q_a, or allowable pile capacity. The allowable soil pressure or pile capacity is selected with an appropriate factor of safety, using principles of soil mechanics considering both ultimate load and settlement measured in short-time load tests. Any long-time settlement records available on existing older structures in the vicinity, as well as short-time test data filed with the local Building Authority, are used to establish the safe bearing pressures for various soils specified in statutory local building codes.

SOIL PRESSURE AND PILE CAPACITY

The ACI Building Code is prepared for inclusion in a general statutory building code, and it is expected that the safe allowable bearing capacity for soils or piles will be specified therein or established by test, according to the local building ordinances. The provisions of the ACI Building Code regulate design of footings for column design loads, usually, $U = 1.4\ D + 1.7\ L$ (Eq. 9-1) or $U = 0.75\ (1.4\ D + 1.7\ L + 1.7\ W)$ (Eq. 9-2). The proportions of the bearing area are selected for the allowable soil pressure, q_a, under actual (unfactored) D, L, W, and E specified in the General Code (Section 15.2.4).

FOOTING DESIGN

The engineer is thus required to select the area of the footing for specified loads (unfactored) and to perform the structural design for factored loads. An assumed soil pressure which results from the factored loads will be used for the design of the footings to resist factored loads. This design soil pressure, q_s, or the design pile load used to determine shears and moments in the strength design of footings is determined most simply by dividing the column design load, U, by the area of the footing or by the number of piles. If the engineer does not wish to accumulate both factored and unfactored loads for multistory columns or when the ratio of D/L is not known, such as in the preparation of footing tables for gravity loads, an average load factor such as 1.6 can be used. The allowable soil pressure or the allowable pile load multiplied by this factor is used to determine the design load that the footing can support.

MINIMUM THICKNESS

The minimum thickness above the bottom reinforcement prescribed for concrete footings is as follows:

Plain concrete footings on soil	8 in. (Section 15.9.1)
Reinforced concrete footings on soil	6 in. (Section 15.9.2)
Reinforced concrete footings on piles	12 in. (Section 15.9.2)

Plain concrete footings on piles are not permitted (Section 15.7.3).

SHEAR

The nominal shear stresses for one-way (beam) shear $v_u = V_u/\phi b_w d$ (Section 11.10.1a) and two-way (punching) shear $v_u = V_u/\phi b_o d$ (Section 11.10.1b) are calculated in the conventional manner. In computing the external shear on any section through a footing supported on piles, the portion of the pile reaction assumed as producing shear on the section is based on a straight line interpolation between full value for center of piles located at $d_p/2$ outside the section and zero value at $d_p/2$ inside the section (Section 15.5.5). d_p is the diameter of the pile at the footing. Design shear, V_u, is increased by the capacity reduction factor, ϕ, for shear.

The critical sections for reinforced and plain concrete footings for shear are shown in Fig. 12-1. For an isolated rectangular footing, one-way shear should be checked in each direction to determine the maximum value. Two-way shear for square and rectangular footings is checked on the total shear periphery *abcd* in Fig. 12-1.

Unless shear reinforcement is provided, the average nominal shear stress v_u must be equal to or less than the shear stress, v_c, carried by the concrete.

$v_c = 2\sqrt{f_c'}$ for one-way shear (Section 11.4.1) or for reinforced sections, the alternate "long" formula may be used

$v_c = 1.9\sqrt{f_c'} + 2500\ \rho_w V_u d/M_u$ for one-way shear (Section 11.4.2)

$v_c = 4\sqrt{f_c'}$ for two-way shear (Section 11.10.3)

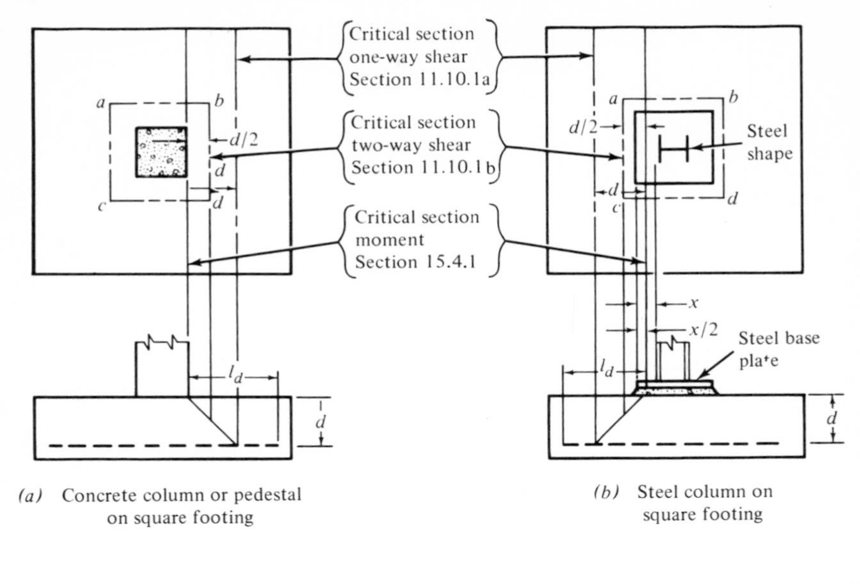

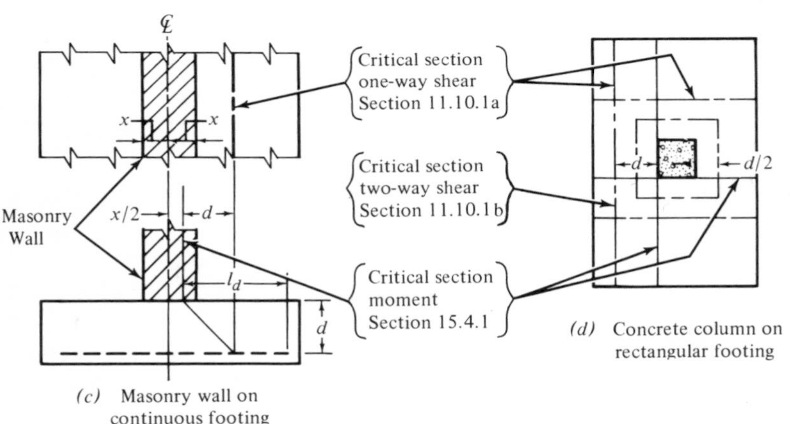

Fig. 12–1 Critical sections for shear and moment.

SHEAR REINFORCEMENT

Shear reinforcement is seldom used in individual column footings or in two-way mat footings. It is usually more practicable and economical to select footing depths so that the shear stress is less than the shear allowed on concrete (see Fig. 12-2). In combined footings and individual column footings connected by beam-like "straps," shear reinforcement is commonly used, and for special conditions, reinforcement for punching shear may be desirable.

If shear reinforcement, consisting of bars or wires, is provided for one-way beam shear, the value of $(v_u - v_c)$ is limited to a maximum value of $8\sqrt{f_c'}$ (Section 11.6.4).

If shear reinforcement, consisting of bars or wires, is provided for two-way shear the shear carried by concrete, v_c, must be reduced from $4\sqrt{f_c'}$ to $2\sqrt{f_c'}$. Nominal shear stress, v_u, is limited to a maximum value of $6\sqrt{f_c'}$ (Section 11.10.3).

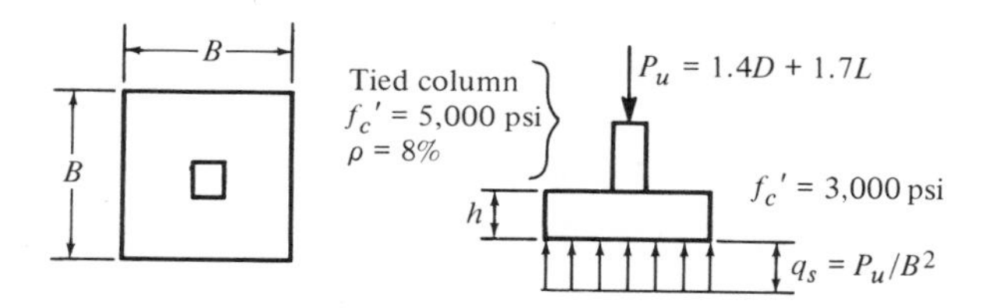

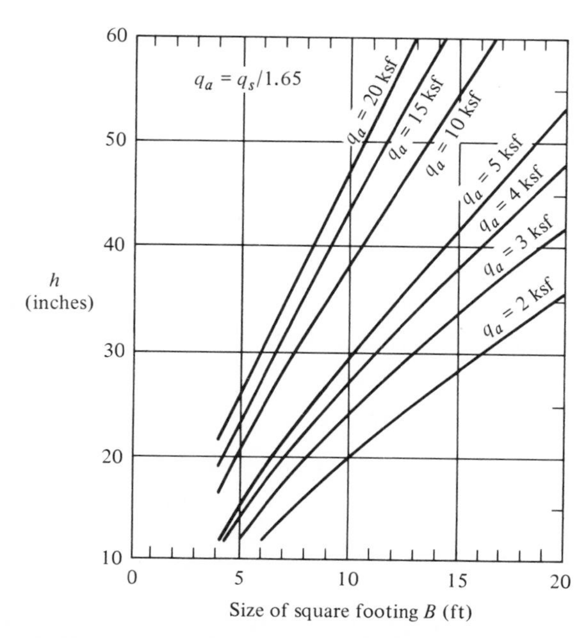

Fig. 12–2 Minimum depth without shear reinforcement.

For plain concrete footings, the Code limits the average one-way beam shear stress ($v_u = V_u/\phi b_w d$) to a maximum value of $2\sqrt{f_c'}$ and the average two-way shear stress to a maximum value $4\sqrt{f_c'}$ (Section 15.7.2). It can be shown that the shear stress is not critical. The tension stress due to flexure controls the thickness of plain concrete square or continuous wall footings on soil.

MOMENT

The critical sections for moment for footings are shown in Fig. 12-1.

For plain concrete footings the maximum permissible flexural tension stress in the concrete is limited to $5\phi\sqrt{f_c'}$ (Section 15.7.2). For a ϕ factor of 0.65 (Section 9.2.1.5) the permissible flexural tension stress (f_t) varies with f_c' as shown in Fig. 12-3.

Based on the Code criteria for bending (Section 15.7.2), the thickness of plain concrete continuous wall footings can be determined from the following expression.

h = total thickness of wall footing in inches

$$h = (0.08)(x)\sqrt{\frac{q_s}{\sqrt{f_c'}}}$$

q_s = design soil pressure in lb per sq ft

x = distance in inches between the face of a concrete wall and the edge of the footing or the distance from a point halfway between the centerline of a masonry wall and face of the wall to the edge of the footing.

Figure 12-4 shows how the thickness of plain concrete wall footings must be increased as the projecting width is increased. Note that for a design soil pressure of 8400 psf the thickness is approximately equal to the projection.

DEVELOPMENT OF MOMENT REINFORCEMENT

Tension in the reinforcement must be developed on each side of the critical section for moment (Section 12.1.1). For isolated column footings and continuous wall footings, sufficient length between the critical section (Section 12.1.3) and the nearest end of the bar to develop the bar in tension must be provided (see Fig. 12-1). This development length (l_d) in tension varies directly as the area and the yield strength of the bar and inversely as $\sqrt{f_c'}$ (Section 12.5). The length required (l_d) to develop various

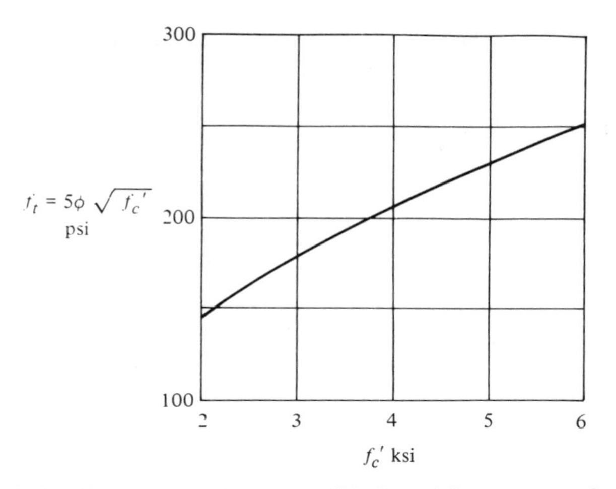

Fig. 12–3 Flexural tensile stress (f_t) for plain concrete footings.

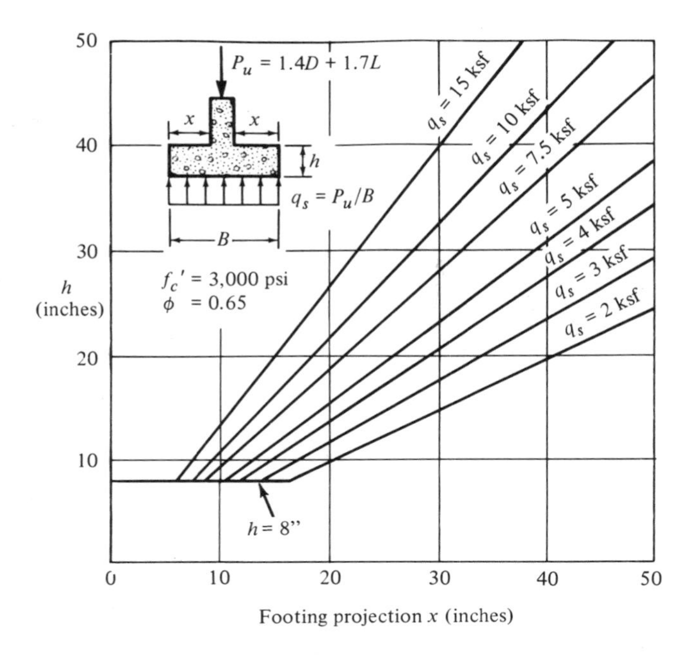

Fig. 12–4 Continuous plain concrete wall footings.

size Grade 60 straight bars in tension for different concrete strengths is shown in Table 13-4. If sufficient length is not available to develop a bar size selected for flexure, the required development length may be reduced by using a larger bar size at a larger spacing (if the spacing can be increased from less to more than 6 in.), smaller sizes or bars can be furnished, or standard hooks can be used to decrease the required l_d (Section 12.5 d). The amount of equivalent straight bar length provided by a standard hook is shown in Table 13-6 (Section 12.8). Note that 20 percent less development length is required if bars are spaced more than 6 in. on centers.

Combined footings with short cantilevers and deep beam footings sometimes do not provide sufficient length between the critical section and the end of the bar for development of the tension reinforcement. In these cases additional steel, hooks, or other means of anchorage must be used, such as welding a bar of equal size at right angles to the end of the tension bars.

BEARING

When the loaded area (area of column pier or base plate) and the supporting area (area at the top of the footing) are equal, bearing stress on the loaded area of the footing must be equal to or less than 0.85 $\phi f_c'$ (Section 10.14.1) (see Fig. 12-5). When the supporting area (*efgh*) is larger than the loaded area (*abcd*) on all sides, the bearing stress must be equal to or less than 0.85 $\phi f_c' \sqrt{A_2/A_1}$ (Section 10.14.2). A_1 is the loaded area. A_2, the supporting area, is the lower base of the largest frustum of a right pyramid or cone contained wholly within the footing with upper base the loaded area, A_1, and with side slopes of 2 horizontal to 1 vertical, as shown in Fig. 12-4. Figure 12-5 also shows in graphical form how the bearing stress on the loaded area varies with the ratio A_2/A_1.

When the bearing stress on the loaded area exceeds 0.85 $\phi f_c' \sqrt{A_2/A_1}$ reinforcement must be provided for the *excess* by extending the longitudinal bars into the footing or by dowels (Section 15.6.3). One additional code requirement concerning column dowels *only* is that the minimum tensile capacity in each face of the column must equal one-fourth of vertical reinforcement in that face times f_y (Section 7.10.5). Extended longitudinal bars or dowels from a column or pedestal, if required, must be four or more in number and have a minimum area equal to 0.005 times the loaded area of the column or pedestal (Section 15.6.5). There are no limitations on the maximum number of dowels that can be used in a footing other than those due to spacing and cover requirements (Sections 7.4.4 and 7.4.5). Dowels can be the same size as, or smaller than, #5 through #18 column compression bars, but cannot be more than one size larger than #5 through #11 column bars (Sections 15.6.5 and 15.6.8).

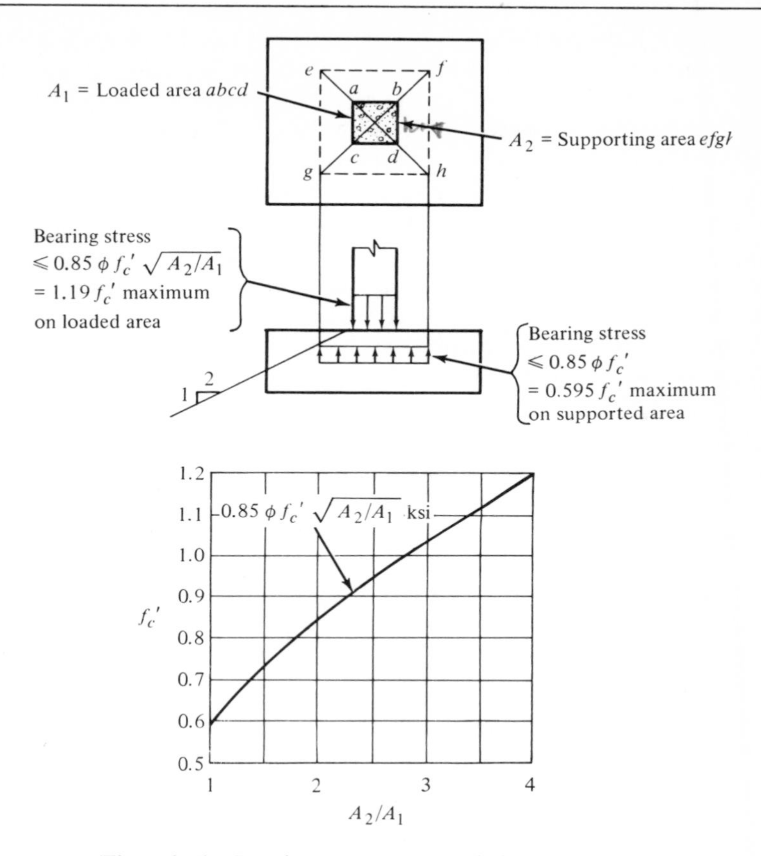

Fig. 12–5 Bearing stress on loaded area.

If the maximum bearing stress on the loaded area of a column does not exceed 0.85 $\phi f_c'$, and there is no tension on the section ($e < h/6$), column bars or dowels do not need to be extended into the footing for compression if they can be developed in compression (Section 12.6) within the pedestal height (three times the minimum column dimension) above the footing. (See Table 13-1. for Grade 60 bars.) The authors recommend that a minimum of one #5 dowel be provided in each corner of a column and pedestal for protection during construction. The required embedment depth for dowels into the footing must develop the larger of

1. excess compression to be transferred by dowels, or for columns only
2. one-fourth the tensile capacity of the steel in each face of the column

Thus, the minimum calculated embedment for the minimum size (#5) dowels could be (1) 0 in. for compression, or (2) 0.25 l_d for tension = (0.25) (15 in.) = 3.75 in. The authors recommend a minimum embedment of 12 in. with footing depths > 15 in. For shallow footings where tension embedment controls, a 90° end hook can be employed to reduce total embedment depth (Section 12.8) (see Table 13-6).

Critical compression dowel embedment cannot be reduced by end hooks (Section 12.8.3) nor with $f_c' > 4444$ psi. If the extended longitudinal bars or dowels of the same size cannot be developed in compression within the footing depth required by strength design, the footing can be made thicker, an equivalent area of smaller size bars can be furnished as shown, or a concrete pedestal as shown in Fig. 12-6 can be provided.

SPLICES

Lap splices for #14 and #18 bars are not permitted (Section 7.5.2) except as column-to-footing compression dowels when spliced with bars of a

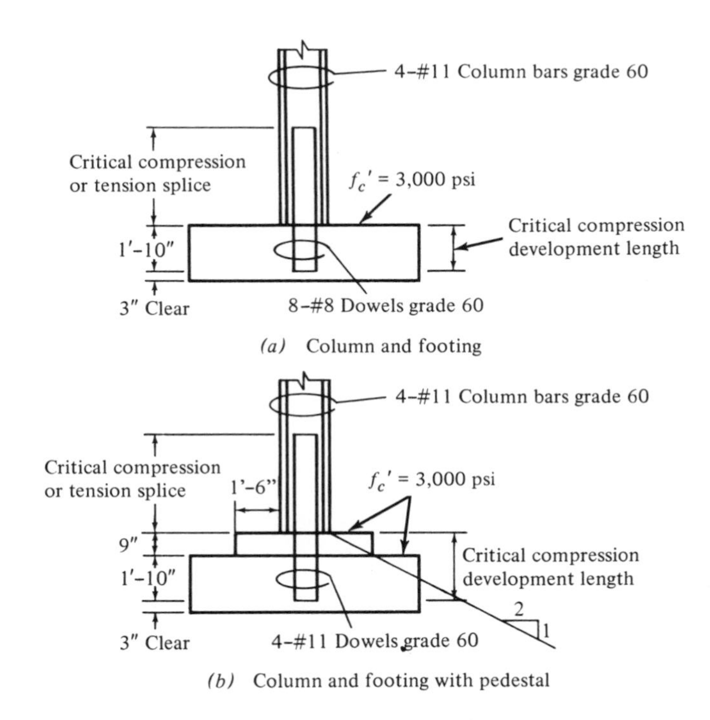

Fig. 12–6 Dowel embedment examples.

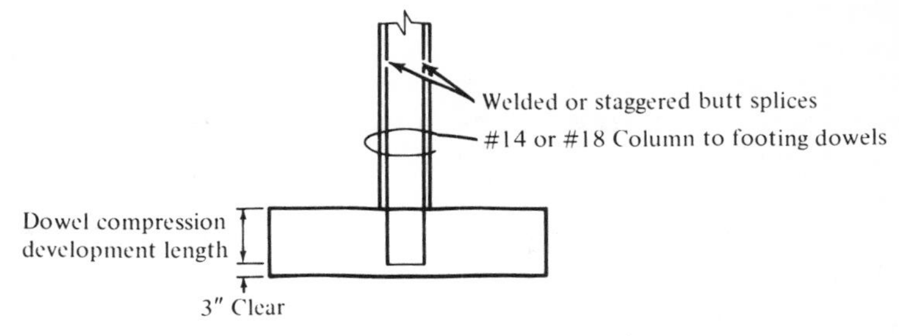

(a) Column to footing compression dowels

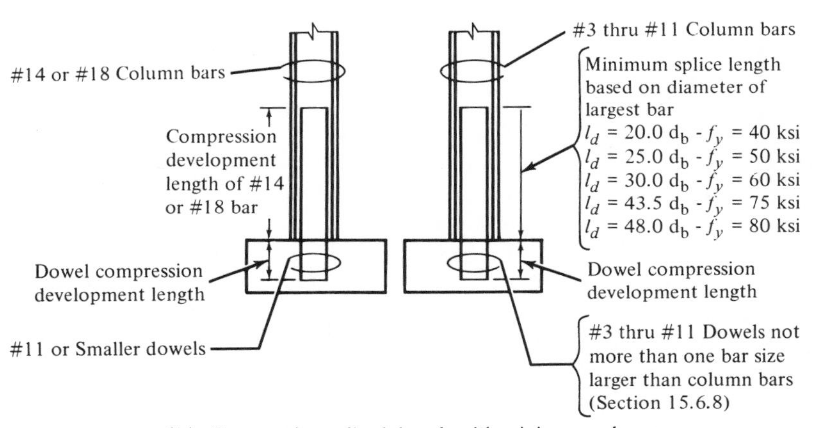

(b) Compression spliced dowels with minimum column ties ($A_v < 0.0015\ bs$ & $f_c' \geqslant 3{,}000$ psi)

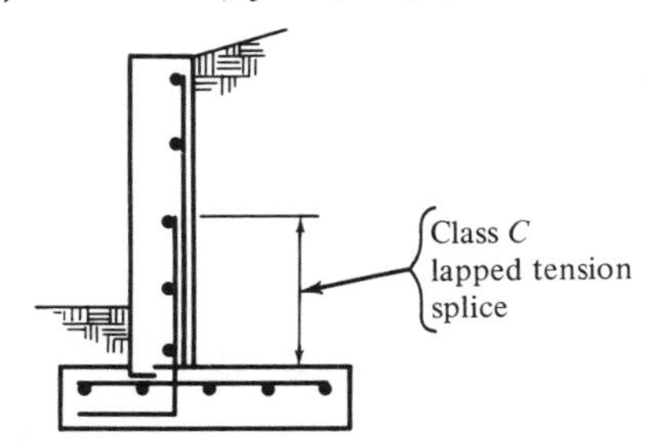

(c) Tension spliced dowels

Fig. 12–7 Footing dowell splices.

smaller size (Section 15.6.8) (see Fig. 12-7). Note the absence of hooks which cannot be considered effective in compression (Section 12.8.3).

#3 through #11 footing dowels that are spliced in compression with column bars must project into the column the minimum compression splice distance (Section 7.7.1) unless this distance can be reduced because the area of the column ties is greater than 0.0015 *bs* (Section 7.7.1.2) or the splice is made in a spiral column (Section 7.7.1.3). When the size of spliced compression footing dowels and column bars is different, but the total area of each is the same, the splice length must be based on the diameter of the larger bar.

When footing dowels are designed for critical tension, and spliced with column bars or with wall bars of a retaining wall above, the splice lengths depend on the class of splice required (Section 7.6.1). In the usual

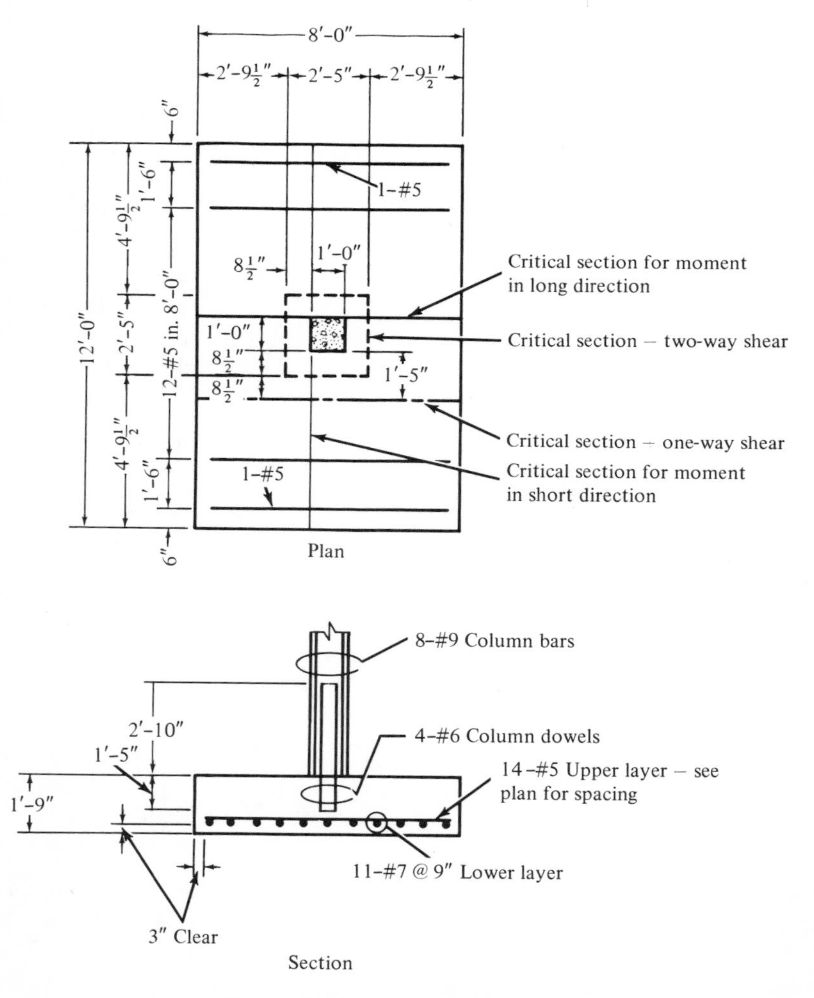

Fig. 12–8 Footing design example.

case, except for columns controlled by tension or high retaining walls, all footing dowels are spliced at one location at a point of maximum stress (Fig. 12-7) and a Class C splice is required.

For Class C splices with $f_c' = 4{,}000$ psi, the splice length for Grade 60 bars varies between 42 and 72 diam. (See Table 13-7 for Class A, B, and C splices with various concrete strengths.)

EXAMPLE. Consider the simple case of an isolated rectangular footing supporting a 12 in. square tied concrete column with eight #9, as shown in Fig. 12-8. Because of site conditions, the footing is restricted to a maximum width of 8′–0″ and is supported on soil with a safe allowable soil pressure, q_a, of 3,000 psf. The actual (unfactored) column dead load is 163 kips and the actual column live load is 100 kips. Grade 60 bars and concrete with $f_c' = 3{,}000$ psi have been selected for footing and column.

Determine required area of footing

Actual column dead load	$D = 163$ kips
Actual column live load	$L = 100$ kips
Estimated weight of footing	$= 25$ kips
$D + L$	$= 288$ kips

$$\text{Area of footing } A_f = \frac{D + L}{q_a} = 288/3 = 96.0 \text{ sq ft}$$

A rectangular footing 8′–0″ × 12′–0″ provides the required area without exceeding the allowable soil pressure, and we can now determine the design soil pressure, q_s, to be used in the strength design of the footing.

$$1.4\,D = (1.4)(163) = 228 \text{ kips}$$

$$1.7\,L = (1.7)(100) = 170 \text{ kips}$$

$$1.4\,D + 1.7L = 398 \text{ kips}$$

$$\text{Design soil pressure, } q_s = 398/96 = 4.15 \text{ ksf}$$

Thickness

Footing depth is usually determined by two-way shear and for this example a trial size depth of 1′–9″ will be assumed and checked. Two-way shear is checked on the shear periphery, *abcd* (Section 11.10.3), with an assumed average depth equal to 4 in. less than the total thickness to allow for cover and one bar diameter.

Two-way shear

$$V_u = \frac{(4150)}{0.85}\left\{(8.00)(12.00) - \left[\frac{(12 + 17)}{12}\right]^2\right\}$$

$$= \frac{(4150)}{0.85}(96 - 5.84)$$

$$= 440{,}000 \text{ lb}$$

$$v_u = \frac{440{,}000}{(4)(29)(17)} = 223 \text{ psi} \approx 4\sqrt{f_c'} = 219 \text{ psi} \quad \text{OK}$$

One-way shear is maximum in long direction by inspection.

$$d = 17.50 \text{ and } b_w = 96 \text{ in.}$$

$$V_u = \frac{(4150)(8.00)(4.04)}{0.85} = 157{,}800 \text{ lb}$$

$$v_u = \frac{157{,}800}{(96)(17.50)} = 94 \text{ psi OK} < 2\sqrt{f_c'} = 109 \text{ psi}$$

Moment Reinforcement

Determine the area of steel in the long direction with an assumed d of 17.62″ based on #6 bars in the bottom layer.

$$A_s = \frac{(4.15)(8)(5.5)(2.75)(12)}{(0.90)(60)(0.95)(17.62)} \text{ (}j\text{ assumed at 0.95)}$$

$$A_s = 6.67 \text{ sq in. (long way)}$$

Determine the maximum diameter of bar that can be developed in the long direction in a length of 5′–3″. From Table 13-4 for $f_c' = 3{,}000$ psi a #10 bar can be developed in a distance of 4′–8″ and a #11 bar in 5′–8″ or the distance can be calculated (Section 12.5).

$$\#10\text{: } l_d = 0.04\, A_b f_y/\sqrt{f_c'}$$

$$= (0.04)(1.27)(60{,}000)/\sqrt{3000} = 56 \text{ in.}$$

$$\#11\text{: } l_d = (0.04)(1.56)(60{,}000)/\sqrt{3000} = 68 \text{ in.}$$

Select the reinforcing bars in the long direction, the minimum number of bars being based on a maximum spacing of 18 in. or $(96 - 6)/18 = 5$ bars.

6-#10 $A_s = 7.62$ sq. in. (Minimum number)

7-#9 $A_s = 7.00$ sq. in.

9-#8 $A_s = 7.11$ sq. in.

11-#7 $A_s = 6.60$ sq. in. (Minimum area)

15-#6 $A_s = 6.60$ sq. in.

22-#5 $A_s = 6.83$ sq. in.

Determine the area of steel in the short direction with an assumed d of 16.75 based on #7 bars in the lower layer and #6 bars in the upper layer.

$$A_s = \frac{(4.15)(12)(3.5)(1.75)(12)}{(0.90)(0.60)(0.95)(16.75)} = 4.26 \text{ in.}^2 \text{ (short way)}$$

Determine the maximum diameter of bar that can be developed in the short direction in a length of 3′–3″. From Table 13-4 for $f_c' = 3{,}000$ psi a #8 bar can be developed in a distance of 3′–1″ and a #9 bar in 3′–8″.

Select the reinforcing bars in the short direction, based on a maximum spacing of 18 in. $= (144 - 6)/18 = 7.7$, say, 8 bars.

8-#7 $A_s = 4.80$ sq. in.

10-#6 $A_s = 4.40$ sq. in.

14-#5 $A_s = 4.34$ sq. in. (Minimum area)

22-#4 $A_s = 4.40$

In a rectangular footing, a part of the bars in the short direction equal in amount to $\frac{2}{\beta + 1}$ times those required for flexure must be placed uniformly across a band width, B, equal to the short dimension of the footing (Section 15.4.4). S is equal to the ratio of the long side to the short side of the footing, or in this case equal to

$$\frac{2}{\beta + 1} = \frac{2}{\frac{12}{8} + 1} = 0.80$$

for fourteen #5 in the short direction, the number to be placed in bandwidth, B, equals (14)(0.8) = 11.2, say, 12 bars. This leaves one bar to be placed on each side of bandwidth, B. If this bar is placed 18 in. from bandwidth, B, it will be located within 6 in. of the edge of the footing, which is satisfactory.

Bearing Stresses

Determine the bearing stress on the loaded area of the footing.

$$f_c = 398{,}000/12^2 = 2760 \text{ psi} \quad \text{OK} < 0.85\phi f_c' \sqrt{\frac{A_2}{A_1}} = 3570 \text{ psi}$$

Dowels are not required for compression, but 4 dowels of the maximum size that can be developed in a depth of 21 − 3 = 18 inches will be furnished to provide 25 percent tensile capacity in each face. Referring to Table 13-1, a #6 compression dowel can be fully developed in 17 in.

Maximum tensile area in one column face above = three #9 = 3.00 in.^{2}

Required dowel area for tension in each face,

$$A_s = (0.25)(3.00) = 0.75 \text{ in.}^2$$

Area of dowels provided in each face,

$$A_s = (2)(0.44) = 0.88 \text{ in.}^2 > 0.75 \text{ in.}^2$$

(See Table 13-4). Read l_d = 19 in. for #6 bars, f_c' = 3,000 psi, other than top bars. The embedment required for tension development is

$$E = \frac{0.75}{0.88}(19 \text{ in.}) = 16.3 \text{ in.} < 18 \text{ in. available}$$

13 splices and details of reinforcement

GENERAL

Chapter 7 of the 1971 Code is titled "Details of Reinforcement," but the formulas and most of the basic design requirements controlling splice lengths, anchorage, hook effectiveness, and stirrup anchorage are located in Chapter 12. Special design requirements for details in shear and torsion are given in Chapter 11; for crack control in Chapter 10; for flat slabs in Chapter 13; walls in Chapters 10 and 14; footings in Chapter 15; and seismic design in Appendix A. In this Guide, the design requirements of Chapter 12 and applications as in Chapter 7 are presented in this chapter, together with cross references to other chapters herein on various different elements of design. Seismic details are outside the scope of this Guide.

BOND

The bond of concrete to reinforcement is a complex phenomenon. It involves adherence and friction on smooth surfaces and bearing against the deformations (lugs) of rebars. Both effects produce tensile secondary stresses at right angles to the rebars which may cause the concrete to split. The state of stress at right angles to the bars due to external loads is also an important factor; confinement by external stress or lateral enclosing reinforcement also deters splitting.

Most of the research on bond has involved at least several of these effects, making it impossible to separate them and so to write a rational set of formulas from pue theory. All codes, including the 1971 Code, deal with bond as an average value over the length of a bar in question. Earlier codes differentiated between two kinds of bond: "flexural bond" as a function of shear, and "anchorage bond" or end embedment as a function of the bar stress, using the same average bond value for each. Recent research has indicated that the "flexural bond" is unimportant to the ultimate strength; generally that sufficient embedment, "development length" on either side of a point of peak stress will ensure against failure. The 1971 Code does not, therefore, use the term, "bond," but does frame all requirements in terms of development length. The only remnant of the earlier Code limits on flexural bond is a limitation on the *net* development length after deduction of anchorage length for positive moment tension bars at simple supports and points of inflection (Section 12.2.3). *Net* development length for the bar size selected may not exceed M_t/V_u.

LAP SPLICES

Lap splices are treated as a special condition of stress transfer. A natural discontinuity develops at each cutoff end, somewhat helpful in a compression lap splice and a definitely detrimental stress raiser in tension lap splices. The splitting effect of transferring stress through shear to the concrete is aggravated in a lap splice, as two opposing tensile bar forces are transferred (see Fig. 13-1). Where tensile lap splices are located on a line along which splitting could occur, the spacing of the splices limits their effectiveness (Section 12.5-d) (see Fig. 13-2).

EMBEDMENT

The term "embedment" is used, more or less interchangeably in practice, with "anchorage." Both terms represent the length of bar past the point of maximum stress to the end of the bar, in which the total stress in the bar is transferred to or from the concrete. The term "anchorage," often more specifically described as "end anchorage," connotes a transfer of tensile stress to the concrete. Ordinarily, both terms imply a bar length in a region of no stress or decreasing stress in concrete parallel to the bar, as in end anchorage or embedment past the face of a support. Embedment can refer to a tensile or compressive stress transfer at any point in a structure, and is, therefore, the more general term. Older codes called the stress transfer process "bond"; the 1971 Code does not use this now-

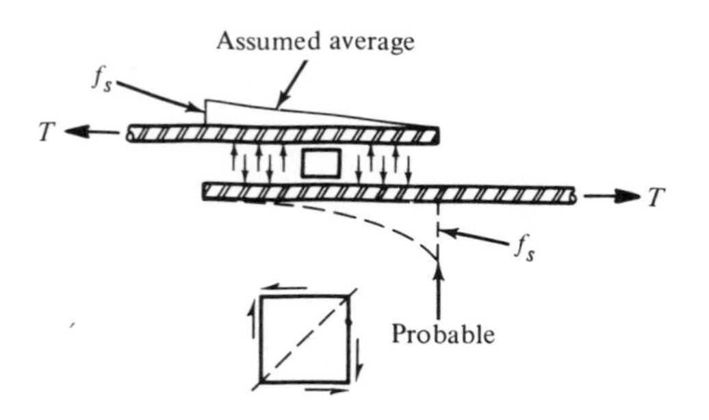

Fig. 13–1 Two-dimensional splitting at tension splices.

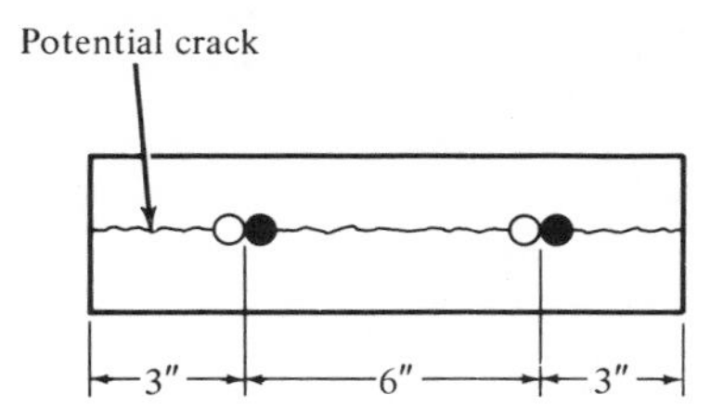

Fig. 13–2 Minimum spacings for use of 0.8 l_d in tension lap splices.

outmoded term. In that Code, the stress transfer is called "development"; and the length in which this transfer occurs is "development length," l_d.

Compression Embedment

Compression embedment involves a length of bar in which compressive stress is transferred to the concrete. The Code term, compression development length, is that length required to develop from, or transfer to, the concrete the full compression capacity of a given size bar. Compression development requirements in the Code include an allowance for the transfer of compressive stress by simple end bearing on the concrete (regardless of end condition, squared or sheared). Tests* have shown as much as 15,000 psi so transferred where splitting is somewhat restrained. The Code specifically forbids consideration of the bar length in an end hook as effective for transfer of compression for this reason (Sections 12.1.1 and 12.8.3). The basic Code requirement is that the stress at any section of all bars shall be developed on either side of that section. Compression development is commonly required in (1) compression splices and (2) bars carrying compression to a zone of no computed parallel stress such as column dowels to footings, extensions of bottom flexural bars into an interior support, etc.

For single bars or 2-bar bundles, standard Grade 60 sizes #3 through #18, the compression development length required for fully stressed bars is stated quite simply as (1) 1200 $d_b/\sqrt{f_c'}$, or (2) 18 d_b, whichever is greater (Section 12.6.1), but shall not be less than 8 in. (Section 12.6.2). For bundled bars, these development lengths shall be increased by 20 percent for bars in a 3-bar bundle and by 33 percent for bars in a 4-bar bundle (Section 12.7.1). For dowels from column to footing, only the compressive stress in the column verticals *in excess* of that permitted to be transferred by bearing of the gross concrete column section area

* "Column Tests with Offset Lap Splices of Zero Lap," PCA.

need be transferred by compression dowels (Section 15.6.3). See Table 13-1 for minimum compression dowel embedment in inches for basic embedments lengths (without spiral enclosure) in the usual range of strengths with normal weight concrete. Note that Table 13-1 gives embedment length into *footing* for full development of Grade 60 bars, or extension of bottom bars in flexural members past the face of a support for full development.

Footing Dowels

For footing dowels, the size of the dowel and the lap length for the splice in the column depend upon the stress to be transferred. Consider the (hypothetical) columns described below as cases A and B, pictured in Fig. 13-3.

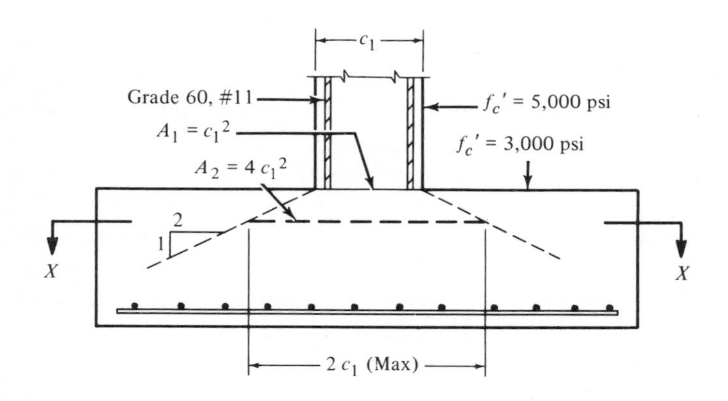

Case A: $\rho = 0.04\ A_g$

A_g per bar $= 39.0$ in.^{2}

At $e = 0.10\ h$, $P_u/A_g = 3{,}500$ psi (See Fig. 10-5, Chapter 10.)

Case B: $\rho = 0.08\ A_g$

A_g per bar $= 19.5$ in.^{2}

At $e = 0.10\ h$, $P_u/A_g = 4{,}500$ psi

$A_2 = 4C_1^2$; $A_1 = C_1^2$; $A_2/A_1 = 4$; $\sqrt{A_2/A_1} < 2$ (Section 10.14)

Fig. 13–3. Dowels in compression.

TABLE 13-1 Minimum Compression Dowel Embedment*

(Basic Compression Development Lengths, ℓ_d)

Bar Size	f'_c (Normal Weight Concrete)			
	3,000 psi	3,750 psi	4,000 psi	Over 4,444 psi
#3	8″	8″	8″	8″
#4	11″	10″	10″	9″
#5	14″	12″	12″	11″
#6	17″	15″	14″	14″
#7	19″	17″	17″	16″
#8	22″	20″	19″	18″
#9	25″	22″	22″	20″
#10	28″	25″	24″	23″
#11	31″	28″	27″	25″
#14	37″	33″	32″	31″
#18	50″	44″	43″	41″

* For $f'_c > 4{,}444$ psi, minimum embedment $= 18d_b$

Note: For embedments enclosed by spirals use 0.75 length shown but not less than 8 in.

TABLE 13-2 Compression Lap Splices for Grade 60 Bars*

Bar Size	Minimum Lap Length All concretes with $f'_c \geq 3{,}000$ psi		
	Standard Lap (1)	Within Ties (2)(4) $A_v \geq 0.0015\ b_s$	Within Column Spirals (3)(4)
#3	12″	12″	12″
#4	15″	13″	12″
#5	19″	16″	14″
#6	23″	19″	17″
#7	26″	22″	20″
#8	30″	25″	23″
#9	34″	28″	25″
#10	38″	32″	29″
#11	42″	35″	32″
#14 & #18	←— LAP SPLICES NOT PERMITTED (5) —→		

(1) Standard lap is minimum for use in all members, including tied columns, except as provided below.

(2) See Table III for minimum size and spacing of tie arrangements to meet the requirement $A_v \geq 0.0015\ bs$.

(3) For use in spirally reinforced columns with spirals conforming to Section 7.12.2.

(4) Reduced laps apply to beams containing stirrups meeting limits in Table III or spirals meeting Section 7.12.2.

(5) Except as column-to-footing dowels.

* Courtesy Concrete Reinforcing Steel Institute.

In case A, bearing allowed at the top of the footing = (2)(0.85) ϕ f_c' (2)(0.85)(0.70)(3,000) = 3,570 psi. At section *X-X*, the allowable bearing = (0.85) (0.70) (3,000) = 1,785 psi. For this condition the Code does not require dowels (Sections 15.6.1, 15.6.2, 15.6.3, and 15.6.4). The authors would recommend at least four dowels, one in each corner of such a column, of at least the basic development length (Table 13-1), and at least #5 in size, to prevent overturning in case of construction collisions, and to assist bar placers in locating and fastening down the column reinforcement cage. See Example, Chapter 12, for *tensile* requirements at column footings (Section 7.10.5).

In case B, the excess compression for transfer by dowels = (19.5) (4,500 − 3,570) = 18,100 lb per vertical bar. Dowels are required for this excess, and must comprise at least 0.5 percent of the column area. A minimum of four dowels are required (Section 15.6.5). Minimum area = 0.005 × 19.5 = 0.0975 in.^{2} per #11 vertical. Since each dowel must transfer 18.1 kips, the area required for compression is (18.1/60) in. = 0.302 in.^{2}. One #5 dowel per vertical will suffice. The dowels should be embedded 14 in. into the footing as noted in Table 13-1. The dowel must be lap spliced into the column above the footing. If suitable ties are available or provided in the splice length, the lap length can be 16 in.; with lesser ties, the dowel must be lapped 19 in. (see Table 13-2; see also Table 13-3 for minimum tie requirements to qualify for the reduction in compression splice lengths).

To satisfy the requirement of 25 percent tensile capacity in each face of the column (Section 7.10.5) the compression dowels selected will not suffice. Minimum area per dowel = (0.25)(1.56) = 0.39 in.^{2}. Use #6 dowels. Embedment into footing = 19 in. (Table 13-4). Lap slice into column 31 in. for a Class C splice, Table 13-7 (Section 7.6.3.2.2).

Tension Embedment

Tension embedment is a length of bar in which the stress transferred to the concrete is tensile. Tensile development length, l_d, is the length of bar required to develop, or to transfer to the concrete, the full tensile capacity of the bar. The length, l_d, is made a function of the yield strength of the bar; $\sqrt{f_c'}$; the concrete splitting strength, f_{ct}; depth of concrete below a horizontal bar (causing water rise); bar spacings and cover; actual design stress in the bar; and lateral confinement, spirals (Section 12.5). Use of the splitting strength, f_{ct}, involves separate provisions for lightweight aggregate concrete (see Chapter 14). Increased l_d is required for bundled bars: in 3-bar bundles, 20 percent; in 4-bar bundles, 33 percent (Section 12.7; same as for compression development). See Chapter 10 for applications to bundles.

TABLE 13-3 Minimum Size and Spacing of Ties Enclosing Reduced Length Lap Splices for Compression

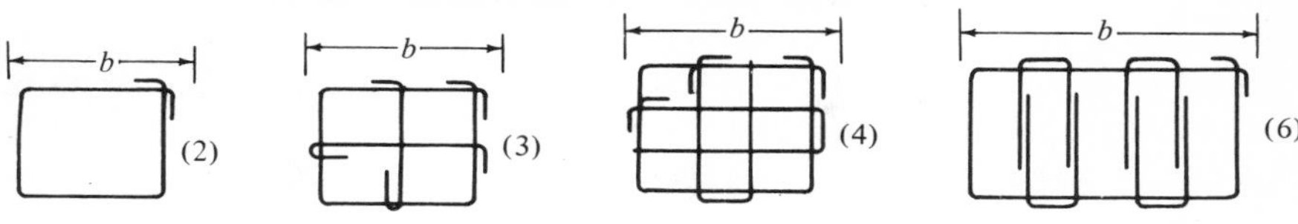

Column Width b	MAXIMUM SPACING, s, FOR MIN. A_v (in.)											
	#3 Ties				#4 Ties				#5 Ties			
	No. of Legs				No. of Legs				No. of Legs			
	2	3	4	6	2	3	4	6	2	3	4	6
12"	12	—	—	—	—	—	—	—	—	—	—	—
14"	11	16	—	—	—	—	—	—	—	—	—	—
16"	9	14	18	—	17	—	—	—	—	—	—	—
18"	8	12	16	27	15	22	—	—	—	—	—	—
20"	7	11	15	22	13	20	—	—	20	—	—	—
22"	6	10	13	20	12	18	24	—	19	28	—	—
24"	—	9	12	18	11	17	22	33	17	24	—	—
26"	—	8	11	17	10	15	20	31	16	24	32	—
28"	—	—	10	16	10	14	19	28	15	22	29	—
30"	—	—	10	15	—	13	18	27	14	20	27	41

Note: b = column width perpendicular to effective tie legs

TABLE 13-4 Tension Development Length (ℓ_d), (inches)

(Minimum Straight Embedmeht Lengths *)

Bar Size	f'_c = 3,000 psi		f'_c = 3,750 psi		f'_c = 4,000 psi		f'_c = 5,000 psi		f'_c = 6,000 psi	
	Top Bars	Other Bars	Top Bars	Other Bars	Top Bars	Other Bars	Top Bars	Other Bars	Top Bars	Other Bars
#3	13	12	13	12	13	12	13	12	13	12
#4	17	12	17	12	17	12	17	12	17	12
#5	21	15	21	15	21	15	21	15	21	15
#6	27	19	25	18	25	18	25	18	25	18
#7	37	26	33	24	32	23	29	21	29	21
#8	48	35	43	31	42	30	38	27	34	25
#9	61	44	55	39	53	38	48	34	43	31
#10	78	56	70	50	67	48	60	43	55	39
#11	96	68	86	61	83	59	74	53	68	48
#14	130	93	117	83	113	81	101	72	92	66
#18	169	120	151	108	146	104	131	93	119	85

* 1. For bars enclosed in standard column spirals, use 0.75 ℓd.
2. For bars spaced 6 in. or more, use 0.8 ℓ_d. (Such as usual temperature bars.)
3. Longer embedments for lightweight aggregate concrete are generally required, depending upon the tensile splitting strength f_{ct}.
4. Standard 90° or 180° end hooks may be used to replace part of the required embedment. See TABLE 5 and TABLE 6

* Courtesy Concrete Reinforcing Steel Institute.

Application of all the various interdependent tensile development requirements to each element in design would be extremely tedious and could be a confusing waste of design time. For practical design the authors recommend instead that the Code user merely check the actual dimensions available for tensile development in the connection (or from a bend point or a cutoff point established as a fraction of the span on his typical details), compare to a table of development lengths required for each bar size, and select the bar size allowable. Such a table is easily constructed for any special set of conditions. Table 13-4 presents the basic l_d-values for each size bar over the usual range of concrete strengths used. The footnotes to Table 13-4 should be consulted if modifications for special conditions are required. Note that separate values are tabulated for "top bars" and "other bars," and that top bars are defined as horizontal bars with more than 12 in. of concrete cast below them (Section 12.5-b). This definition *excludes* from the "top" bars, the top layer of bars in a slab reinforced top and bottom, if thickness is less than the sum (12 in. + d_b + cover). It generally *includes* all horizontal reinforcement in walls and all column ties as "top" bars.

EXAMPLES OF DEVELOPMENT LENGTH. Some simplified applications of the Code requirements for tension and compression development are shown in Fig. 13-4.

CANTILEVERED SLABS AND FOOTINGS

In Fig. 13-4 (*a*) the simplest possible application of the development concept is shown, the end anchorage for cantilever reinforcing in a uniformly loaded cantilever. In Fig. 13-4 (*b*) note that the dimension available for development is (l − 3 in.) where l is the overall clear span from the critical section at the face of the wall or column, and 3 in. is the cover required. The numbers in parentheses indicate Code sections regulating items so marked.

One-Way Slabs

Figure 13-5 shows the tensile development length requirements for the end anchorage of positive moment steel in slabs. The maximum bar size is established here by a computed dimension involving the moment/shear ratio taken at the point where the steel is to be fully stressed. For the cantilever end in Fig. 13-4, uniformly loaded, the ratio (M_t/V_u) becomes 0.5l where l = clear cantilever span. For the simple end support in Fig. 13-5, the ratio for uniform loading (M_t/V_u) becomes 0.250 l where l = clear span.

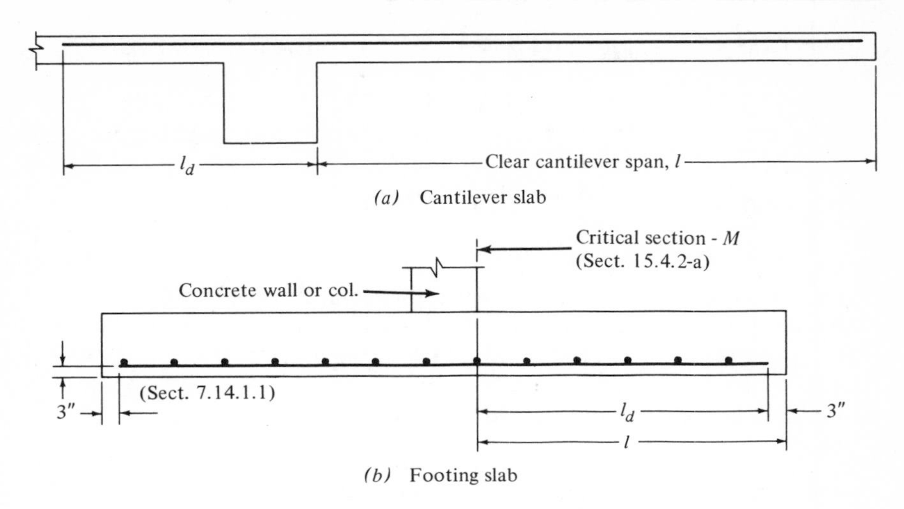

Fig. 13–4. Basic tension development lengths.

If some of the bars are to be cut off for savings in steel, note how the cutoff point locations must be computed. For practical design within the ordinary range of spans, the authors suggest that no attempt be made to use varying lengths of bars. The savings in material will often be less than the additional costs of added design time, detailing, fabricating, placing, and inspection required. For longer spans or where a great number of identical spans are repeated, the savings in material will be worthwhile, but the variations in shear from concentrated or partial span loading would have to be considered (Sections 12.1.6.1, 12.1.6.2, and 12.1.6.3).

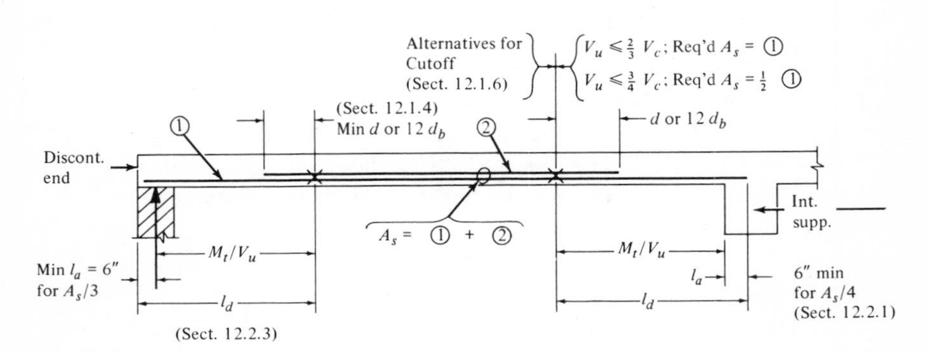

Fig. 13–5 Tensile development lengths for positive moment steel in one-way slabs.

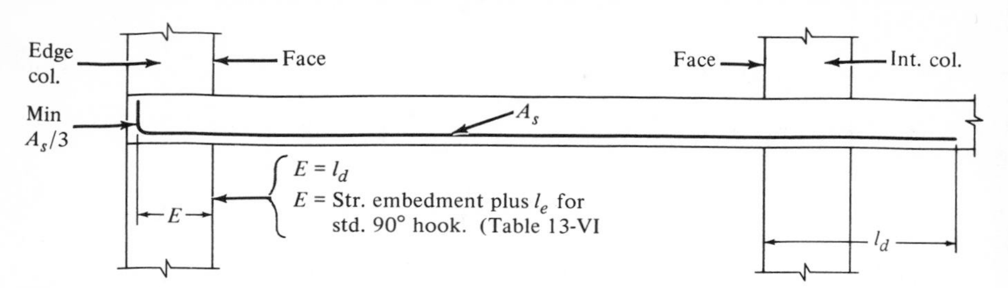

Fig. 13–6 Tensile anchorage of positive moment steel in beams (when beams are part of the primary lateral load-resisting system).

PRIMARY LATERAL LOAD RESISTING FRAMES

When the flexural member is a part of the "primary lateral load resisting system," the same percentages of positive reinforcement extended into the support as shown in Fig. 13-5 must be anchored for full development of f_y in tension at the face of the support (Section 12.2.2) (see Fig. 13-6). "A primary lateral load resisting system" is usually defined as the beam-column frames without shear walls or in combination with shear walls designed for only a part of the lateral load resistance. Note that the end hook, if required, will become an effective part of the total anchorage for tension (see Table 13-4). Although the use of standard end hooks is recommended for economy, the additional straight length in a longer 90° bend is also considered part of the effective tension anchorage. When this detail is used, only the straight embedment from the face of the support to the tangent of the 90° bend can be considered effective for compression when the lateral loads reverse in direction (see Table 13-1).

TENSION DEVELOPMENT FOR NEGATIVE MOMENT STEEL

A similar detailing problem is shown in Fig. 13-7. In this case note that full development in tension is required for the top bars in both directions from the face of the support (Section 12.1.1). The additional requirements are measured from the point of inflection (Sections 12.1.4 and 12.3.3). The requirements for anchorage of negative moment reinforcement apply to all connections, and are not limited only to those which are part of primary lateral load resisting systems. Where the connection involves principal framing elements, such as beams and columns, enclosure (to prevent splitting) must be provided for splices of continuing reinforcement and end anchorage of reinforcement terminating in the connection (Section 7.11). Such enclosure may consist of external concrete, where other mem-

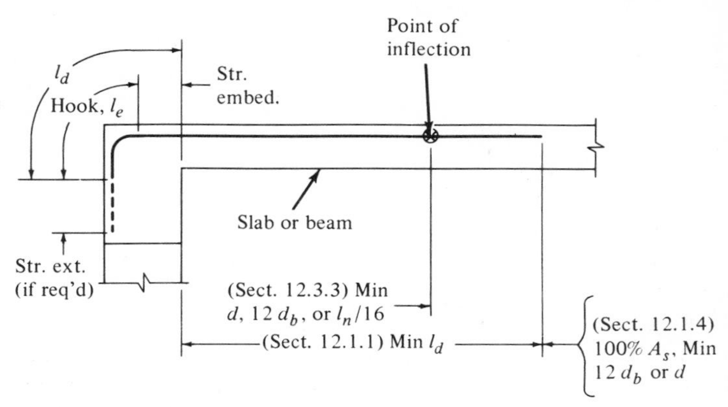

Fig. 13–7 Tension development lengths for negative moment steel at discontinuous ends.

bers frame into the connection at right angles, or internal closure as provided by ties, stirrups, or spirals.

Figure 13-8 shows the development requirements for negative moment steel at interior supports, applicable to both slabs and beams. Note that use of varying bar lengths requires that the designer satisfy a number of minimum development length requirements, whichever is larger controlling.

END ANCHORAGES IN LIMITED DISTANCES

The end anchorage requirements for top bars at discontinuous ends of any principal member (beam) can present a practical problem of fitting into

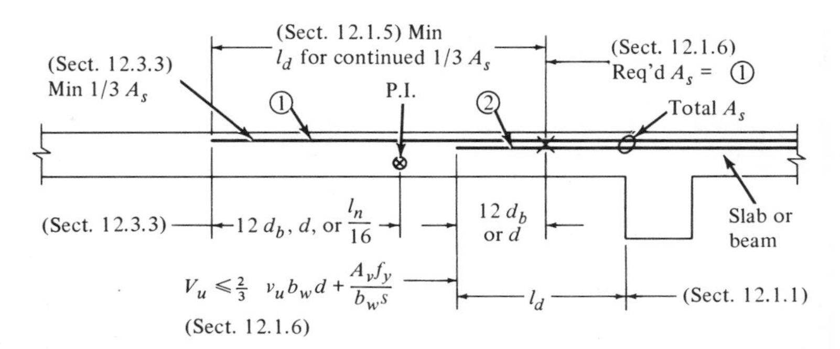

Fig. 13–8 Tension development lengths for negative moment steel at interior supports.

available form space. The same problem arises for the bottom bars in a primary lateral load resisting system. The problem becomes critical when the column depth is limited and an end hook is required, but the beam is too shallow for the standard size vertical end hook. To provide access for placing concrete in the column, the floor system steel is frequently placed after the column is cast. For the end anchorage shown in Fig. 13-7, one solution is to employ L-shaped dowels with one end set into the column and the other spliced to the main beam bars. If set before concreting, these bars interfere with the entry of concrete; if placed into the plastic concrete after concreting the column, they may easily be overlooked. Several alternative details will satisfy the Code and will in some cases facilitate bar placing and concrete placing. Since confining concrete is not present on all sides of the column, confining closed ties are required (Section 7.11). It will at times be possible to combine the dowel requirement for the bottom bars with a partial U-shaped tie with wide beams (see Fig. 13-9). This tie (maximum size for usual columns #5) will be fully developed at the face of the column and can be combined with the development possible by straight extension of, and end hooks on, the bottom bars (see Table 13-7 for tensile splice requirements, Class C). Since the Code requires anchorage sufficient to develop 33 percent of the bottom steel fully, the

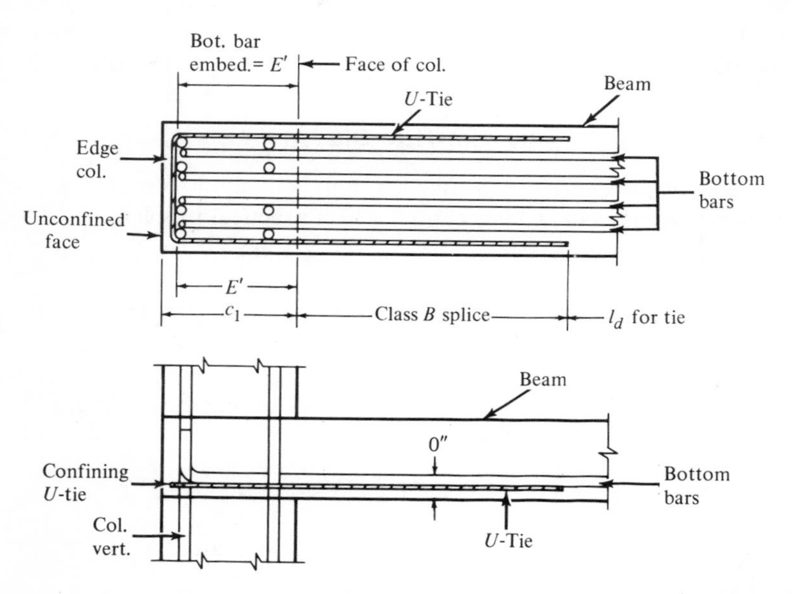

Fig. 13–9 Anchorage in limited space for bottom bars in a primary lateral load-resisting system.

same total tension can be developed by providing 1/2 l_d for 66 percent. Although this connection would not develop the same ductility, if the anchorage consisted of some straight embedment plus a standard 90° end hook, recent tests show that the full f_y will be developed, but at larger slip. If the modified tie of Fig. 13-9 is used and a #5 bar capacity is inadequate, and bending radii for a larger size bar will not fit, a 2-bar bundle of #5 bars, one atop the other, can be used.

EXAMPLE. BOTTOM BAR ANCHORAGE. (Figure 13-9) $f_c' = 4{,}000$ psi; assume four #9 Grade 60 bottom bars $A_s = 4.00$ in.^{2}) are the required one-third midspan bottom A_s to be fully developed at the face of the column. $c_1 = 16$ in.; $E' = 13.5$ in.; required $E = 16$ in. (Table 13-6); total l_d for #9 bottom bars = 38 in.; $E - E' = 2.5$ in. The additional stress per bar to be developed is $\frac{2.5}{38}(60) = 4$ ksi; total additional force = $(4)(1.00)(4) = 16$ kips. Use one U-tie, #4. Force = $(2)(0.20)(60) = 24$ kips.

TOP BARS IN LIMITED ANCHORAGE DISTANCE

When space is limited the top bar anchorages are even more difficult to achieve since the Code requires full development of 100 percent of the top steel required for flexure. Under some conditions where column is wider than the beam framing into it or where a deep spandrel beam is available for extension of top steel, the top beam bars may be spliced to a larger number of smaller bars spread into the adjacent slab flanges of the beam (see Fig. 13-10(*a*)).

Under very restricted conditions, such as with heavy top bars, no spandrel beam, dimension $c < E$, dimension $h < A$, the designer often has no choice other than to extend the top bars, or preferably long L-shaped dowels spliced to the top bars, below the construction joint into the column below. Fig. 13-10 (*a*) and (*b*) shows two alternatives that can sometimes be used to supplement the development required sufficiently to permit all floor system steel to be placed after the lower portion of the column is cast.

OTHER EXAMPLES. Other examples showing practical applications of development requirements to particular structural elements will be found in other chapters: one-way slabs, Figs. 3-3 and 3-4; one-way joist construction, Fig. 4-4; two-way flat plates, Figs. 5-13 and 5-14; waffle flat slabs, Fig. 7-9; footings and dowels, Figs. 12-6 and 12-7.

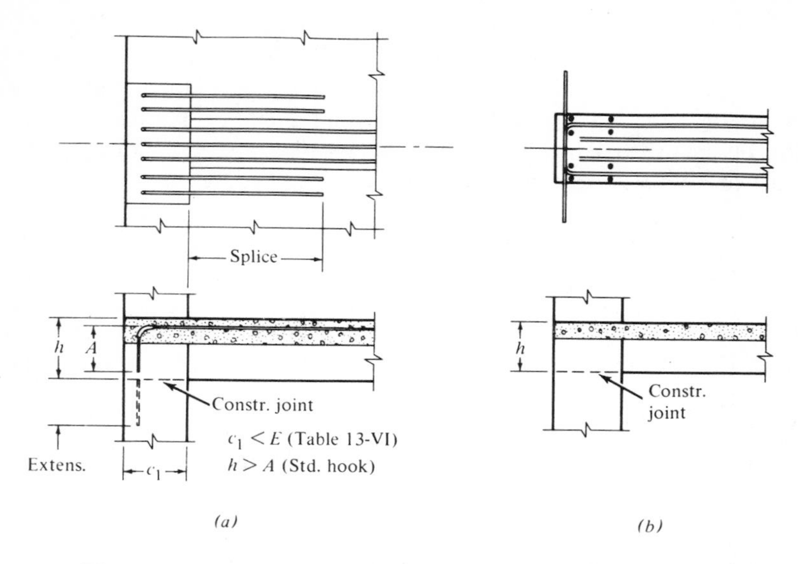

Fig. 13–10 Top bar anchorages in limited dimensions.

HOOKS

End hooks are a vital part of tension anchorage. The Code provides a procedure whereby the straight tensile embedment length, l_e, equivalent to that developed by a standard end hook, can be computed (Sections 12.8.1 and 12.8.2). Briefly, two steps are involved:

1. The stress developed by an end hook, $f_h = \xi\sqrt{f_c'}$, where ξ is a constant prescribed for standard hooks of top bars and other bars in each size (Section 12.8.1). Values of ξ may be increased 30 percent where confinement perpendicular to the plane of the hook is present. This confinement may be provided by external means (concrete member or mass) or internal means (lateral reinforcement).

2. The equivalent length, l_e, is computed by solving the formula for l_d, substituting l_e for l_d, and f_h for f_y (Section 12.5-a). Note that $\sqrt{f_c'}$ will cancel out for #14 and #18 bar sizes, for bars #8 through #11 with $\sqrt{f_c'} \leq$ 6,000 psi, and for bars #3 through #7 with $f_c' \geq$ 3,000 psi Equivalent hook lengths are *not* subject to the various adjustments for other conditions affecting embedment (Section 12.5-b, -c, and -d).

Since this procedure is too tedius to perform routinely, the designer will require a table of l_e values (see Table 13-5).

Even with the value l_e in hand, the main design dimension is yet to be

TABLE 13-5 ℓ For Standard End Hooks for Grade 60*

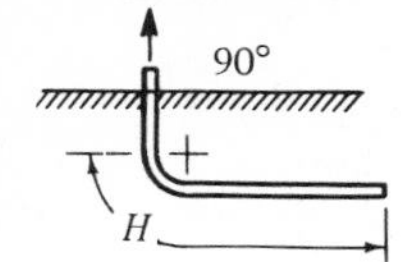

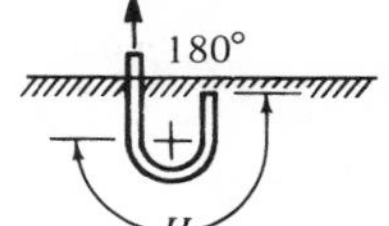

Bars #14 — #18 All f'_c
Bars #8 — #11 $f'_c \leq 6{,}000$ psi
Bars #3 — #7 $f'_c \geq 3{,}000$ psi

ℓ_e = straight embedment equivalent in tension to H.

Bar Size	For General Use in Enclosed Members *		In narrow members with no enclosure	
	Top Bars	Other Bars	Top Bars	Other Bars
#3	6″	6″	4″	4″
#4	8″	8″	6″	6″
#5	10″	10″	7″	7″
#6	11″	12″	8″	10″
#7	12″	17″	9″	13″
#8	15″	22″	11″	17″
#9	19″	28″	14″	22″
#10	24″	32″	18″	24″
#11	29″	34″	23″	26″
#14	37″	37″	28″	28″
#18	32″	32″	24″	24″

* General Use—In massive elements or connections of principal members conforming to Section 7.11 (restrained against splitting).

TABLE 13-6 Minimum Tension Embedment, *E*, with Standard End Hook for Grade 60 Bars*

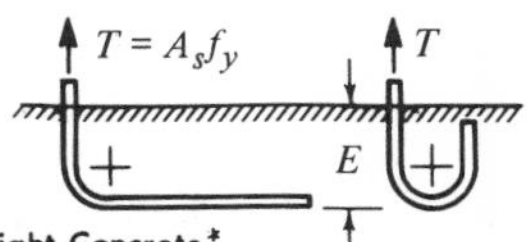

General Use, Enclosed Members, Normal Weight Concrete*

Bar Size	f'_c = 3,000 psi		f'_c = 3,750 psi		f'_c = 4,000 psi		f'_c = 5,000 psi		f'_c = 6,000 psi	
	Top Bars	Other Bars	Top Bars	Other Bars	Top Bars	Other Bars	Top Bars	Other Bars	Top Bars	Other Bars
#3	9″	8″	8″	7″	8″	7″	7″	5″	6″	5″
#4	11″	8″	10″	7″	10″	7″	9″	5″	8″	5″
#5	14″	8″	13″	7″	13″	7″	11″	5″	10″	5″
#6	20″	10″	18″	8″	17″	8″	16″	6″	15″	5″
#7	29″	13″	25″	10″	24″	10″	22″	7″	20″	6″
#8	38″	17″	33″	13″	31″	12″	27″	9″	24″	7″
#9	48″	22″	42″	17″	40″	16″	35″	12″	30″	9″
#10	60″	31″	52″	25″	50″	23″	43″	18″	38″	14″
#11	74″	41″	63″	34″	61″	32″	52″	26″	46″	21″
#14	104″	67″	90″	57″	87″	54″	75″	46″	66″	40″
#18	151″	103″	133″	90″	128″	87″	113″	76″	101″	67″

* (Concrete Weight ± 145)

Notes: 1. For use in unconfined narrow sections subject to splitting, reduce value of hook embedment (Table V) and increase E, tabulated above.

2. For light-weight aggregate concrete, increase ℓ_d (Table IV) by aggregate factor and add increase to *E*, tabulated above.

* Courtesy Concrete Reinforcing Steel Institute.

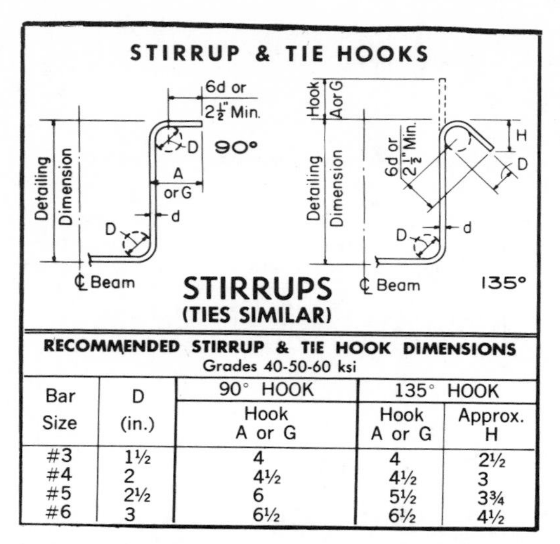

RECOMMENDED STIRRUP & TIE HOOK DIMENSIONS
Grades 40-50-60 ksi

Bar Size	D (in.)	90° HOOK	135° HOOK	
		Hook A or G	Hook A or G	Approx. H
#3	1½	4	4	2½
#4	2	4½	4½	3
#5	2½	6	5½	3¾
#6	3	6½	6½	4½

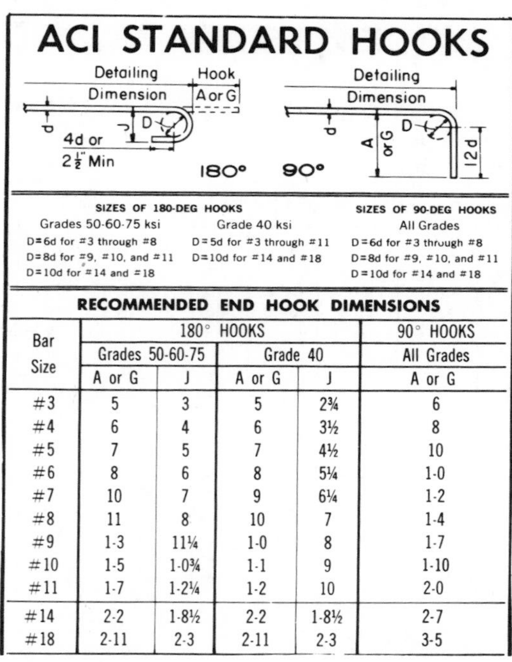

SIZES OF 180-DEG HOOKS

Grades 50-60-75 ksi
D = 6d for #3 through #8
D = 8d for #9, #10, and #11
D = 10d for #14 and #18

Grade 40 ksi
D = 5d for #3 through #11
D = 10d for #14 and #18

SIZES OF 90-DEG HOOKS

All Grades
D = 6d for #3 through #8
D = 8d for #9, #10, and #11
D = 10d for #14 and #18

RECOMMENDED END HOOK DIMENSIONS

Bar Size	180° HOOKS				90° HOOKS
	Grades 50-60-75		Grade 40		All Grades
	A or G	J	A or G	J	A or G
#3	5	3	5	2¾	6
#4	6	4	6	3½	8
#5	7	5	7	4½	10
#6	8	6	8	5¼	1-0
#7	10	7	9	6¼	1-2
#8	11	8	10	7	1-4
#9	1-3	11¼	1-0	8	1-7
#10	1-5	1-0¾	1-1	9	1-10
#11	1-7	1-2¼	1-2	10	2-0
#14	2-2	1-8½	2-2	1-8½	2-7
#18	2-11	2-3	2-11	2-3	3-5

Fig. 13–11 Standard hooks (*Courtesy of the Concrete Reinforcing Steel Institute*).

resolved—the total embedment dimension, including the end hook which will produce the development required for full utilization of the bar. Simply, $l_d = l_e +$ the straight embedment from tangent of the hook to the critical section. To determine the minimum total embedment, E, the outer radius of the hook must be added.

$E = l_d - l_e + \mathrm{R} + d_b$, where R = the specified minimum (standard) radius of bend for a given bar size and d_b = bar diameter.

See Table 13-6 for minimum tension embedment, E, with standard end hooks for Grade 60 bars. E is the dimension required on drawings.

The hook dimensions shown in Fig. 13-11 are recommended as standard by the reinforcing steel fabricating industry for hooks to meet the minimum requirements of the 1971 Code.

COMPRESSION LAP SPLICES

Compression lap splices for all values of $f_c' \geq 3{,}000$ psi are specified to vary as a direct function of the bar diameter, d_b, and the grade of steel, f_y (Section 7.7.1.1) (see Table 13-2). Three different lap lengths in terms of bar diameters for compression splices are specified for full development under common conditions of use:

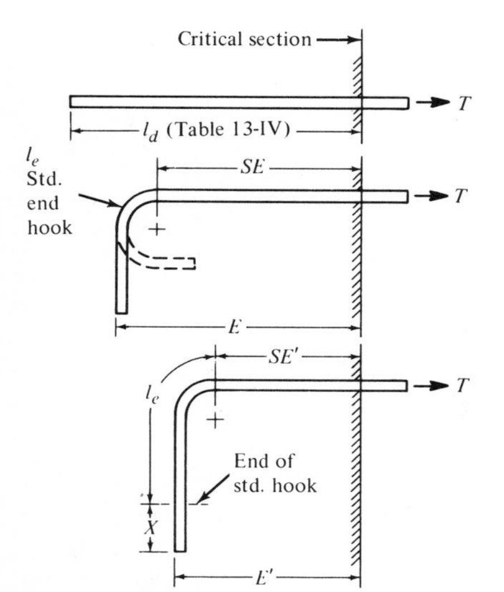

Full embedment is such that the tensile design stress, $T = f_y A_s$. Straight embedment for $T = f_y A_s$ for each size of bar, l_d, is given in Table 13-4 (Section 12.5).

The value for embedment of a standard end hook is l_e, in terms of equivalent straight development distance. Let full embedment with an end hook, $E = l_d$ = straight embedment $SE + l_e$. See Table 13-5 for l_e and Table 13-6 for E. Reference (Section 12.8).

Where available straight embedment before, hook $SE' < SE$, an extension of straight embedment may be added. If $SE' = SE - X$,

$$X = E - E'$$

required extension beyond hook where embedment depth available = E' and E = embedment for full $f_y A_s$. See Table 13-6.

Fig. 13–12 End anchorages for full development of bars.

1. 30 d_b for general use without regard to confinement (Section 7.7.1.1)
2. 25 d_b for use when confined by ties or stirrups with an effective area of at least 0.0015 *hs* (Section 7.7.1.2) where *h* is the width of the member at right angles to the legs of the ties and *s* is the spacing of the ties. See Table 13-3 for minimum tie sizes, maximum tie spacings, and a graphic definition of width, *h*.
3. 22.5 d_b for use only when confined within spirals (Section 7.7.1.3). Standard column spirals (Section 7.12.2) will satisfy this requirement.

Compression lap splices are commonly used for the bottom bars at the interior supports of continuous flexural members or for the vertical bars in multistory columns. In slabs and joists, it is not generally practicable nor required that confining ties or stirrups be provided, and so, if required, compression lap splice lengths of $30d_b$ are used. For compression reinforcement in beams and girders, enclosing stirrups are required (Section 7.12.5); these must meet the minimum size and maximum spacing limits for column ties (Section 7.12.3). Since these limits will usually supply an effective area equal to the required minimum 0.0015 *hs,* compression lap lengths of 25 d_b can be used. The minimum length of lap for compression lap splices is 12 in. (Section 7.7.1). This limit will control laps for #3 and #4 bar sizes only, usually only in lightly reinforced slabs see Table 13-2).

Special Requirements for Compression Lap Splices —Other Strengths

There are a number of additional provisions for special conditions (Section 7.7). For concrete strengths, $f_c' < 3{,}000$ psi, all required lap lengths must be increased by one-third. For $f_y > 60{,}000$ psi, all required lap lengths must be increased by the ratio $\frac{(0.9\ f_y - 24)}{30}$, where f_y is in ksi (Section 7.7.1). For lower yield strength grades of bars, the required lap length is reduced directly in the ratio $f_y/60$, where f_y is in ksi.

BUNDLED BARS. For bars in 3-bar bundles, the above lap lengths must be increased 20 percent; in 4-bar bundles, 33 percent (Section 7.5.3). Note that these reductions in effectiveness prescribed for bundled bars are based on the exposed bar surface areas and, therefore, are intended to include any added dowels or "splice bars" as part of the number in the bundle. A bundle consisting of three through bars with an added splice splice bar past past butted cutoff points becomes a 4-bar bundle at the

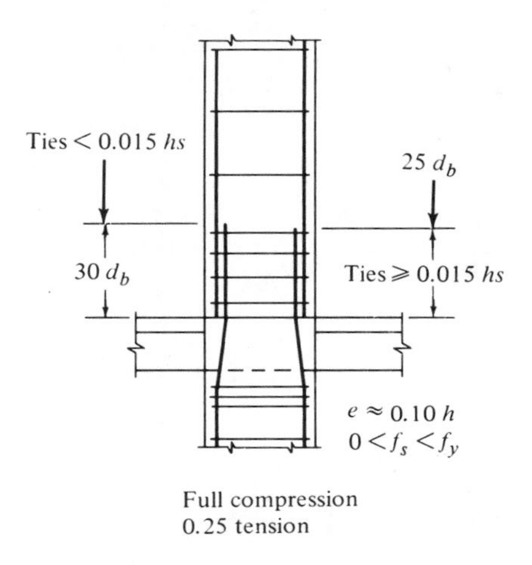

Fig. 13–13 Typical interior column.

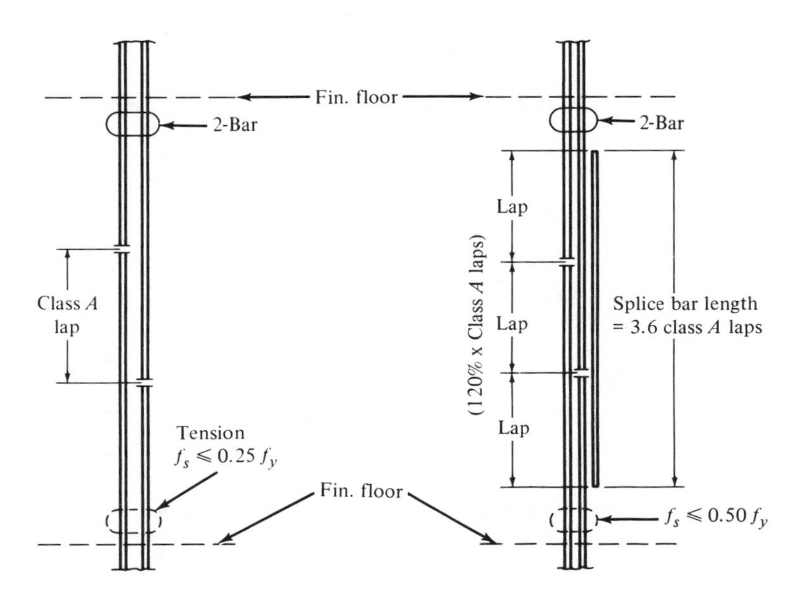

Fig. 13–14 End bearing + lap splices in bundled column verticals.

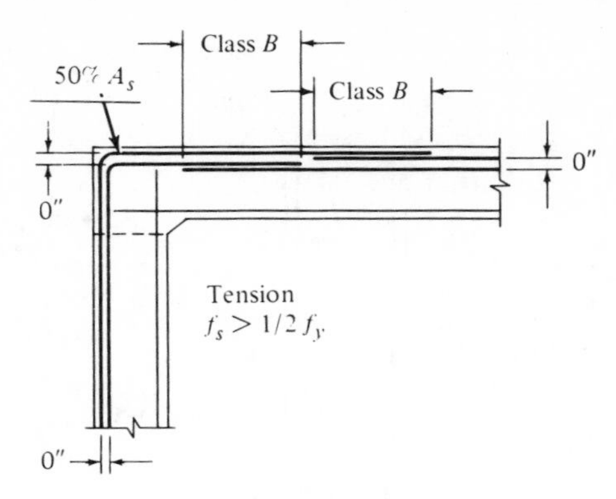

Fig. 13–15 Class B lap splices.

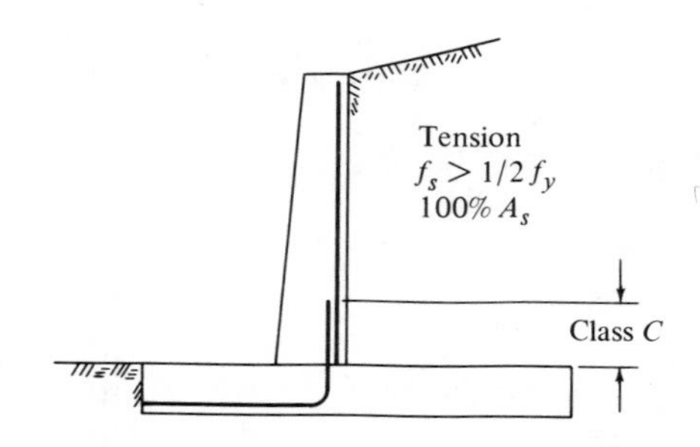

Fig. 13–16 Class C lap splices.

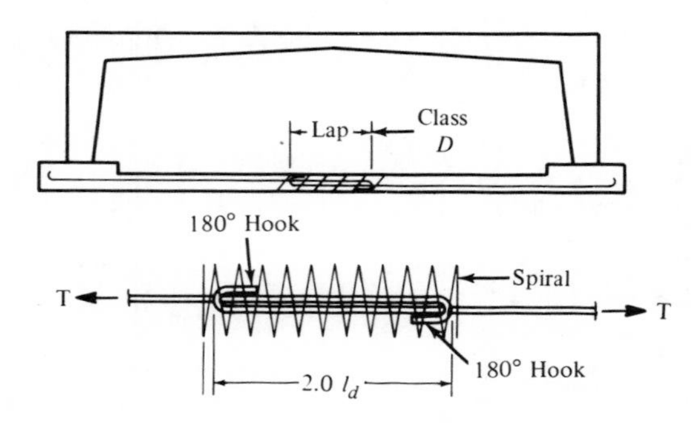

Fig. 13–17 Tension Tie—Class D splice.

splice point for the purpose of dimensioning length of the splice bar (see Chapter 10 for examples).

LARGE BARS. Lap splices are not permitted for bars larger than #11 (Section 7.5.2) except that #14 and #18 column verticals may be lap spliced to smaller size bars used as dowels for column-to-footing connections where the vertical bars are required for design compression only (Section 15.6.8) (see Chapter 12 for examples).

COLUMN VERTICALS MINIMUM TENSILE CAPACITY. For compression lap splices of column verticals, a number of special detailing requirements apply (Section 7.10). Where the column faces are offset 3 in. or more, separate dowels must be used (Section 7.10.2). A minimum tensile capacity equal to 1/4 A_{sf} must be maintained in each face of a column, where A_{sf} is the area of vertical bars in that face (Section 7.10.5). It will be noted from a comparison of the required lap lengths for compression in Table 13-2 to those for tension in Table 13-7 that the compression lap lengths are more than adequate to develop 25 percent of the tensile capacity.* Where various conditions of loading may result in stress reversals, so that the tensile design stress exceeds about 20 percent of f_y in the column verticals, tensile requirements may govern. The total tensile capacity provided in each face of a column must be at least twice the maximum calculated tension (Section 7.10.3) (see the following section, "Tension Splices," and Chapter 10).

Positive Connections in Compression Splicing

In addition to compression lap splices the Code permits the use of positive connections or welding for compression splices of all bar sizes and requires same for #14 and #18.

Compression splices for bars required to transmit compressive stress only may consist of end bearing on square cut ends (Section 7.7.2). When conditions permit their use, end bearing splices will be the most economical type for #14 and #18 bars, since material costs, field time, and labor, and inspection requirements are minimal. Lateral confining reinforcement in the region of end bearing splices must be provided (Sections 7.7.2 and 7.11). The tolerance on squareness of square-cut ends, 1-1/2°, prescribed,

---

* The reason for the minimum tensile capacity in each column face of 25 percent is to approximate the ductility of column splices upon which most American experience in actual construction is based. Most existing buildings designed under previous codes employed 20-bar diameter laps of Grade 40 standard size bars, equal to approximately 25 per cent tensile capacity for Grade 60 bars under the 1971 Code.

is based upon full-size tests in which access for the concrete grout to the bearing area was not prevented (Commentary 7.7.2). Proprietary clamping devices are available which will hold the square-cut ends of the bars in close concentric contact even for changing bar sizes and also provide openings for inspection of fit that permit access by the grout. Note: when specifying, it is usual to specify "saw-cut" to the required tolerance for maximum economy.

For compression splices of large bars where large stress reversals from compression to tension are expected under some conditions of loading, welded splices or proprietary sleeve coupler splice devices are usually used (Section 7.5.5) (see the following section.)

TENSION SPLICES

Types

Tension splices for all bars #11 or less in size may consist of (1) lap splices, (2) full welded splices, (3) full positive connections, or (4) welded splices or positive connections designed for lesser values than "full" tension. *Note:* "full" tension design requires a splice capacity 125 percent of the minimum specified f_y (Section 7.5.5).

Classes of Tensile Lap Splices

Tension lap splices are specified in Class A, B, C, or D (Section 7.6.1). See Tables 13-7 and 13-8 for the lap lengths required with the commonly used range of concrete strengths. Lap lengths required are specified as a function of the tensile development length, l_d, which varies with f_y, f_c', bar size, bar spacing, confinement condition, bar position (top bar or other bar), and, for lightweight concretes, the splitting strength, f_{ct}, (Section 12.5). For tensile development lengths, see Table 13-4. The location of the splice, maximum design stress at the location, confinement available, and the percentage of the total steel area spliced at the one point determine which class of lap splice may be used (Sections 7.6.2 and 7.6.3). These requirements may be summarized as follows:

Class A—at points away from the maximum tensile stress where the computed maximum design stress required in all bars at the section is less than 0.5 f_y, and no more than 75 percent of the total A_s is spliced within one Class A lap length of the section (Section 7.6.3.2.1).

Class B—(1) where design stress in all bars is less than 0.5 f_y and more than 75 percent of the total A_s is spliced (Section 7.6.3.2.2), or (2) at

TABLE 13-7 Basic* Tension Lap Splices for Grade 60 Bars*

(Definitions and usage of Class A, B, C, splices—pp.15–4, 15–5)

Bar Size	CLASS A, B & C LAP SPLICE LENGTHS (inches) f'_c = 3,000			f'_c = 3,750			f'_c = 4,000			f'_c = 5,000			f'_c = 6,000		
	A	B	C	A	B	C	A	B	C	A	B	C	A	B	C
#3	12	16	20	12	16	20	12	16	20	12	16	20	12	16	20
#4	12	16	20	12	16	20	12	16	20	12	16	20	12	16	20
#5	15	20	26	15	20	26	15	20	26	15	20	26	15	20	26
#6	19	25	33	18	24	31	18	24	31	18	24	31	18	24	31
#7	26	34	45	24	31	40	23	30	39	21	27	36	21	27	36
#8	35	45	59	31	40	53	30	39	51	27	35	46	25	32	42
#9	44	57	74	39	51	67	38	49	65	34	44	58	31	40	53
#10	56	72	95	50	65	85	48	63	82	43	56	73	39	51	67
#11	68	89	116	61	80	104	59	77	101	53	69	90	48	63	82

* Normal weight concrete; bars other than top bars.

1. For vertical bars centered in walls, slab bars and temperature bars in slabs or footings with less than 12 in. or concrete below, etc., spaced 6 in. or more, use 0.8 basic lap lengths shown, but not less than 12 in.
2. In standard spiral columns, use 0.75 basic lap lengths shown, but not less than 12 in.

TABLE 13-8 Tension Lap Splice for Top Bars, Grade 60*

Bar Size	CLASS A, B & C LAP SPLICE LENGTHS (inches) f'_c = 3,000			f'_c = 3,750			f'_c = 4,000			f'_c = 5,000			f'_c = 6,000		
	A	B	C	A	B	C	A	B	C	A	B	C	A	B	C
#3	12	16	21	12	16	21	12	16	21	12	16	21	12	16	21
#4	17	22	29	17	22	29	17	22	29	17	22	29	17	22	29
#5	21	27	36	21	27	36	21	27	36	21	27	36	21	27	36
#6	27	35	46	25	33	43	25	33	43	25	33	43	25	33	43
#7	38	48	63	33	43	56	32	41	54	29	38	50	29	38	50
#8	48	63	82	43	56	74	42	55	71	38	49	64	34	45	58
#9	61	80	104	55	71	93	53	69	90	48	62	81	43	56	74
#10	78	101	132	70	91	118	67	88	115	60	78	103	55	72	94
#11	96	124	163	86	111	146	83	108	141	74	96	126	68	88	115

Note: 1. For normal weight concrete.

2. For horizontal bars centered in walls and slab bars and temperature bars in slabs or footings with more than 12 in. of concrete below, spaced 6 in. or more, use 0.8 lap lengths shown, but not less than 12 in.

* Courtesy Concrete Reinforcing Steel Institute.

points of maximum tensile stress, where no more than 50 percent of A_s is spliced (Section 7.6.3.1.1).

Class C—at points of maximum tensile stress where more than 50 percent of A_s is spliced (Section 7.6.3.1.1).

Class D—in tension tie members (Section 7.6.2). Note that the Class D splices must be enclosed in a spiral, and for bars larger than #4, the ends of the bars must be hooked 180°. No reduction in development length is allowed for the effect of the spirals or the end hooks (Section 7.6.1).

Except for compression only at footings (Section 15.6.8), welded splices or other positive connections must be used for #14 and #18 bars (Section 7.5.2) and may be used for the smaller bars (Section 7.5.5). A "full" welded splice is a butt splice capable of developing in tension 125 percent of the (specified) f_y; a "full" positive connection, capable of developing in tension or compression or both, as required, 125 per cent of f_y (Sections 7.5.5.1 and 7.5.5.2). Lesser value welds, including lap or groove types, and lesser value positive connections may be used at points of known less than maximum tensile stress, where splices are staggered at least 24 in. and where, together with unspliced bars, at least twice the maximum calculated tensile capacity is provided, but not less than 20,000 psi on the total steel area through the section containing splices (Section 7.6.3.2.3). In spliced sections of columns, the minimum tensile capacity must also be at least twice the maximum calculated tension (Section 7.10.3) but not less than 25 percent of f_y on the total steel area in each face (Section 7.10.5).

WELDED WIRE FABRIC (WWF)

Smooth Wires

(Wire, ASTM A-82; WWF, ASTM A-185). Embedment of smooth wire WWF is dependent entirely upon the anchorage of the welds to the cross-wires. Full development (embedment for full strength) requires two cross-wires embedded, with the closer wire at least 2 in. beyond the critical section. Embedment of one cross-wire at least 2 in. beyond the critical section is considered half effective, for $1/2\ f_y$ (Section 12.10.1) (see Figs. 13-18 and 13-19).

Lap splice lengths are similarly established by the cross-wire embedment. Except for temperature reinforcement, lap splices of smooth WWF at points of maximum stress should be avoided. With larger sizes of smooth wire (W 31 = 5/8 in. ϕ) at close spacings if lap splices must be employed at points of maximum design stress, the minimum embedment should be increased. Calculations for such embedment may be based upon the shearing resistance of the wedge of concrete from the nearest embedded cross-

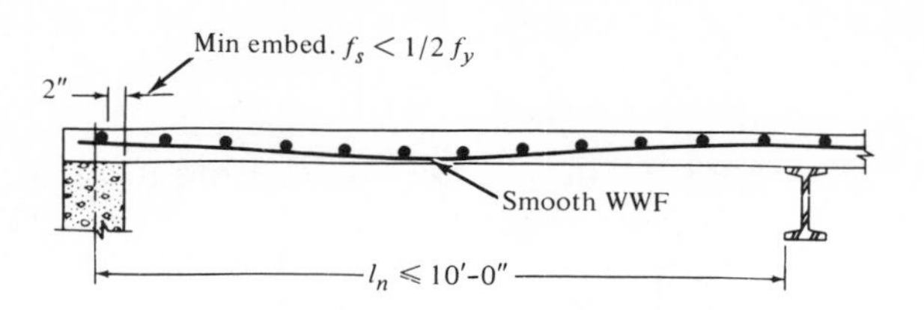

Fig. 13–18 Half-strength embedment.

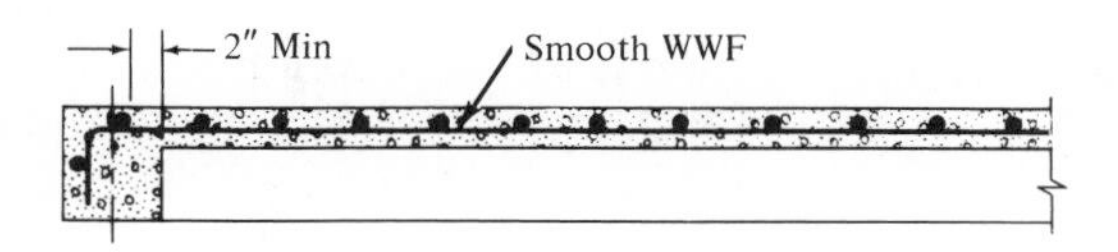

Fig. 13–19 Full-strength embedment.

wire to the potential crack. Where calculated maximum tensile stress exceeds $1/2\ f_y$, the lap length required is one spacing of cross-wires plus 2 in. measured from the end cross-wires (Section 7.8.1). For splices where the maximum calculated tensile stress is always less than $1/2\ f_y$, the outermost cross-wires must overlap 2 in. (Section 7.8.2). (see Figs. 13-20 and 13-21).

Deformed Wires

(Wire, ASTM A-496; WWF, ASTM A-497). Deformed wire has a development capacity recognized in the Code. The basic development length,

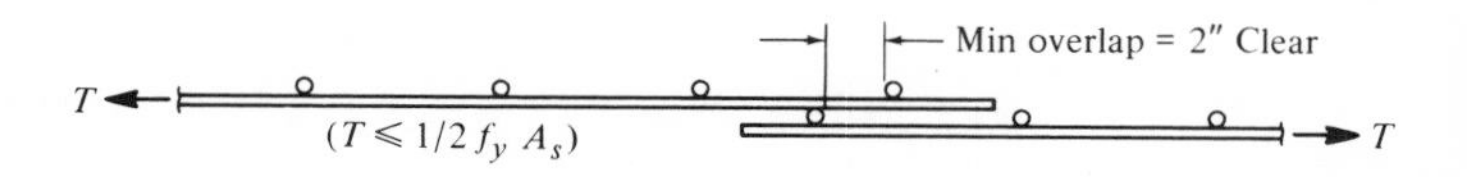

Fig. 13–20 Smooth WWF—lap splice for $\frac{1}{2}f_y$.

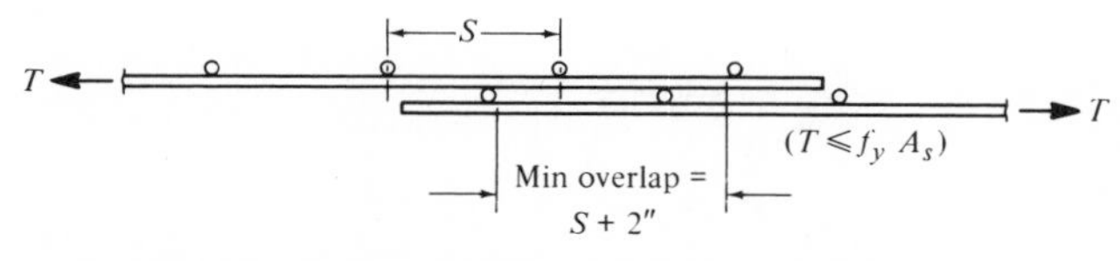

Fig. 13–21 Smooth WWF—full lap splice.

$$l_d = \frac{0.03\ d_b f_y}{\sqrt{f_c'}} \quad \text{(Section 12.5-a)}$$

This basic development length is subject to adjustment for various conditions. As top reinforcement, with more than 12 in. of concrete cast below it, the development length becomes 1.4 × the basic l_d (Section 12.5-b); in lightweight aggregate concretes, 1.33 × basic l_d for all light-weight and 1.18 for sand-lightweight (Section 12.5-c). See Chapter 14 for special provisions concerning lightweight aggregate concrete.

Deformed WWF is developed by both the deformations along the wire and the welded cross-wire anchorages. The development length,

$$l_d = \frac{0.03\ d_b(f_y - 20{,}000\ n)}{\sqrt{f_c'}}$$

where f_y is in psi, and n = the number of cross-wires embedded 2 in. or more from the critical section. The minimum $l_d = 250\ A_w/s_w$, where A_w is the area of the wire in tension and s_w is the spacing of the wires being spliced (Section 12.10.2 (see Fig. 13-22).

For lap splices of the deformed wires, the minimum lap length is 1.5 l_d or 12 in., whichever is greater. For lap splices of deformed WWF, the minimum lap length is 1.5 l_d, but not less than $A_w/s_w\ (360 - 8l_w/d_b)$ in which l_w is the extension of the wires in the direction of the splice past the last cross-wire and d_b is the wire diameter in inches of the wires spliced (Section 7.9.1) (see Fig. 13-23).

It will be noted that all splice and embedment requirements for wire are based upon the full f_y, except the smooth WWF half-strength splice. In most building elements using WWF as the principal reinforcement, of necessity the splices occur at the edges of the fabric, whether in flat sheets

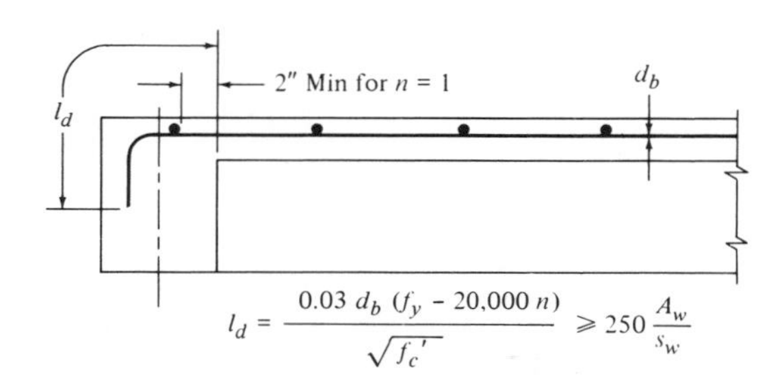

Fig. 13–22 Embedment for deformed WWF.

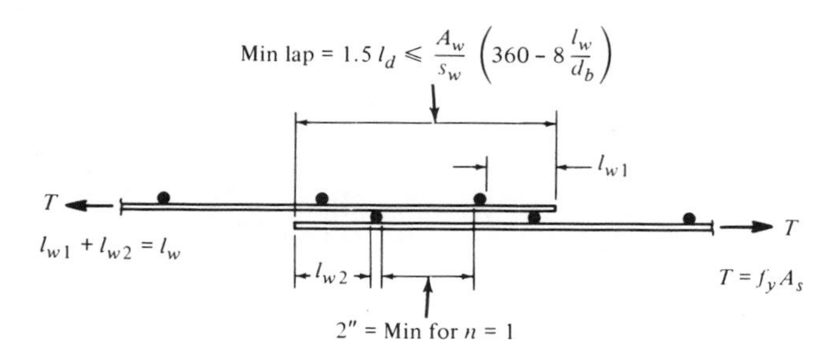

Fig. 13–23 Lap splices for deformer WWF.

or in rolls, and at ends of the rolls; it is seldom feasible to plan the location of all splices away from points of maximum stress. Also, it should be noted that temperature and shrinkage reinforcement in rigid concrete buildings is designed as fully stressed from end anchorage to end anchorage at each end of the element or structure, unlike temperature steel in slabs on ground where stress is a function of the friction and weight of slab between the critical section and nearest free joint.

SPACING OF REINFORCEMENT

The most significant new Code provisions regarding the spacing of reinforcement in the last decade concern bundled bars and noncontact lap splices. Where spacing and cover limitations are based on the bar size, a bundle of bars is treated as a single bar of a diameter derived from the equivalent total area of the bundle (Section 7.4.2). Contact lap splices are encouraged, and noncontact lap splices are limited so that the maximum transverse spacing between the single bars is not to exceed one-fifth of the required lap length or 6 in. (Section 7.5.4) (see Fig. 13-24).

In flexural members, the minimum clear spacing between bars is one bar diameter, but not less than 1 in. For parallel layers, as in beams, bars in the upper layer must be positioned directly above those in the layer below, but not less than 1 in. clear distance above (Section 7.4.1). In walls and one-way slabs, the maximum bar spacing is $3h$ (where h = thickness of wall or slab) but not more than 18 in. (Section 7.4.3). For two-way slabs maximum spacing of bars is $2\ h \leq 18$ in. (Section 13.5.1). For temperature steel only, maximum spacing is $5h \leq 18$ in. (Section 7.13). Temperature steel at right angles to the ribs, in the slabs of one-way slab-joist systems, must be supplied (Section 8.8.6) and the maximum spacing is $5\ h \leq 18$ in., where h = thickness of top slab. In columns, vertical bars (or

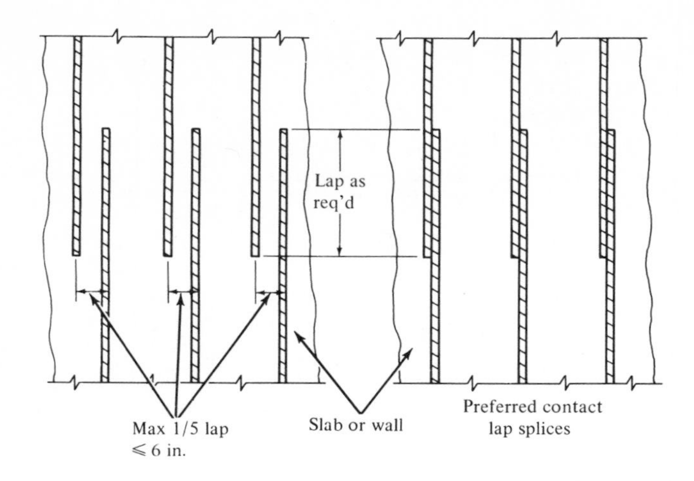

Fig. 13–24 Spaced and contact lap splices.

bundles) must have a clear spacing not less than $1\frac{1}{2}$ bar diameters or $1\frac{1}{2}$ in., whichever is larger (Section 7.4.4).

It will be noted that all of the above minimum bar spacings are contingent upon the use of concrete with a nominal maximum size of coarse aggregate not larger than three-fourths the minimum clear spacing (Section 3.3.2). Normal practice for design and detailing of reinforcement is to employ the minimum bar spacings with the regulation of the aggregate size prescribed in the concrete section of the project specifications to suit the bar spacings. The engineer can waive the maximum limits on aggregate size if proper workability and consolidation of the concrete is achieved (Section 3.3.2). If not, adjustments in the concrete are required in any event (Section 5.4.3).

CONCRETE COVER

The Code requirements for concrete cover, intended to provide protection to the reinforcement for ordinary exposure conditions, have been consolidated into one code section (Section 7.14). Only minor changes have been made from past codes. Except with earth forms, 2 in. is the maximum cover required. With earth forms a 3 in. cover is required to allow for a tolerance of − 1 in. at each side of the excavation. For cast-in-place beams, girders, and columns the specified clear cover is measured to the outermost reinforcement (Section 7.14.1.1), be it ties, stirrups, or spirals if present; or to main bars if not. For precast concrete (Section 7.14.1.2) or prestressed concrete (Section 7.14.1.3) separate requirements, applicable simultaneously, either of which may control, are given.

In the design of precast beams, girders, or columns, it is not sufficient to allow the minimum cover to the outermost steel plus the diameter of same as cover to the main reinforcement. For example, if a precast column contains #8 vertical bars and #4 ties, cover to the vertical bars will control. Minimum cover to the #8 bar equals one bar diameter, 1 in.; minimum cover to the #4 tie equals 3/8 in., but 1/2 in. must be furnished in order to provide the 1 in. cover to the vertical bars (Section 7.14.2).

It will be noted that the required cover for precast or prestressed concrete is less than that for nonprestressed cast-in-place concrete. These reductions recognize the greater *accuracy* possible in the positioning of reinforcement under plant control when precast; and the greater *effectiveness* of cover when prestressed to keep cracks closed under dead load (tension $\leqq$ zero).

It is intended that the minimum covers prescribed be increased (1) where local codes require additional fire protection, or (2) where members are exposed to highly corrosive environments (Sections 7.14.3 and 7.14.5). Such increases are not expected to be proportional to all specified minimum cover limits given. Beyond about 2 in., increases in cover do not provide proportional increases in protection against penetration of corrosive liquids or gases and improvement in the quality of the concrete itself or other protection is required (Section 7.14.3). As an example, for sewage retention, slab and wall cover prescribed for cast-in-place concrete (3/4 and $1\frac{1}{2}$ in.) might be increased to 2 in. together with use of a strength of concrete higher than that required structurally, use of sulfate-resisting cement (Sections 4.2.6 and 4.2.7), and air entrainment (Section 4.2.5). Proprietary "shrinkage controlled" cements have produced some construction with marked reduction in random shrinkage cracking. The use of concrete made with such cements is not specifically regulated by the Code, but it is acceptable where performance requirements govern (Section 1.1.3 and 1.4).

14 prestressed concrete

GENERAL

Prestressed concrete is reinforced concrete in which there have been introduced internal stresses of such magnitude and distribution that the stresses resulting from loads are counteracted to a desired degree. For design under the Code, the internal stresses can be introduced by pretensioning (a method of prestressing in which the tendons are tensioned before the concrete is placed) or by posttensioning (a method of prestressing in which the tendons are tensioned after the concrete has hardened (Section 2.1).

Although limited in no way in the Code, the provisions of the latter for prestressed concrete were developed primarily for slabs, beams, and columns, and contain no specific provisions for application to nonlinear prestressing for elements such as pipes, circular tanks, etc. (Code Commentary 18.1.1). Code provisions apply only to structural members prestressed with high strength steel tendons made from wire—seven wire strands—or from high-strength alloy steel bars (Section 18.1.1).

EXCEPTIONS TO THE GENERAL CODE

Prestressed concrete is subject to the provisions of the entire Code, except the following specifically excluded portions:

(Section 8.6)	Redistribution of moments on a limit design basis for prestressed concrete is covered in Section 18.12.
(Sections 8.7.2, 8.7.3 and 8.7.4)	The width and thickness of the flange of prestressed concrete T-beams is left to the judgment of the engineer. The empirical provisions for conventional reinforced concrete, if applied to prestressed concrete, would exclude many prestressed products in satisfactory use (Code Commentary Sections 18.1.2 and 18.1.3).
(Section 8.8)	Concrete joists. Empirical limits are justified for

conventional reinforced concrete joist construction but not for prestressed concrete joist construction.

(Sections 10.3.2 and 10.3.3) Limiting steel percentage. For prestressed concrete separate limits are prescribed (Section 18.8).

(Section 10.3.6) Minimum eccentricity for members subject to a compression load—not intended for prestressed concrete.

(Section 10.5) Minimum steel requirements. Prestressed concrete minimums are prescribed separately (Section 18.8).

(Section 10.9.1) Limits for reinforcement in compression members. Compression members with an average prestress of 225 psi or higher are excluded from these limits. Compression members with an average prestress less than 225 psi for purposes of these requirements are not considered as prestressed (Section 18.14).

(Chapter 13) Two-way slabs. Prestressed two-way slab design is separate (Section 18.13).

(Chapter 14) Walls. Empirical requirements for walls are not suitable for prestressed concrete.

Note: Empirical limits on reinforcement for walls designed by the rational method (Sections 10.16.2, 10.16.3, 10.16.4 and 10.16.5) have *not* been excluded from application to prestressed concrete (see Chapter 11).

HIGH-STRENGTH PRESTRESSING STEEL

High-strength steel tendons in prestressed concrete must conform to "Specifications for Uncoated Seven-Wire Stress-Relieved Strand for Prestressed Concrete" (ASTM A 146) or the "Specifications for Uncoated Stress-Relieved Wire for Prestressed Concrete" (ASTM A 421). Other steel strands or wire may be used providing that they conform to the minimum requirements of A 416 or A 421 and have no properties that make them less satisfactory (Section 3.5.9).

The prestressing strand is made by stranding six wires to form helices around a straight seventh wire. The straight wire has a slightly larger diameter to insure that the outer wires, when under stress, will grip the center one. The properties of Grades 250 and 270 seven-wire uncoated stress-relieved steel strands conforming to ASTM A 416, intended for use in prestressed concrete construction, are shown in Table 14-1. Note that the Grade 270 strand has a larger area, weight, and breaking strength than

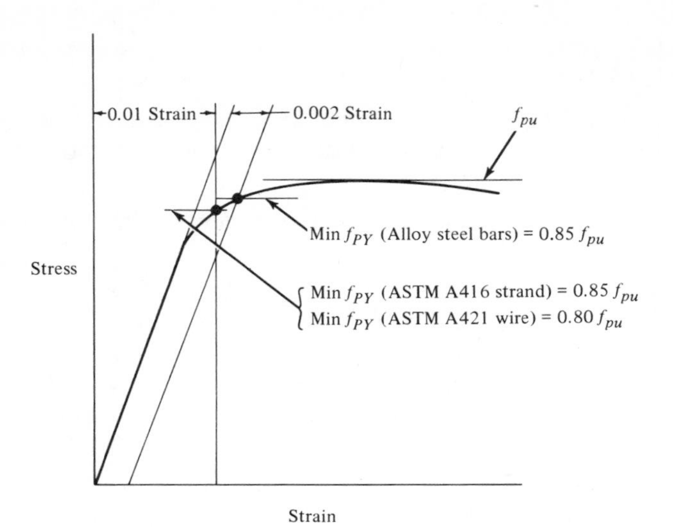

Fig. 14–1 High-strength prestressing steel.

TABLE 14-1 Properties of ASTM A 416 High-Strength Steel Strand

Strand Nominal Diameter (inches)	STRAND AREA SQUARE INCHES		STRAND WEIGHT (Lb/Ft)		STRAND BREAKING STRENGTH—KIPS	
	Grade 250	Grade 270	Grade 250	Grade 270	Grade 250	Grade 270
1/4 (0.255)	0.036	—	0.122	—	9.0	—
5/16 (0.313)	0.058	—	0.197	—	14.5	—
3/8 (0.375)	0.080	0.085	0.272	0.290	20.0	23.0
7/16 (0.438)	0.108	0.115	0.367	0.390	27.0	31.0
1/2 (0.500)	0.144	0.153	0.490	0.520	36.0	41.3

Grade 250. The minimum yield strength of ASTM A 416 strand based on 1 percent total strain is 0.85 f_{pu}, as shown in Figure 14-1.

The properties of uncoated stress-relieved wire conforming to ASTM A 421, commonly used in prestressed linear concrete construction, are shown in Table 14-2. The minimum yield strength for ASTM A 421 wire based on 1 percent, total strain is equal to 0.80 f_{pu}, as shown in Figure 14-1.

High-strength alloy steel bars are used principally for posttensioned prestressed concrete construction. They are available in diameters ranging from 1/2 to 1-3/8 in. These bars are required to be proof-stressed to 85 percent of the guaranteed tensile strength, as shown in Fig. 14-1. After proof-stressing they must have a yield strength, f_{py}, not less than 0.85 f_{pu}, an elongation at rupture (0.2 percent offset) not less than 4 percent and a reduction in area at rupture not less than 20 percent (Section 3.5.10. High-strength alloy steel bars are produced with a minimum tensile strength as high as 160,000 psi.

Large-diameter high-strength wire strand made of 7, 19, 37, or more galvanized or uncoated hard-drawn wires, spirally wound, is also available for prestressing steel, but must meet the minimum requirements of ASTM A 416 to conform to Code requirements (Section 3.5.9).

The modulus of elasticity of high-strength prestressing steel strand is usually less than wire and it should be determined by tests or supplied by the manufacturer (Section 8.3.2).

Cover

The minimum concrete cover for prestressed and nonprestressed reinforcement, ducts, and end fittings of cast-in-place and precast prestressed concrete members varies with exposure and manufacturing conditions. If members exposed to a corrosive environment and members requiring greater cover for fire protection (Section 7.14.5) are excluded, the minimum concrete cover in inches for prestressed reinforcement and nonprestressed reinforcement in prestressed members, other than in members manufactured under plant control where cover can be the same as that for precast concrete (Section 7.14.1.2), is as follows:

	Tension in precompressed tension zone less than $6\sqrt{f'_c}$	Tension in precompressed tension zone greater than $6\sqrt{f'_c}$
Members cast against and permanently exposed to earth	3″	4½″
Members exposed to earth or weather:		
Wall panels, slabs, and joists	1″	2″
Other members	1½″	3″
Members not exposed to weather or in contact with ground:		
Slabs, walls, joists	¾″	¾″
Beams, girders, columns: Principal reinforcement	1½″	1½″
Beams, girders, columns: Ties, stirrups, spirals	1″	1″
Shells and folded plate members:		
Reinforcement ⅝ in. and smaller	⅜″	⅜″
Other reinforcement	d_b but not less than ¾″	

TABLE 14-2 Properties of ASTM A 421 High-Strength Steel Wire

Wire Diameter Inches	Wire Area Square Inches	WIRE TENSILE STRENGTH f_{pu} (ksi)	
		Button Anchorage *BA*	Wedge Anchorage *WA*
0.192	0.0289	—	250.0
0.196	0.0302	240.0	250.0
0.250	0.0491	240.0	240.0
0.276	0.0598	—	235.0

When prestressed concrete members are subject to corrosive atmospheres, the amount of concrete cover must be suitably increased or other protection must be provided (Section 7.14.3).

Loss of Prestress

The amount of prestress in a prestressed concrete member varies with time. The initial prestressing force can be measured with reasonable accuracy by reading a pressure gauge on the jack and checked by measuring tendon elongation. The final prestressing force is more difficult to determine and can only be estimated because it varies with time and is affected by slip at anchorage, elastic shortening of concrete, creep of concrete, shrinkage of concrete, and relaxation of steel stress, and includes frictional losses due to intended or unintended curvature in the tendons. The difference between the initial prestressing force and the effective prestressing force at any later time is defined herein as the loss of prestress.

The loss of prestress is difficult to predict and can only be estimated. T. Y. Lin,* suggests that for average steel and concrete properties and curing conditions encountered with prestressed concrete members that the average percentage of loss of prestress for losses other than frictional losses can be taken as follows:

	Pretensioning, %	Postensioning, %
Elastic shortening	3	1[a]
Creep of concrete	6	5
Shrinkage of concrete	7	6
Creep in steel	2	3
TOTAL	18	15

[a] Elastic shortening loss is impossible; this allowance probably represents loss during securing the anchorage.

* *Prestressed Concrete Structures,* 2d ed. (New York: John Wiley Sons, Inc., 1960).

When specific loss data are lacking, the loss in prestress—not including friction loss—can be assumed as 35,000 psi for pretensioning and 25,000 psi for posttensioning.*

Frictional losses can be calculated from experimentally determined wobble and curvature coefficients which must be controlled during stressing operations by initial overstressing, measured elongation, and, if necessary, jacking from each end (Section 18.6.2). The Code Commentary sets forth a range of values for wobble and curvature coefficients for grouted tendons in metal sheathing and unbonded mastic coated and pregreased tendons.

The calculated design strength (ultimate strength) of a prestressed concrete member with bonded tendons is little affected by the magnitude of the loss of prestress, but any error in the provisions for the loss of prestress could result in a false prediction of the behavior of the member under service loads and cause calculations for stresses, cracking load, and deflection or camber to be in error.

Flexural Strength

The flexural strength of prestressed concrete members must be computed by the strength design methods of the Code (Section 18.7). In designing for strength, the assumptions provided for nonprestressed members apply (Section 10.2), except the assumption of "flat yielding" (Section 10.2.4) which applies only for nonprestressed reinforcing steel that conforms to Sections 3.5.1, 3.5.6 and 3.5.8 of the Code (Section 18.3.1).

In the calculation of the design strength, the Code requires that f_{ps}, calculated stress in the prestressing steel at design load, be substituted for f_{py}, specified yield strength, (Section 18.7). Different types of high strength prestressing steels have different stress-strain characteristics and each lacks a sharp and distinct yield point (Fig. 14-1). It is therefore difficult to predict the actual stress at design load closely unless the stress-strain curve for the steel being used is known.

When the information to determine f_{ps} is not available, the Code allows approximate values to be used, provided that the effective prestressing force after losses at service loads is not less than 50 percent of the ultimate strength f_{pu} (Section 18.7.1).

For unbonded tendons the approximate value given for f_{ps} is

$$f_{ps} = f_{se} + 10{,}000 + f_c'/100\rho_p \quad \text{(Section 18.7.1, Eq. 18-4)}$$

where f_{ps} must not be greater than f_{py} or greater than $f_{se} + 60{,}000$.

---

* Tentative Recommendations for Prestressed Concrete," *ACI Journal, Proceedings,* **54** (7) (Jan. 1958): 545-578 and the Code Commentary.

The flexural design strength of members with unbonded tendons is generally less than, and occurs at, a lower steel stress than in members with equivalent bonded tendons. Large variations in the steel stress of unbonded tendons at design strength have been reported by different investigators for different members and are due to differences in these members such as profile of prestressing steel, moment diagram, length : depth ratio, friction coefficient between presressing steel and duct and amount of bonded nonprestressed reinforcement. The rational method for prediction of design strength of unbonded members requires: (1) an accurate estimate of losses and (2) an accurate deflection analysis to predict increase in prestress from service load to design (ultimate) load based on known stress-strain properties of the steel.

For bonded tendons the approximate value given for f_{ps} is

$$f_{ps} = f_{pu}(1 - 0.5\rho_p \, f_{pu}/f_c') \quad \text{(Section 18.7.1 Eq. 18-3)}$$

The expression for f_{ps} for members with bonded tendons is shown graphically in Fig. 14-2 where high-strength prestressing steels with f_{pu} =

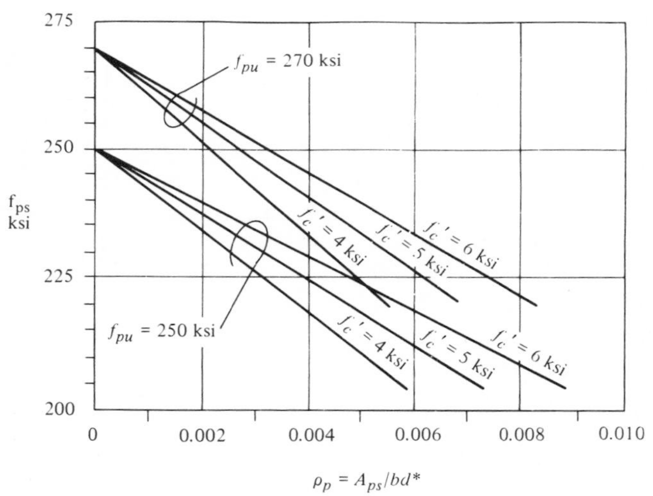

*For flanged sections, where the neutral axis or the depth of the equivalent rectangular stress block at design strength falls below the flange, p is calculated using the width of the web and the steel area is taken as only that amount that is required to develop the design compressive strength of the web. (Section 18.0).

Fig. 14–2 Approximate values of f_{pu} for bonded tendons.

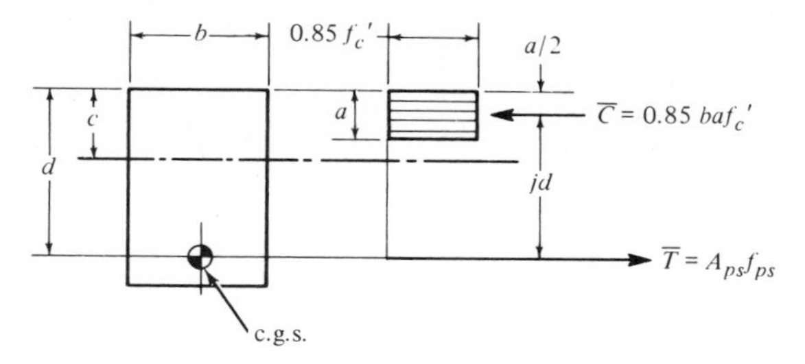

Fig. 14–3 M_t for rectangular prestressed sections.

250 and 270 ksi are used in combination with concrete with $f_c' = 4$, 5 and 6 ksi.

Note in Fig. 14-2 that values for f_{ps} for bonded tendons are discontinued at steel percentages where $\omega_p = \rho_p f_{ps}/f_c'$ becomes greater than 0.30, which is the maximum value permitted to avoid the condition of an over-reinforced beam which will fail in compression and whose design strength will be controlled by the strength of the concrete (Section 18.8.2). For values of ω_p less than 0.3, a flexural member will fail in tension and the design strength will be controlled by the strength of the prestressing steel.

For rectangular sections the theoretical design moment strength, M_t, for flexural members reinforced only with high strength prestressing steel can be derived as shown in Fig. 14-3.

When $\rho_p f_{ps}/f_c' < 0.3$ tension controls and the flexural member is under-reinforced and the theoretical design moment strength is

$$a = \frac{A_{ps}f_{ps}}{0.85f_c'b} = \frac{\rho_p f_{ps} d}{0.85f_c'}$$

$$M_t = \phi\, Tjd = \phi A_{ps}f_{ps}jd$$

$$= \phi A_{ps}f_{ps}(d - a/2)$$

$$= \phi A_{ps}f_{ps}(d - 0.59\rho_p\, f_{ps}d/f_c')$$

$$= \phi A_{ps}f_{ps}d(1 - 0.59\omega_p)$$

This equation is the same as that in the Code Commentary (Section 18.7). When $\rho_p f_{ps}/f_c' = 0.3$ compression starts to control and the flexural member starts to become overreinforced and the theoretical design moment strength approaches

$$M_t = \phi A_{ps} f_{ps} d(1 - 0.59\omega_p)$$
$$= \phi \rho_p bd\, f_{ps} d(1 - 0.59\omega_p)$$
$$= \phi \rho_p \frac{f_{ps}}{f_c'} f_c' bd^2(1 - 0.59\omega_p)$$
$$= \phi \omega_p f_c' bd^2(1 - 0.59\omega_p)$$
$$= \phi(0.30bd^2)[1 - (0.59)(0.30)]$$
$$= \phi(0.25 f_c' bd^2)$$

This equation is the same as that shown in the Code Commentary (Section 18.8.2). The authors recommend the use of a ϕ-factor of 0.70 for this brittle type of concrete compression failure.

For flanged sections such as T-beams, where the neutral axis and the depth of the equivalent rectangular stress block fall below the flange, the theoretical design moment strength, M_t, for flexural members reinforced only with high-strength prestressing steel can be derived as shown in Fig. 14-4.

Let $A_{psw} = A_{ps} - A_{psf}$ = the steel area required to develop the ultimate compressive strength of the web of a flanged section. When $A_{psw} f_{ps}/b_w d f_c' < 0.03$ tension controls and the flexural member is underreinforced and the theoretical design moment strength is

$$\overline{T}_f = \overline{C}_f = A_{psf} f_{ps} = 0.85 f_c'(b - b_w)h_f$$
$$M_t = \phi(\overline{T}_f j_f d + T_w j_w d)$$
$$= \phi[A_{psf} f_{ps}(d - h_f/2) + A_{psw} f_{ps}(d - a/2)]$$
$$= \phi[0.85 f_c'(b - b_w)h_f(d - h_f/2) + A_{psw} f_{ps}(d - a/2)]$$

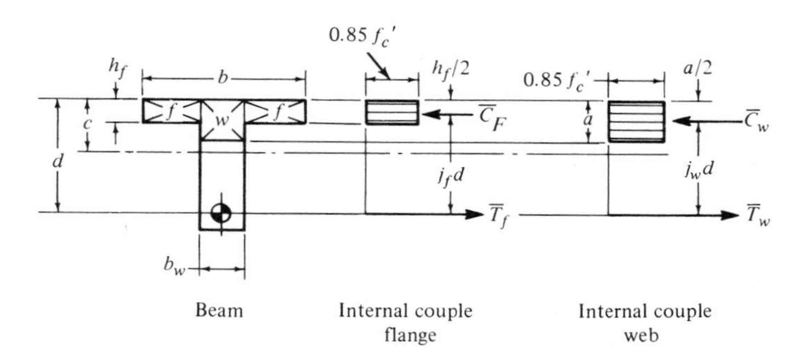

Fig. 14–4 M_t for T-beam prestressed sections.

This equation is shown in the Code Commentary (Section 18.7). When $A_{psw}f_{ps}/b_w d f_c' = 0.3$ compression starts to control and the flexural member starts to become overreinforced and the theoretical design moment strength approaches

$$a = \frac{A_{psw}f_{ps}}{0.85f_c'b_w} = \frac{0.3b_w df_c'}{0.85f_c'b_w} = 0.353d$$

$$M_t = \phi[0.85f_c'(b - b_w)h_f(d - h_f/2) + 0.30b_w df_c'(d - a/2)]$$

$$= \phi\left[0.85f_c'(b - b_w)h_f(d - h_f/2) + 0.30b_w df_c'\left(d - \frac{0.353d}{2}\right)\right]$$

$$= \phi[0.85f_c'(b - b_w)h_f(d - h_f/2) + 0.25b_w d^2 f_c']$$

This equation is shown in the Code Commentary (Section 18.8.2). Again the authors recommend the use of a ϕ-factor of 0.70 here for a brittle concrete compression failure.

As stated before, the fact that bonded reinforcement is prestressed does not affect the calculated flexural strength of the member and nonprestressed reinforcement, in combination with prestressed steel, can be considered to contribute to the tension force in a member at design load an amount equal to its area times its yield strength (Section 18.7.2).

To avoid the sudden flexural failure of an underreinforced prestressed concrete member due to an overload, such as when the flexural capacity is reached immediately after cracking, the total amount of prestressed and nonprestressed reinforcement cannot be less than that required to develop a design load in flexure at least 20 percent greater than the cracking load (Section 18.8.3).

Behavior at Service Conditions

In addition to having adequate flexural strength, prestressed concrete members must behave properly at service conditions at all loads that the member will be subjected to from the time the prestress is applied (Section 18.2.1). Service conditions requiring proper behavior include transfer of the prestressing force (when the tensile force in the prestressed steel is transferred to the concrete usually combined with all or part of the dead load, and compression stress in the concrete is usually high relative to ultimate concrete compressive strength). Service dead and live loads after long-time volume changes have occurred represent a condition that must also be considered. Other service conditions that must be considered for proper behavior include: stress concentrations due to prestressing (Section

18.2.2); effects on the adjoining structure of elastic and plastic deformations, deflections, changes in length, and rotations caused by the prestressing (Section 18.2.3); temperature and shrinkage (Section 18.2.3); possibility of buckling in a member between points where the concrete and the prestressing steel are in contact and of buckling in their webs and flanges (Section 18.2.4); and deflections (Section 9.5.4).

Permissible stresses in the concrete and the prestressing steel at service loads are given to insure serviceability (Code Commentary Section 18.4). They do not automatically guarantee adequate flexural strength which must satisfy design load combinations (Section 9.3). Stresses in the concrete and the prestressing steel can be calculated using the basic assumptions (Section 18.3.2).

Permissible flexural stresses in the concrete at service loads must be within prescribed limits shown in Table 14-3, unless it can be shown experimentally or analytically that performance will not be impaired (Sections 18.4.1, 18.4.2, and 18.4.3).

When the calculated flexural tension stress after transfer but before losses exceeds $3\sqrt{f_c'}$, reinforcement must be provided to resist the total tension force in the concrete computed on the assumption of an uncracked section (Section 18.4.1).

TABLE 14-3 Allowable Stresses in Concrete at Service Loads

	f'_{ci} 4 ksi	f'_{ci} 5 ksi	f'_{ci} 6 ksi
After transfer but before losses			
Compression, $f_c \leq 0.6\ f'_{ci}$	2,400 psi	3,000 psi	3,600 psi
Tension in members without prestressed or unprestressed bonded auxiliary reinforcement in the tension zone, f_t, $\leq 3\ \sqrt{f'_{ci}}$	184 psi	212 psi	232 psi
	f'_c 4 ksi	f'_c 5 ksi	f'_c 6 ksi
Service load stresses after losses			
Compression, $f_c \leq 0.45\ f'_c$	1,800 psi	2,250 psi	2,700 psi
Tension in precompressed tensile zone, $f_t \leq 6\ \sqrt{f'_c}$	379 psi	424 psi	465 psi
Tension in precompressed tensile zone where immediate and long-time deflection calculated on the basis of the transformed cracked section and on biiineal moment deflection comply with limits (Section 9.5), $f_t \leq 12\ \sqrt{f'_c}$	758 psi	848 psi	930 psi

TABLE 14-4 Allowable Stresses in Prestressing Steel at Service Loads

	$f_{pu} = 145$ ksi	$f_{pu} = 250$ ksi	$f_{pu} = 270$ ksi
Tension due to jacking $\leqq 0.8\ f_{pu}$	116 ksi	200 ksi	216 ksi
Tension in pretensioning tendons immediately after transfer, or in post-tensioning tendons immediately after anchoring $\leqq 0.7\ f_{pu}$	101 ksi	175 ksi	189 ksi

Permissible flexural stresses in the high-strength prestressing steel at service loads must be within prescribed limits (Section 18.5) shown in Table 14.4.

Shear

Concrete is strong in *pure* shear. The Code limits the design shear where it is inappropriate to use it as a measure of diagonal tension to a maximum value of $0.2\ f_c'$ but not more than 800 psi (Section 11.15.3).

For design of prestressed concrete flexural members, the average or nominal shear stress, $V_u/\phi b_w d$, is used as a measure of the diagonal tension. In this expression the distance, d, is taken as that from the extreme compression fiber to the centroid of the tension reinforcement, but not less than $0.8\ h$ (Section 11.2.1). Sections located less than a distance $h/2$ from the face of the support can be designed for the shear, V_u, at $h/2$ (Section 11.2.2), when the reaction in the direction of the applied shear introduces compression into the end of the member.

There are two types of diagonal tension cracks that occur in flexural members; web-shear cracks caused by principal tensile stresses that exceed the tensile strength of the concrete and flexural-shear cracks initiated by flexural cracks. When flexural cracks occur, the shear stresses above the crack increase resulting in increased principal tensile stresses that cause flexural-shear cracks when the principal tensile stresses exceed the tensile strength of the concrete. Both types of cracks may be visible in prestressed concrete members with loads greater than the service loads (Code Commentary Sections 11.2.3 and 11.2.4).

To restrain the growth of both types of diagonal tension cracks, and to increase ductility (Code Commentary Section 11.1), the Code requires that a minimum area of shear reinforcement be provided in all prestressed concrete flexural members, except: (1) where v_u is less than one-half of v_c; or (2) where the depth of the members (h) is less than 10 in., 2.5 times the thickness of the compression flange, or one-half the thickness of the web;

or (3) where tests show that the required ultimate flexural and shear capacity can be developed with the shear reinforcement omitted (Section 11.1).

Where shear reinforcement is required for prestressed concrete members the minimum area, A_v, must be equal to or greater than

$$A_v = 50\, b_w s/f_y \quad \text{(Section 11.1.2, Code Eq. 11-1).}$$

Alternatively, a minimum area

$$A_v = \frac{A_{ps} f_{su} s}{80\, f_y d}\sqrt{\frac{d}{b_w}} \quad \text{(Section 11.1.2, Code Eq. 11-2)}$$

can be used if the effective prestress force is at least equal to 40 percent of the tensile strength of the flexural reinforcement (Section 11.1.2).

The design yield strength of the shear reinforcement for prestressed members may not be assumed greater than 60,000 psi (Section 11.1.3).

Shear reinforcement is required to consist of stirrups placed perpendicular to the axis of the member (Section 11.1.4a). Such stirrups may consist of welded wire fabric with wires placed perpendicular to the axis of the member (Section 11.1.4b). Vertical legs of shear reinforcement cannot be spaced further apart than 0.75 h or 24 in. (Section 11.1.4).

Figure 14-5 shows in graphical form the minimum area of shear reinforcement for prestressed concrete members of two different widths and varying depths (Section 11.1).

Minimum shear reinforcement required (Code Eq. 11-1) is shown in Fig. 14-5 for flexural members with minimum d ($d = 0.8\, h$), $f_y = 60{,}000$ psi and a stirrup spacing equal to 0.75 h, but not greater than 24 in. The alternative minimum shear reinforcement required (Code Eq. 11-2) is also shown in Fig. 14-5 for $f_{pu} = 250{,}000$ psi, an effective prestressing force equal to or greater than 40 percent of the tensile strength of the flexural reinforcement, minimum d ($d = 0.8$ h), $f_y = 60{,}000$ psi and a stirrup spacing equal to 0.75 h, but not greater than 24 in. It is obvious from Fig. 14-5 that, when minimum stirrups only are required, Code Eq. 11-2 will require less shear reinforcement.

In prestressed concrete members without shear reinforcement the shear carried by the concrete is assumed equal to the shear causing significant inclined cracking and must not exceed

$$v_c = 0.6\sqrt{f_c'} + 700\, V_u d/M_u \quad \text{(Section 11.5.1, Eq. 11-10)}$$

unless a more detailed analysis is made (Section 11.5.2). Allowable shear

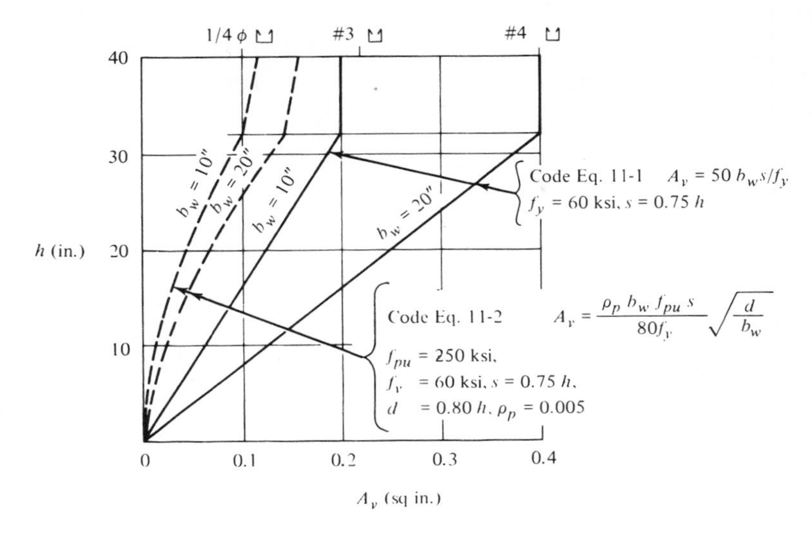

Fig. 14–5 Minimum shear reinforcement for prestressed concrete members.

on the concrete, v_c, need not be less than $2\sqrt{f_c'}$ and cannot be greater than $5\sqrt{f_c'}$ (Section 11.5.1). The design moment, M_u, occurs simultaneously with shear, V_u, at a particular section and d is the distance from the extreme compression fiber to the centroid of the prestressing tendons at this section. Note that d is not limited to a minimum value of 0.80 h. $V_u d/M_u$ cannot be taken greater than 1.0.

For a simple span beam, supporting a uniform load w, the quantity $V_u d/M_u$ in Code Eq. 11-10 at a distance x from the support can be expressed as follows (Code Commentary Section 11.5.1):

$$\frac{V_u d}{M_u} = \frac{w(l/2 - x)d}{(w/2)(l - x)x} = \frac{d(l - 2x)}{x(l - x)}$$

Using this expression to calculate v_c

$$v_c = 0.6\sqrt{f_c'} + \frac{700d(l - 2x)}{x(l - x)}$$

which is shown graphically in Fig. 14-6 for concretes with $f_c' = 4{,}000$ and 6,000 psi and for d/l ratios between 0.01 and 0.06.

There is less than 4 percent difference between values of v_c for $f_c' = 4{,}000$

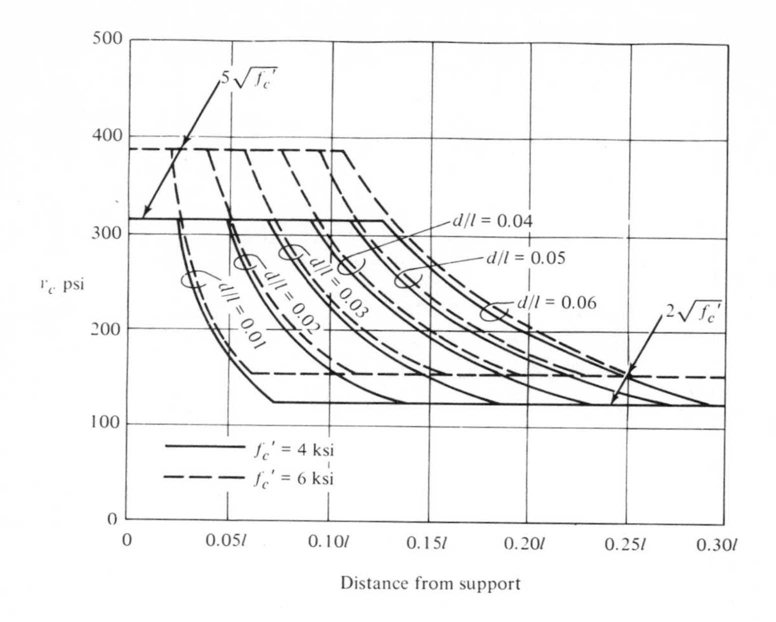

Fig. 14–6 v_c simple beam with uniform load.

psi and for $f_c' = 6{,}000$ psi, except where v_c approaches a maximum permitted value of $5\sqrt{f_c'}$ near the support or a minimum value of $2\sqrt{f_c'}$ near the center two-thirds of the span.

When it is desired to use a more accurate method to determine the shear carried by the concrete or when the effective prestress force is less than 40 percent of the tensile strength of the flexural reinforcement, it will be necessary to use the lesser of the shear stress causing inclined flexure-shear cracking, v_{ci}, or web-shear cracking, v_{cw}.

$$v_{ci} = 0.6\sqrt{f_c'} + \frac{V_d + (V_l M_{cr}/M_{max})}{b_w d} \quad \text{(Section 11.5.2, Eq. 11-11)}$$

$$v_{cw} = 3.5\sqrt{f_c'} + 0.3f_{pc} + V_p/b_w d \quad \text{(Section 11.5.2, Eq. 11-12)}$$

Web-shear cracking is predicted as the shear stress, v_{cw}, causing a principal tension stress becomes approximately $4\sqrt{f_c'}$ at the centroidal axis of the cross section (Code Commentary Section 11.5.2). In general v_{cw} will be the maximum value for the shear stress carried by the concrete similar to the maximum value of $5\sqrt{f_c'}$ in the approximate equation. The computed shear stress carried by the concrete need not be less than $1.7\sqrt{f_c'}$, which is smaller than the minimum value of $2\sqrt{f_c'}$ for an approximate equation due to the possibility of a smaller prestress force.

For uniformly loaded noncomposite beams, the equation for v_{ci} reduces to the following expression (Code Commentary Section 11.5.2):

$$v_{ci} = 0.6\sqrt{f_c'} + \frac{V_u M_{cr}}{M_u b_w d}$$

It is important to note that with equations for v_{ci} and v_{cw}, d is the distance from the extreme compression fiber to the centroid of the prestressing tendons, and need not be taken less than 0.8 h (Section 11.5.2.1). The approximate equation for v_c requires that d be the actual distance from the extreme compression fiber to the centroid of the prestressing steel with no minimum limit (Section 11.5.1).

When the nominal design shear stress ($v_u = V_u/\phi b_w d$) is greater than the shear stress carried by the concrete (Sections 11.5.1 or 11.5.2) shear reinforcement must be provided (Section 11.2.4). Proper shear reinforcement will insure that the prestressed concrete member will develop its design flexural strength without serious shear distress.

Shear reinforcement must be placed perpendicular to the axis of the member and spaced not farther apart than 0.75 h, or not more than 24 in. (Section 11.1.4). When the nominal shear stress minus the shear stress carried by the concrete ($v_u - v_c$) is greater than $4\sqrt{f_c'}$ the maximum spacing of shear reinforcement must be reduced to 0.375 h but not more than 12 in. (Section 11.6.3).

The maximum shear in excess of that carried by the concrete, ($v_u - v_c$), must be equal to or less than $8\sqrt{f_c'}$ (Section 11.6.4).

The shear reinforcement required to carry the shear in excess of that carried by the concrete can be determined as

$$A_v = (v_u - v_c)b_w s/f_y \text{ (Section 11.6.1, Eq. 11-13)}$$

The design yield strength of the shear reinforcement, f_y, must not be greater than 60,000 psi (Section 11.1.3). This limit provides a control on diagonal crack width (Code Commentary Section 11.1.3).

Shear reinforcement can also be determined from design aids such as are shown in Figs. 14-7 and 14-8 which give the shear stress capacity ($v_u - v_c$) of #3 and #4 stirrups, respectively, for beam web widths from 2.5 to 30 inches. As an example of the use of these design aids, let it be required to select stirrups for a 10 in. wide by 20 in. deep prestressed concrete beam with a value of ($v_u - v_c$) equal to 300 psi at a distance of

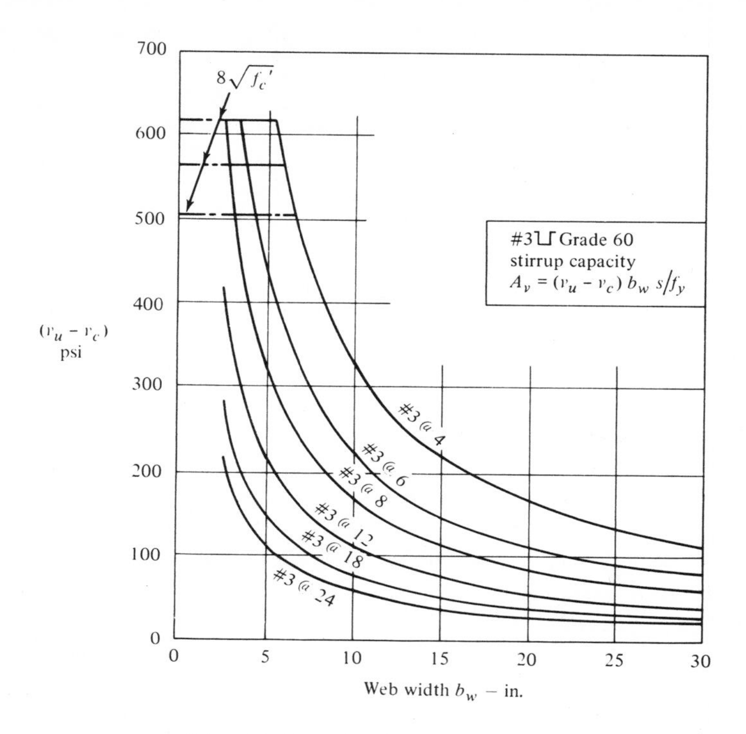

Fig. 14–7 #3 stirrup capacity.

$h/2$ from the support. From Figs. 14-5 and 14-6 it is evident that #3 stirrups at a 4 in. spacing or #4 stirrups at an 8 in. spacing are satisfactory.

The spacing of the #3 stirrups could also be obtained from Fig. 14-8 for #4 stirrups by simple proportion of the stirrup area as follows:

$$(\text{Spacing of } \#3\ \sqcup) = \frac{(\text{Spacing of } \#4\ \sqcup)(\text{Area of } \#3\ \sqcup)}{(\text{Area of } \#4\quad)}$$

$$= \frac{(8.00)(0.22)}{(0.40)}$$

$$= 4.4 \text{ in.}$$

The spacing of #5 stirrups or other sizes can be determined from Figs. 14-7 or 14-8 in a similar manner or from the figures in Chapter 9 of this Guide.

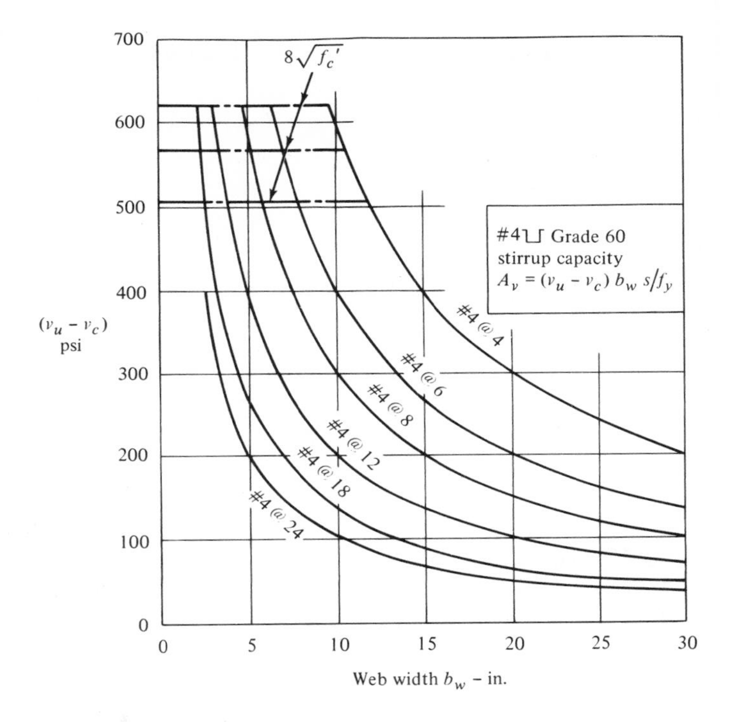

Fig. 14–8 #4 stirrup capacity.

A design example at the close of this chapter shows in detail some of the problems in design for shear. In particular note Fig. 14-14.

The Code does not contain design criteria for torsion and shear for prestressed concrete members (Section 11.7). When the prestressing steel is unbonded, some bonded reinforcement must be provided in the precompressed tensile zone of flexural members and distributed uniformly over the tension zone near the extreme tensile fiber (Section 18.9.1).

The amount of bonded reinforcement provided for beams and one-way slabs must be equal to or greater than

$$A_s = N_c/0.5f_y \quad \text{(Section 18.9.2)}$$

or

$$A_s = 0.004\, A \quad \text{(Section 18.9.2)}$$

where N_c is equal to the tensile force in the concrete under a load equal to

$D + 1.2\ L$; A is equal to the area of that part of the cross section between the flexural tension face and the center of gravity of the gross section; and f_y is the yield strength of the bonded reinforcement which must not be greater than 60,000 psi (Section 18.9.2). The authors find the definition of N_c inapplicable because of its ambiguity, and recommend interpretation thus: "N_c = total tensile force under loads consisting of effective prestress (after losses) plus dead load (D) plus 1.2 times live load (1.2 L)."

For two-way slabs with unbonded prestressing tendons the amount of bonded reinforcement provided must be equal to or greater than one-half of the amount required for one-way slabs (Section 18.9.3). This amount may be reduced where there is no tension in the precompressed tensile zone at service loads, but no lower limit nor specific determination for such reduction is prescribed (Section 18.9.3).

DESIGN EXAMPLE. The final design of a cast-in-place prestressed concrete posttensioned 60′–0″ simple span T-beam will be used to illustrate some of the design decisions required by the Code. An elevation of the beam is shown in Fig. 14-9 and the dimensions of the cross section are shown in Fig. 14-10. A preliminary design has been made and it indicates that 13-1/2 in. round 270k parabolically draped prestressing tendons will meet the code requirements for strength. Two #5 reinforcing steel bars (f_y = 60,000 psi) are placed in the bottom of the beam in the corners of the stirrups for anchorage of web reinforcement (Section 12.13.2). A final design check will now be made to determine if the nonprestressed reinforcement used in combination with the prestressed steel will provide the necessary strength for design load (Section 18.7.2) and also meet the service load stress limitations (Section 18.4 and 18.5).

The final design check will be made for concrete with an ultimate com-

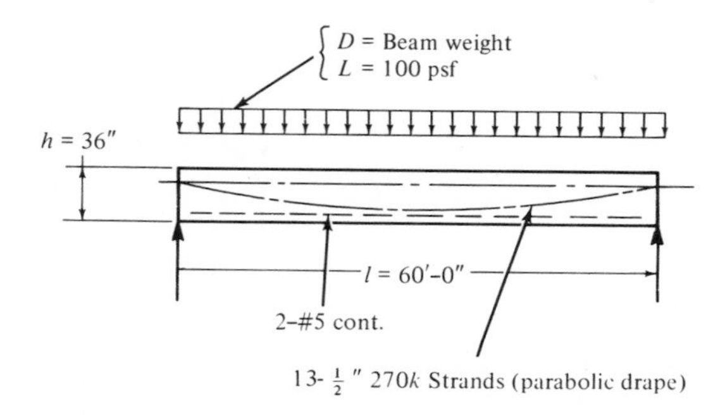

Fig. 14–9 Prestressed concrete beam profile.

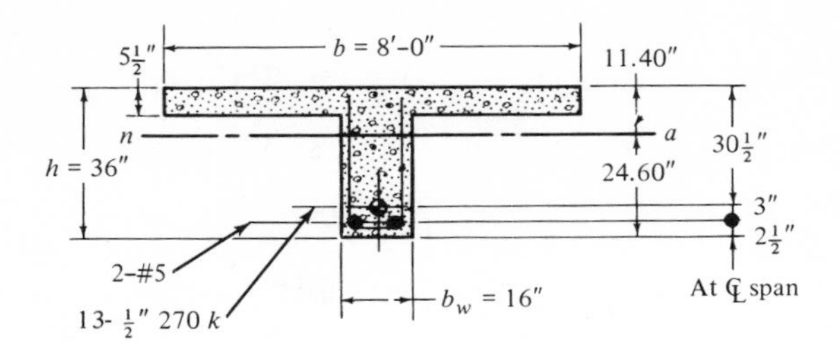

Fig. 14–10 Beam cross section.

pressive strength of 4,000 psi at the time the prestressing force is applied to the concrete (transfer) and 6,000 psi at design load (1.4D + 1.7L).

The geometrical properties of the section are as follows:

Area and centroid of section

$$(80.0)(5.5) = 440 \times 33.25 = 14{,}630$$
$$(16.0)(36.0) = 576 \times 18.00 = 10{,}368$$
$$\Sigma A_c = 1{,}016 \qquad \Sigma A_y = 24{,}998$$

$$y_t = \Sigma A_y / \Sigma A = 24.60 \text{ in.}$$

Moment of inertia

$$(16)(24.60)^3/3 = 79{,}397$$
$$(96)(11.4)^3/3 = 47{,}409$$
$$126{,}806$$
$$-(80)(5.9)^3/3 = -5{,}476$$
$$I_g = 121{,}330 \text{ in.}^4$$

The beam loads, moments, and shear are as follows:

Service loads, moments and shears

$$D = (1016)(150)/(144)(1000) = 1.058 \text{ kips/ft}$$
$$L = (100)(8)/(1000) = .800 \text{ kips/ft}$$
$$D + L = 1.858 \text{ kips/ft}$$

$$V_{(D+L)} \text{ at support} = (1.858)(30) = 55.74 \text{ kips}$$

$$M_D \text{ at center of span} = (1.058)(60)^2/8 = 476 \text{ ft-kips}$$

$$M_{(D+L)} \text{ at center of span} = (1.858)(60)^2/8 = 836 \text{ ft-kips}$$

Design loads, moments and shares

$$w_d = (1.058)(1.4) = 1.481 \text{ kips/ft}$$

$$w_l = (0.800)(1.7) = 1.360 \text{ kips/ft}$$

$$w_d + w_l = 2.841 \text{ kips/ft}$$

$$V_u = (\text{at support}) = (2.841)(30) = 85.23 \text{ kips}$$

$$M_u = (\text{at center of span}) = (2.841)(60)^2/8 = 1278 \text{ ft-kips}$$

The behavior of the beam under service conditions will be determined by calculating the flexural stresses due to service dead load and the prestressing force that will be acting immediately after transfer and before prestress losses. This condition will produce the maximum compression stress in the concrete in the bottom of the beam and the maximum tension stress in the concrete in the top of the beam, if any. The flexural stresses due to service dead plus service live load and the prestressing force that will be acting after all prestress losses will also be calculated to determine the maximum compression stress in the concrete in the top of the beam and the maximum tension stress in the concrete in the bottom of the beam. The stress in the steel at transfer will be assumed equal to the maximum value of 0.7 f_{pu} (Section 18.5.2) and the effective prestress force after losses will be based on an effective stress of 0.6 f_{pu}. The locations of the prestressing strands and the reinforcing bars are shown in Figs. 14-9 and 14-10.

Behavior at service loads

Stresses at jacking with service dead load (refer to Fig. 14-11):

$$f_{ci}' = 4{,}000 \text{ psi}$$

$$f_{pu} = 270 \text{ ksi}$$

$$M_D = 476 \text{ ft kips}$$

$$A_{ps} = (13)\ 1/2'' \text{ round } 270\ k \text{ strands} = (13)(0.153) = 1.989 \text{ in.}^2$$

$$P_s = 0.70\ f_{pu}A_{ps} = (0.70)(270)(1.989) = 375.9 \text{ kips}$$

where P_s = the prestressing force at jacking end. Summing moments

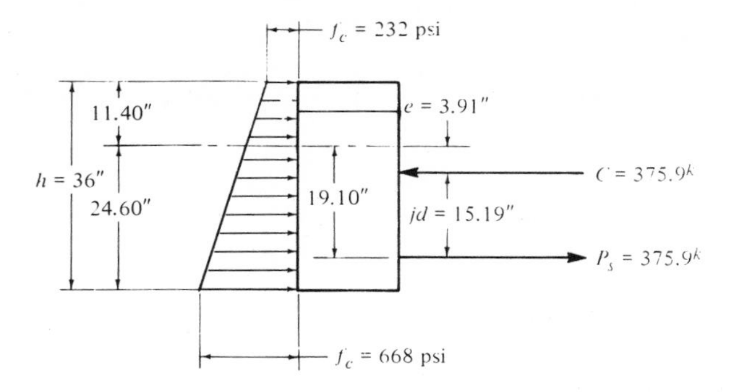

Fig. 14–11 Service load stresses at Transfer.

about the centroid of the prestressing force

$$Cjd = M_D$$

and

$$jd = M_D/C = (476)(12)/375.9 = 15.19 \text{ in.}$$

letting e equal the eccentricity of the compression force about the centroid of the gross concrete section,

$$e = 19.10 - 15.19 = 3.91 \text{ in. and}$$

$$f_c(\text{top fiber}) = -\frac{P_s}{A_c} + \frac{P_s ey}{I_g}$$

$$= -\frac{375.9}{1016} + \frac{(375.9)(3.91)(11.40)}{121{,}330}$$

$$= -0.375 + 0.138$$

$$= -0.232 \text{ ksi compression OK} < 0.60 f_{ci}' = 2.40 \text{ ksi}$$

$$f_c(\text{bottom fiber}) = -\frac{P_s}{A_c} - \frac{P_s ey_t}{I_g}$$

$$= -\frac{375.9}{1016} - \frac{(375.9)(3.91)(24.60)}{121{,}330}$$

$$= -0.370 - 0.298$$

$$= -0.668 \text{ ksi compression OK} < -2.40 \text{ ksi}$$

Service load stresses after losses (refer to Fig. 14-12):

$$f_c' = 6{,}000 \text{ psi}$$

$$f_{pu} = 270 \text{ ksi}$$

$$M_{(D+L)} = 836 \text{ ft-kips}$$

$$A_{ps} = 1.989 \text{ in.}^2$$

$$P_s = 0.60\, f_{pu}A_{ps} = (0.60)(270)(1.989) = 322.2 \text{ kips}$$

Summing moments about the centroid of the prestressing force

$$Cjd = M_{(D+L)}$$

and

$$jd = M_{(D+L)}/C = (836)(12)/322.2 = 31.14 \text{ in.}$$

letting e equal the eccentricity of the compression force about the centroid of the gross concrete section

$$e = 31.14 - 19.10 = 12.04 \text{ in. and}$$

$$f_c(\text{top fiber}) = -\frac{P_s}{A_c} - \frac{P_s e y}{I_g}$$

$$= -\frac{322.2}{1016} - \frac{(322.2)(12.04)(11.40)}{121{,}330}$$

$$= -0.317 - 0.364$$

$$= -0.681 \text{ ksi compression OK} < 0.45 f_c' = -2.70 \text{ ksi}$$

$$f_t(\text{bottom fiber}) = -\frac{P_s}{A_c} + \frac{P_s e y_t}{I_g}$$

$$= -\frac{322.2}{1016} + \frac{(322.2)(12.04)(24.60)}{121{,}330}$$

$$= -0.317 + 0.787$$

$$= +0.470 \text{ ksi tension OK} \approx 6\sqrt{f_c'} = 0.465 \text{ ksi}$$

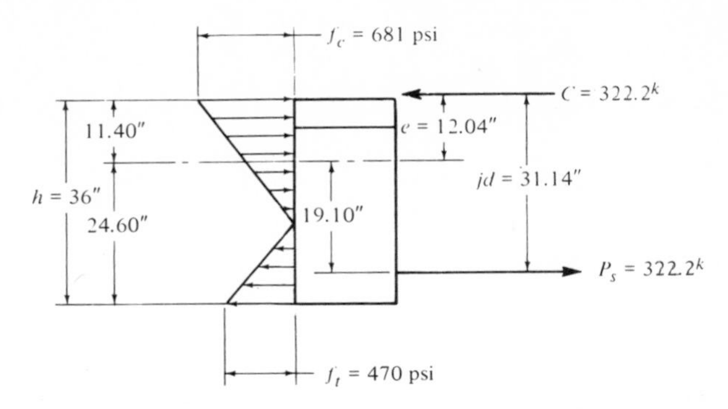

Fig. 14–12 Service load stresses after losses.

It is obvious from an examination of Fig. 14-12 that the tensile stress of 470 psi in the concrete in the bottom of the beam under service dead load and service live load after losses is less than the modulus of rupture of $7.5\sqrt{f_c'} = 7.5\sqrt{6{,}000} = 581$ psi (Section 9.5.2.2). This means that the beam will not crack under service dead and live loads. A service load equal to dead load plus 115 percent service live load will be required to produce a tension stress equal to the modulus of rupture and cause cracking due to flexure. The absence of cracks in any beam will result in a member with less deflection and one that is more corrosion resistant than if cracks were present.

If the tension stress in the concrete due to flexure caused by service loads does not exceed $6\sqrt{f_c'}$, the instantaneous deflection of a prestressed concrete member can be based on the modulus of elasticity of the concrete (Section 8.3.1) and on the gross amount of inertia of the concrete (Section 9.5.4.1). A tension stress greater than $6\sqrt{f_c'}$ but less than $12\sqrt{f_c'}$ is permitted only if it can be shown that the immediate and long-term deflections based on the transformed cracked moment of inertia and on bilinear moment deflection relationships will conform to Code requirements (Section 18.4.3).

The beam of our example will not crack under service loads because the maximum tension stress is less than the modulus of rupture ($470 < 581$) and the immediate long-term deflection can be approximated as follows:

$$E_c = 57{,}000\sqrt{f_c'} \quad \text{(Section 8.3.1)}$$

$$= 57{,}000\sqrt{6{,}000} = 4{,}415{,}220$$

$$\Delta(\text{Gravity loads}) = \frac{5Wl^3}{384E_cI_g}$$

$$\Delta(\text{Prestress force}) = \frac{5P_sel^2}{48E_cI_g}$$

$$\Delta(\text{Service dead load only}) = \frac{(5)(1058)(60)(720)^3}{(384)(4{,}415{,}220)(121{,}330)} = 0.576 \text{ in.} \downarrow$$

$$\Delta(\text{Prestress before losses}) = \frac{(5)(375{,}900)(19.1)(720)^2}{(48)(4{,}415{,}220)(121{,}330)} = 0.724 \text{ in.} \uparrow$$

Δ(Inst. service dead load with prestress before losses $= 0.148$ in. $\uparrow$

Δ(Prestress losses @ 15 percent) $= (0.15)(0.724)$ $= 0.109$ in. $\downarrow$

Δ(Service dead load with prestress after losses) $= 0.039$ in. $\uparrow$

Δ(Creep and shrinkage of concrete) $= (2)(0.39)$ $= 0.078$ in.* $\uparrow$

Δ(Long-time service dead load with prestress after losses) $= 0.117$ in. $\uparrow$

Δ(Inst. service live load) $= (0.576)(800)/(1058)$ $= 0.436$ in. $\downarrow$

The example beam is a floor member that is not connected to and does not support any nonstructural elements that would be damaged by large deflections. The maximum instantaneous service live load deflection permitted for such a member (Code Table 9.5(b); Section 9.5.4.3) is 1/480 of the span or 1.50 in. This is much greater than the calculated instantaneous live load deflection of 0.436 in. The example beam is therefore satisfactory under the criteria of the Code for behavior under service loads.

Flexural Strength

The flexural strength necessary for the design load ($1.4\ D + 1.7\ L$) must be computed in accordance with the strength design methods (Section 10.2 and 18.7). The theoretical design strength moment as shown in Fig. 14-13 is equal to the internal design couple $\overline{T}jd$, where d is the distance to the combined centroid of the prestressing steel and the nonprestressed steel and $\overline{T}$ is equal to the tension force in the prestressed and nonprestressed reinforcement. This tension force is equal to the area of the prestressing steel multiplied by the calculated stress in the prestressing steel at design load ($A_{ps}\ f_{ps}$) (Sections 18.7 and 18.7.1) plus the area of the nonprestressed reinforcement times its yield strength (A_sf_y)(Section 18.7.2).

---

* The additional long-time deflection due to creep and shrinkage (Section 9.5.4.2) has been assumed equal to twice the instantaneous deflection due to service dead load with prestress after losses.

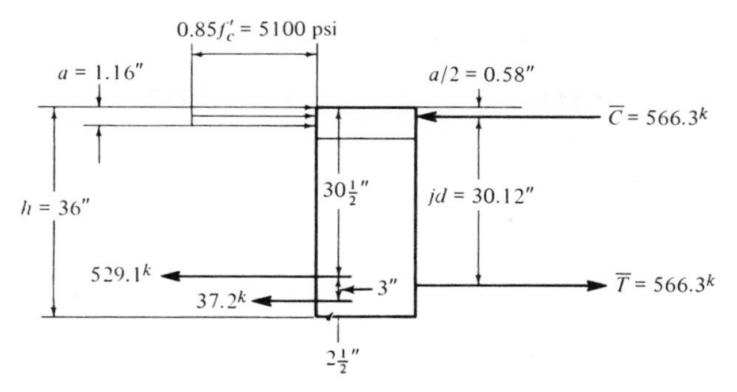

Fig. 14–13 Theoretical moment strength.

$$\overline{T} = (A_{ps}f_{ps}) + A_sf_y)$$

In order to determine $\overline{T}$ and then jd the following steps can be taken:

Step 1. Assume neutral axis will fall in flange.

Step 2. Assume $f_{ps} = 266$ ksi (Fig. 14-2).

Step 3. Determine $\overline{T}$ and d.

14-1/2 in. round strands

14-1/2 in. round strands	$(1.989)(266) =$	$529.1 \times 30.5 =$	16,140
two #5	$(0.62)(60) =$	$37.2 \times 33.5 =$	1246
	$\overline{T} =$	566.3	17,386

$d = (17{,}386)/566.3 = 30.70$ in.

Step 4. Determine jd.

$$\overline{C} = \overline{T} = 0.85\ f_c'ba$$

$$a = \frac{\overline{T}}{0.85\ f_c'b} = \frac{566{,}300}{(0.85)(6000)(96)}$$

$$= 1.16 \text{ in.}$$

$$jd = d - a/2$$

$$= 30.70 - 1.16/2 = 30.70 - 0.58$$

$$= 30.12 \text{ in.}$$

Step 5. Determine percentage of prestressing steel

$$\rho = A_{ps}/bd = (1.989)/(96)(30.70)$$

$$= 0.000675$$

Step 6. Determine f_{ps}

$$f_{ps} = f_{pu}\left(1 - 0.5\,\frac{\rho_p f_{pu}}{f_c'}\right)$$

$$= (270)[1 - (0.5)(0.000675)(270)/(6)]$$

$$= (207)(0.985)$$

$$= 266 \text{ ksi which checks the assumed value in Step 2.}$$

Step 7. Calculate the design moment strength

$$M_u = \phi \overline{T} jd = (0.90)(566.3)(30.12)/(12)$$

$$= 1279 \text{ ft-kips OK} > \text{design moment of 1278 ft-kips}$$

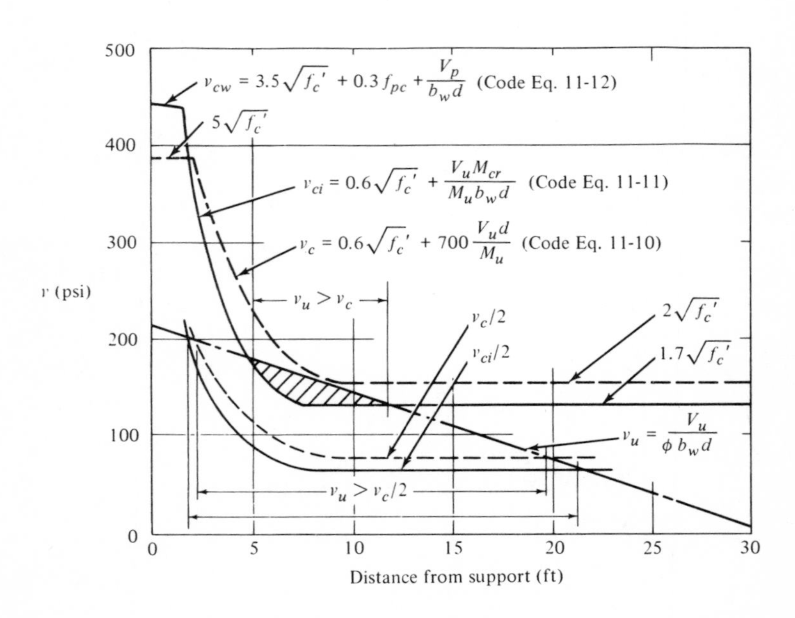

Fig. 14–14 Shear stresses.

Shear Reinforcement

The maximum value of the design shear to be carried by the shear reinforcement ($v_u - v_c$) for a uniformly loaded simple span beam is limited to a maximum value of $8\sqrt{f_s'}$ (Section 11.6.4). It will usually occur between one-eighth and one-quarter of the span from the support for a uniformly loaded simple span beam.

Sample calculations for the shear stresses will be made at one-eighth of the span or 7′–6″ from the support. The results of these calculations and similar calculations for other points are shown in Fig. 14-14.

Calculations for the shear stresses at 7′–6″ from the support are as follows:

$$V_u = 85.2 - (2.841)(7.5) = 85.2 - 21.3 = 63.9 \text{ kips}$$

$v_u = V_u/\phi b_w d = 63.9/(0.85)(16)(0.80)(36) = 163$ psi (In calculating v_u, d need not be taken less than $0.80h$ – Section 11.2.1).

$$v_c \quad \text{(Code Eq. 11-10)} = 0.6\sqrt{f_c'} + 700\ V_u d/M_u$$

$$v_c = 0.6\sqrt{6000} + (700)(63.9)(19.75)/(560)(12)$$

$$= 46 + (700)(0.188) \text{ OK } V_u d/M_u < 1.0 \text{ (Section 11.5.1)}$$

$$= 46 + 132$$

$$= 178 \text{ psi}$$

In calculating v_c, d is taken as the distance from the extreme compression fiber to the centroid of the prestressing tendons (Section 11.5.1).

It is obvious that v_c, the shear stress carried by the concrete is greater than v_u, the average or nominal shear stress. However, v_u is greater than $v_c/2$ (Section 11.1-d), and minimum stirrups are required. The stirrups can be calculated by the less conservative requirement (Code Eq. 11-2) (see Fig. 14-5).

$$A_v \text{ (min)} = \frac{A_{ps} f_{pu} s\sqrt{d}}{80\ f_y d\sqrt{b_w}} = \frac{(1.989)(270)(24)\sqrt{(0.80)(36)}}{(80)(60)(0.80)(36)(\sqrt{16})}$$

$$= 0.125 \text{ in.}^2$$

#3 ⅂⅃⌈ stirrups at a maximum spacing of 24 in. are more than adequate for minimum shear reinforcement area. Fig. 14-14 indicates that minimum stirrups are required from 2′–3″ from the support to a location 19′–9″ from the support (Code Eq. 11-10). #3 ⊔⊓ stirrups spaced one @ 12 in. and nine @ 24 in. from each support will meet this requirement.

More than minimum stirrups would have been required for the length of the span shown shaded in Fig. 14-14 (5′–0″ from the support to 11′–9″ from the support), if v_c had been calculated as the lesser value v_{ci} or v_{cw} (Section 11.5.2). Calculations for v_{ci} and v_{cw} at 7′–6″ from the support are as follows:

e (eccentricity of prestressing force about the centroid of the concrete section at a distance 7′-6″ from the support) =8.35 in.

f_{pe} (compressive force in the concrete at the bottom of the beam due only to the prestress force after losses)

$$f_{pe} = P_s/A_c + P_s e y_t/I_g = (322{,}200)/(1016) + (322{,}200(8.35)(24.60)/(121{,}330)$$

$$= 317 + 545$$

$$= 862 \text{ psi}$$

f_d (tension stress in bottom of beam due to dead load)

$$f_d = M_D y_t/I_g = (209{,}000)(12)(24.60/121330)$$

$$= 510 \text{ psi}$$

$$M_{cr} = \text{(cracking moment)} = \frac{I_g}{y_t}\,[6\sqrt{f_c'} + f_{pe} - f_d] \quad \text{(Section 11.5.2)}$$

$$M_{cr} = \frac{(121{,}330)}{(24.60)}\left[\frac{(6\sqrt{6000}}{(12{,}000)} + 862 - 510\right]$$

$$= 336 \text{ ft-kips}$$

v_{ci} = (shear stress at diagonal cracking due to design loads when such cracking is the result of combined shear and moment)

v_{ci} = (Code equation 11-11)

$$= 0.6\sqrt{f_c'} + \frac{V_d + V_l M_{cr}/M_{\max}}{b_w d}$$

$$= 0.6\sqrt{f_c'} + \frac{V_u M_{cr}}{M_u b_w d}$$

$$= 0.6\sqrt{6{,}000} + \frac{(63.9)(336{,}000)}{(560)(16)(0.80)(36)}$$

$$= 47 + 83$$

$$= 130 \text{ psi.}$$

f_{pc} = (compressive stress in concrete at neutral axis due to dead and live load after prestress losses)

$$f_{pc} = P_s/A_c = 322{,}200/1016$$

$$= 317 \text{ psi}$$

V_p = (vertical component of prestress force after losses)

$V_p = P_s$ (Slope of cable profile)

$$= (322{,}200)\,\frac{(19.1)(2)(22.5)}{(360)(30)}$$

$$= (322{,}200)(0.0796)$$

$$= 25{,}600 \text{ lb}$$

v_{cw} = (shear stress at diagonal cracking due to design loads, when such cracking is the result of principal tensile stress in the web)

v_{cw} = (Code Eq. 11-12)

$$v_{cw} = 3.5\sqrt{f_c'} + 0.3\, f_{pc} + V_p/b_w d$$

$$= 3.5\sqrt{6000} + (0.3)(317) + (25{,}600)/(16)(0.80)(36)$$

$$= 271 + 95 + 56$$

$$= 422 \text{ psi}$$

v_c = the smaller of v_{ci} or v_{cw}

$$= v_{ci}$$

= 130 psi (refer to Fig. 14-14.)

It is evident from the calculations that the shear carried by the concrete as v_{ci} or v_{cw} (Section 11.5.2) is conservative for the example beam. Stirrups based on v_c are more economical and utilize simpler manual calculation methods for the example beam (Section 11.5.1). From Fig. 14-14 it is also evident for the example beam that more than minimum stirrups are required when v_c is calculated as v_{ci} or v_{cw} (Section 11.5.2).

15 structural lightweight aggregate concrete

Unless special exceptions are made, the entire Code is applicable to lightweight aggregate concrete. The principal exceptions and special provisions for lightweight aggregate concrete are discussed and illustrated in the following section.

Lightweight concrete is defined for purposes of this book (and implicitly in the Code) as *structural* concrete containing lightweight aggregate and weighing less than 115 pcf (Section 2.1). Lightweight aggregate is defined and limited to aggregates conforming to ASTM C330 (Section 3.3.1), and having a dry, loose weight of 70 pcf or less (Section 2.1). Foamed lightweight concrete or lightweight concrete containing styrofoam aggregate suitable principally for insulation would not conform to the Code definition of structural lightweight aggregate concrete and would require special approval for proposed use under the Code as reinforced concrete (Section 1.4). Lightweight concrete in which only the coarse aggregate is lightweight is termed "sand-lightweight" concrete; where both coarse and fine aggregates are lightweight, "all-lightweight" concrete (Section 2.1).

Proportioning of lightweight aggregate concretes without trial batches or field experience data is not permitted (Section 4.2.4, Table 4.2.4). When lightweight concrete is expected to be subject to freezing and thawing wet, it must be proportioned for a specified strength, f_c', at least 3,000 psi (Section 4.2.5); for watertight concrete (fresh water), 3,750 psi; and for watertight concrete (sea water), 4,000 psi (Section 4.2.6).

It will have been noted that shear resistance and development (bond) are functions of $\sqrt{f_c'}$ to which the pure tensile strength of concrete is considered proportional. Since the tensile strength of lightweight concrete utilizing certain acceptable lightweight aggregates is substantially below that for normal weight concretes of equal compressive strength, reductions for shear resistance (Section 11.3) and development (Section 12.5-c) are provided. Alternate procedures permit higher values to be used, up to the value for normal weight concretes when the "splitting strength," f_{ct} of the lightweight aggregate concrete to be used is known (Section 4.2.9). Such tests are *not* intended to be used as routine acceptance tests on field concrete (Section 4.2.9.1).

Aside from the simply handled lesser weight and lower tensile strengths, the most important property of lightweight aggregate concrete differentiating its design from normal weight concrete is its lower modulus of elasticity (Section 8.3.1). Note that E_c is proportional to $w^{1.5}$, where w is the weight in pcf. The modular ratio, $n = E_s/E_c$ by definition. For design with unity load and capacity reduction factors using the straight-line theory of stress and strain in flexure, the Code provides that, "except in calculations for deflection . . . n for light weight concrete shall be assumed to be the same as (that) for normal weight concrete of the same strength" (Section 8.10.1.4). For normal weight concretes, E_c may be considered as 57,000 $\sqrt{f_c'}$ (Section 8.3.1).

Under design utilizing load factors and capacity reduction (ϕ) factors, the term n is not employed (Section 10.2, Assumptions) in strength design. The computation of deflections is required to follow usual formulas for elastic deflections, using the modulus of elasticity as defined for both normal and lightweight concretes (Section 9.5.2.2). $E_c = w^{1.5}\, 33\sqrt{f_c'}$ in psi; or, for normal weight concretes only, $E_c = 57{,}000\sqrt{f_c'}$ (Section 8.3.1). The effective moment of inertia for a reinforced concrete section is required to be taken as

$$I_{\text{eff}} = \left(\frac{M_{cr}}{M_a}\right)^3 I_g + \left[1 - \left(\frac{M_{cr}}{M_a}\right)^3\right] I_{cr} \quad \text{(Code Eq. 9-4)}$$

where

$$M_{cr} = \frac{f_r I_g}{y_t} \quad \text{(Section 9.5.2.2)}$$

in which when f_{ct} is not specified

$$f_r = 7.5\sqrt{f_c'} \text{ for normal weight concrete}$$

$$f_r = 5.63\sqrt{f_c'} \text{ for all-lightweight concrete}$$

$$f_r = 6.39\sqrt{f_c'} \text{ for sand-lightweight concrete}$$

Due to the use of the lower value for E_c, lightweight concrete will have a higher moment of inertia than normal weight concrete for the cracked section transformed to concrete, I_{cr} (see Fig. 15-1 comparing I_{cr} for lightweight to normal weight concrete). Thus, for low ratios of M_{cr}/M_a (heavily reinforced sections), I_e will be larger for lightweight concrete, offsetting

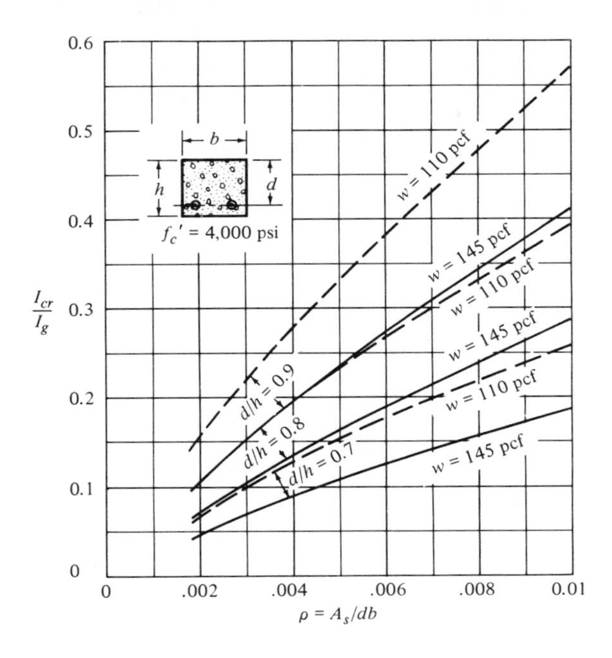

Fig. 15–1 I_{cr} for normal vs lightweight concrete.

considerably the effect of its lower value of E_c in the deflection equations.

EXAMPLE. Compute relative deflections for a reinforced concrete slab in which $f_c' = 4{,}000$ psi; $f_y = 60{,}000$ psi; $h = 5$ in.; $d = 4$ in.; $\rho = 1.0\%$; $d/h = 0.8$; (1) for a normal weight concrete, $w = 145$ pcf and (2) for a lightweight aggregate concrete $w = 110$ pcf.

(1) $E_c = 3.644 \times 10^6$ psi; $\quad n = 8 \quad I_g = \dfrac{(12)(5)^3}{12} = 125$ in.^{4}/ft

From Fig. 15-1, $I_{cr}/I_g = 0.293$. $I_{cr} = (0.293)(125) = 36.6$ in.^{4}/ft

$$M_{cr} = \frac{(7.5)(63.2)(125)}{(2.5)} = 23{,}700 \text{ in.-lb/ft}$$

$$M_u = \phi f_y A_s(d - a/2) \qquad a = \frac{(60)}{(0.85)(4)(12)} = 1.47\ A_s$$

$$= (0.90)(60)(0.48)(4 - 0.35) \qquad = 1.47 \times 0.48 = 0.7 \text{ in.}$$

$$= 94{,}500 \text{ in. lb/ft} \qquad 1/2\ a = 0.35 \text{ in.}$$

Use average load factor for $D + L = 1.6$

$$M_a = \frac{94{,}500}{1.6} = 59{,}000 \text{ in.-lb/ft}$$

$$(M_{cr}/M_a)^3 = (23.7/59.0)^3 = (0.4)^3 = 0.064$$

$$I_e = (0.064)(125) + (0.936)(36.6) = 42.2 \text{ in.}^4\text{/ft}$$

Immediate live load deflection,

$$\Delta_L \approx \frac{100}{E_c I_e} = \frac{100}{(3.644)(42.2)} = 0.65$$

from Fig. 14-1, $I_{cr}/I_g = 0.396$. $I_{cr} = (0.396)(125) = 49.5$ in.^{4}/ft

Since $f_r = 0.75\ f_r$ for normal weight concrete, $\frac{M_{cr}}{M_a} = (0.75)(0.4)$

$$(M_{cr}/M_a)^3 = (0.3)^3 = 0.027$$

$$I_{\text{eff}} = (0.027)(125) + (0.973)(49.5) = 51.6 \text{ in.}^4\text{/ft}$$

Immediate live load deflection,

$$\Delta_L \approx \frac{100}{E_c I_e} = \frac{100}{(2.407)(51.6)} = 0.81$$

To compare relative total deflections, let the ratio of live to dead loads, $L/D = 1/2$ for the lightweight concrete slab, $L = 1.0$; $D = 2.0$; $L + D = 3$; the total immediate deflection is proportional to $\frac{L + D}{E_c I_e}$. Total deflection, $\Delta \approx (0.81)(3) = 2.43$. For the normal weight concrete, the dead load, $D = (2)(145/110) = 2.64$ $D + L = 3.64$. Total deflection, $\Delta \approx (0.65)(3.64) = 2.37$.

(2) $w = 110$ pcf, $E_c = 2.407 \times 10^6$ psi

For the usual lightly reinforced slab, where the ratio of the cracking moment, M_{cr}, to the maximum applied (service load) moment, M_a, approaches or exceeds unity, I_e will become I_g. In this case the relative deflections of the lightweight and normal weight concretes will be computed as proportional to their respective moduli of elasticity. When (Code) Eq. 9-4 gives a value for I_e larger than that of I_g, the requirement that I_e shall not be taken as larger than I_g controls (Section 9.5.2.2).

16 field inspection and construction

GENERAL

The Code is a document intended to become part of a law regulating all construction for the benefit of the public. Thus, there are frequent references outlining obligations (which could be undertaken by the Structural Engineer, or Architect, representing the Owner or the Contractor) for securing approval of the Building Official. The Code covers the entire subject of field inspection with a few short minimum requirements (Section 1.3). It states that inspection of concrete construction shall be performed by a competent person without definition (Section 1.3.1). The inspection shall be performed by an engineer or architect or under their direction. Minimum records which must be kept are prescribed. Since the Code is intended to regulate concrete construction for small and large projects, no requirements for continuous inspection, frequency of intermittent inspection, or inspection procedures can be given.

A full treatment of field inspection or construction is outside the scope of this book. In this chapter important or frequently overlooked code requirements concerning inspection and construction have been collected and are presented with brief explanations and practical precautions. The best references in detail to accepted inspection practice are the ACI 311-64 "Recommended Practice for Concrete Inspection," and ACI *Manual of Concrete Inspection,* SP-2; the authors strongly recommend them to anyone undertaking concrete inspection.

The actual design through construction is a complex procedure with many interdependent obligations among designer, material suppliers, contractors, subcontractors, inspectors acting for the Owner and for the Building Official. Very simply, each party concerned would like his responsibility to be limited; the design Engineer, to correct plans and specifications; the Contractor and subcontractors, to correct execution of same; material suppliers, to material meeting specifications and design dimensions; and inspectors to following the contract documents. When the Code is part of the the law regulating construction, and, as often the case, a part of the contract documents (with a statement that the term "Building Official" means

"Engineer"), it protects the public only. It does not define separate duties of each party involved except in general terms.

If defective form materials were supplied and accepted, put together defectively according to defective form design plans made by a subcontractor and approved by Contractor, design Engineer, and city inspector, and—after approvals by all for concreting—the forms failed, the Code does not assign responsibilities. If a completed structure is found unsafe for design loads, the public is protected by the provision that the Building Official may approve a lower load rating (Section 20.6) and further by standard general code provisions for condemning.

The structural Engineer, through plans and specifications which become "contract documents," establishes terms for material quality and construction procedures. With special provisions, he may require approvals prior to form design, reinforcing bar details and placing plans, construction sequences for casting, schedules for form removal, curing, etc. In general, the Code merely states performance requirements.

MATERIALS

Structural inspection begins with material tests. Compliance with ASTM specifications for concrete materials, aggregate, water, and cement is required (Section 3.1.2). Code requirements supplementing those of ASTM should not be overlooked (see Chapter 1).

FORMWORK

The Code requires that formwork be properly designed and prescribes design considerations[1] (Section 6.1). The contract may or may not require submission of formwork design for approvals by the Engineer. Even when such approval is required and secured, ordinarily the approval is for overall conception only. Where the contract requires approval of the construction before concreting, the inspection is usually limited to dimensions, proper surface preparation, cleanout, etc. For protection of all concerned, the inspector should always check camber, whether specified or not, and record same. Since all formwork design is related to rate and method of casting as well as temperatures, all under control of the Contractor, the Engineer's approval of formwork is not intended to guarantee final safety.

REINFORCEMENT AND PLACING

Ordinarily, the Engineer has required and approved details and placing plans before shipment. Field inspection of this phase should ensure proper placing. Bar supports as shown on approved placing plans should be adequate to carry the load of the bars and *necessary* foot traffic (Section 7.3.1). Usually auxiliary documents prescribing bar support spacings are specified.[2] Tack welding for assembly is prohibited, except by approval of the Engineer (Section 7.3.1). Ordinarily this approval will be given only for tack welds at points of low design stress in each crossing bar. The Code provides tolerances applied simultaneously to cover or effective depth, d (Section 7.3.2). The dimension d is the most structurally important dimension in the inspection. Other important points are proper position of column verticals in plan; end anchorages as detailed; and splices. Dimensions locating bend up points, cutoff points, spacing between slab, footing, or wall bars are relatively unimportant so long as the correct total number of the correct bars are used. $A \pm 2$ in. tolerance is prescribed for bends and ends of bars, except at discontinuous ends of members (Section 7.3.2.2).

Coatings, other than rust, on bars generally should be removed (Section 7.2.1). Rust is generally harmless since "loose" rust possibly detrimental to bond is by definition knocked off in handling. Where rust is considered excessive, the Code provides a specific test to determine if it is excessive (Section 7.2.2). A specimen, any length, is to be wire brushed by hand and subsequently checked for weight and height of deformations. If either is less than the minimum required by the relevant ASTM specification, the inspector may (1) reject the bars, or (2) require cleaning and permit use with sufficient additional material to make up any design deficiencies.

As a practical matter, perhaps the most critical points for inspection of bar placing are (1) dowels for staggered butt splices in column verticals (see Chapter 10) and (2) the first floor reinforcing in multistory projects repeating same. In regard to correcting mislocated dowels, dowels out of plumb, etc., it should be noted that the Code requires bars to be bent cold, and forbids field bending of embedded bars, except as permitted by the Engineer (Section 7.1.4).

PLACING CONCRETE[3]

The Code requires clean forms, clear of ice or debris (Section 5.1.1). Concrete during placing should not be permitted to segregate (Sections 5.3.1, 5.3.2, and 5.4.1). The inspector should not only take (or supervise taking) samples for required cylinder tests, but should observe slump to avoid use of unnecessarily high slump concrete and avoid segregation. If the contract documents or accepted trial batch proportions are based on limited slump, these provisions, of course, should be enforced. Honeycomb is a structural deficiency due to segregation in casting or form leakage with high slump concrete. To prevent it at the base of deep (wall or column) forms, the Code requires placing mortar to a depth of at least 1 in. (Section 5.4.4). This mortar consists simply of the same mix used for concrete with the coarse aggregate omitted and water reduced if necessary.

The location of planned construction joints establishes extent of a single casting. Acceptable locations for joints not indicated on the plans can be specified. Otherwise, casting should be a continuous operation till completion of a panel, section, or element. Generally, joints may be allowed near the center of flexural members, except that joints in a girder should be offset two beam widths from a beam at the center (Section 6.4.3.). Unless a construction joint is located at the top of walls or columns (best practice) a delay in casting (usually 1 to 4 hours) until the concrete in the wall or column takes an initial set is required (Section 6.5.4. Note that column capitals, brackets, beams, girders, haunches, etc. are to be "considered part of the floor system and shall be placed monolithically therewith" (Section 6.4.2).

CURING

Minimum requirements only are given (Section 5.5.1). For all cast-in-place concrete, additional curing is to be expected, since curing is not a formal process ending in 28 days. The reference to high early strength *concrete* (not defined) is usually interpreted as concrete made with high-early strength cement, or equivalent concrete made so by accelerating admixtures or additional cement, generally resulting in achievement of 28 day f_c' in 7 days under laboratory curing. Extremes of temperatures, hot or cold, require special precautions (Section 5.6 and 5.7) including but not limited to curing.[4]

FORM REMOVALS

A number of studies have shown that, in many types of concrete structures lightly loaded in use, the most critical combinations of load-strength occur during construction. Two contradictory considerations are involved. The Contractor is held responsible for demonstrating that the structure is adequately *strong* to support construction loads (Section 6.2.1). Adequate safety (strength) does not insure adequate *stiffness* to prevent permanent deflections considered excessive by the Engineer (or owner). Many Engineers specify arbitrary minimum periods before form removal or reshoring for an arbitrary number of floors or days. The Code supplies performance requirement only, and does not contain any arbitrary time limits which add to cost, except that vertical forms and horizontal forms separate from shoring must remain in place 24 hours or till concrete is strong enough not to be damaged by form removal (Section 6.2.2).

One suggestion that may be helpful to the Engineer in his decision to permit construction loads and in assessing the behavior of proposed shoring or reshoring[1] is to require modulus of elasticity tests (stress-strain) curves for job-cured test specimens, beams or cylinders, as well as strength tests. Such tests will be found also especially valuable for long span shell or folded plate structures where large deflections are anticipated (Section 19.7).[1]

References

1. ACI 347-68; also *Formwork for Concrete,* ACI SP-4.
2. ACI 315 *Manual of Standard Practice for Detailing Reinforced Concrete Structures.* CRSI *Recommended Practice for Placing Reinforcing Bars.*
3. ACI 614-59; *Recommended Practice for Measuring, Mixing, and Placing Concrete.*
4. ACI 605-59; *Recommended Practice for Hot Weather Concreting.*
   ACI 306-66, *Recommended Practice for Cold Weather Concreting.*

17 strength evaluation of existing structures

GENERAL

In outline form, the 1971 Code provides for a comprehensive evaluation of existing structures (Chapter 20). Existing structures include any completed or partially completed construction the safety of which is in doubt, as well as old structures for which the present safety is unknown. In this section, the practical application of Code provisions is briefly explained with some practical precautions, but for an in-depth study, the ACI Committee 437 report* is recommended.

Previous codes provided merely the magnitudes for test loads, term of test loading, and criteria for evaluating flexural load tests. The 1971 Code provides for analytical evaluation, load tests, or a combination of both (Section 20.1). The minimum age of construction for testing and terms of testing are specified (Sections 20.3 and 20.4). Analysis is specifically recommended for evaluation of members other than flexural (Section 20.5), since load tests for footings, columns, walls, etc. are seldom feasible. The Code specifically permits the establishment of a safe load rating, based upon evaluation procedures outlined (Section 20.6). It is *not* intended for use as an acceptance test procedure for new materials or new design or construction methods (Section 1.4), for acceptance of precast elements, nor to resolve controversies in regard to contractural obligations for construction quality (Commentary, Chapter 20).

PROCEDURE

When the safe load rating is doubtful, a five-step procedure is implied:

1. Identify the most doubtful elements or areas.

* "Strength Evaluation of Existing Concrete Buildings," *ACI Journal,* Nov. 1967. *ACI Proc.,* V (64): 705-710.

2. Determine the stress condition of the questionable elements considered critical to the doubtful strength or stiffness of the element.
3. Determine the combination of design live loads and the pattern of loading required in design which will produce the maximum critical stress.
4. Compute the actual amount and extent of test load required, taking into account the live load reductions permitted by both the ACI Code and the general code and taking care to avoid overload of the elements supporting the element to be tested.
5. Execution of the load test (or analysis) and determination of the safe load rating.

DOUBTFUL CONSTRUCTION

Doubtful strength of completed construction may concern the concrete or the steel. The quality of the concrete may be questioned for one or more of the following reasons, among others: (1) low 28-day acceptance test strengths, (2) unanticipated curing and temperature conditions, including freezing or plastic shrinkage (drying) cracking, (3) visibly poor consolidation (honeycomb), and (4) any failure to harden as expected, as evidenced by surface abrasion or impact tests. The effectiveness of the steel, tensile or compressive capacity, can be questioned because of one or more of the following reasons among others: (1) low tests on field samples, (2) concrete strengths less than f_c' (3) use of a steel grade *lower* than specified (higher grade is no problem, obviously), (4) misplaced rebars due to wrong size bar supports, top bars trodden down during concreting traffic, etc., and (5) discovery of overlooked bundles marked for incorporation into elements cast.

The elements or area of elements containing suspected low-strength concrete or reduced steel capacity can usually be located from construction records. At this point all possible methods of nondestructive testing should be employed. These test results may dispel the doubt of quality, or verify the doubt and aid in locating the most doubtful portion, and will provide data on actual conditions for necessary analytical evaluations.

ANALYSIS

Even though the load test may be required and its results will then comprise the final acceptance or the basis for a lower load rating, an analysis is required before the testing. The analysis will determine whether the cause of doubtful strength will critically affect crack widths; capacity in shear, negative moment, positive moment; or deflection (low E_c). Further analysis is necessary to determine the critical required load combination (Sections 9-3, Eq. 9-1, 9-2, and 9-3) and the required pattern of loading to produce the maximum stress on the resisting capacity considered critical (Section 20.3.3).

TEST LOADS

At this point it is necessary to consider the different design loads used for slabs or joist-slabs and the supporting elements, beams or girders, where the loads and spans are such that the live load reductions of the general code are applicable. For example, if no live load reductions are allowed for slabs supported on beams, and slab reinforcement capacity is in question, the test load for the slabs must be applied so as not to "overtest" the beams.

Test Load Patterns, Measurements, and Criteria

If negative moment capacity or shear capacity of a flexural member is considered the critical property to be established by the load test, the Engineer, acting as the representative of the owner, cannot accept the Code limit on deflection only as a criterion of acceptability. The deflection criteria (Section 20.4.6) are intended to represent satisfactory performance under load tests for only the positive moment capacity for end or continuous spans. For simple spans or cantilevers, the deflection criteria can be accepted as evidence of satisfactory total performance. An obvious interpretation (which has been misapplied even in published load test reports) for the term "maximum deflection" (Section 20.4.6) is that it means "maximum deflection at the point which should deflect most under the loading pattern causing maximum deflection at that point." The term "maximum deflection" and the numerical criteria provided in the Code for same are significant only when so interpreted.

It is possible and practically desirable to utilize the deflection measurements during a load test intended to develop critical shear or negative moment (Section 20.3.3). The Code prescribes a combination of analysis and test (Section 20.1). The analysis prescribed (Section 20.2) should include an appropriate reduction factor to be applied to the maximum numerical limits for deflection (Section 20.4.6-a and -b) when the test load pattern is established to test the shear or negative moment capacity in continuous flexural members. See Fig. 17-1 for loading patterns on one-way flexural members.

For two-way construction the ideal loading pattern to test positive moment capacity for direct application of the deflection criteria extends at least two panels in each direction from the panel in which the test measurements are taken (see Fig. 17-2). Where practical considerations make such an extensive loading pattern impracticable, the maximum allowable deflection criterion should be reduced appropriately.

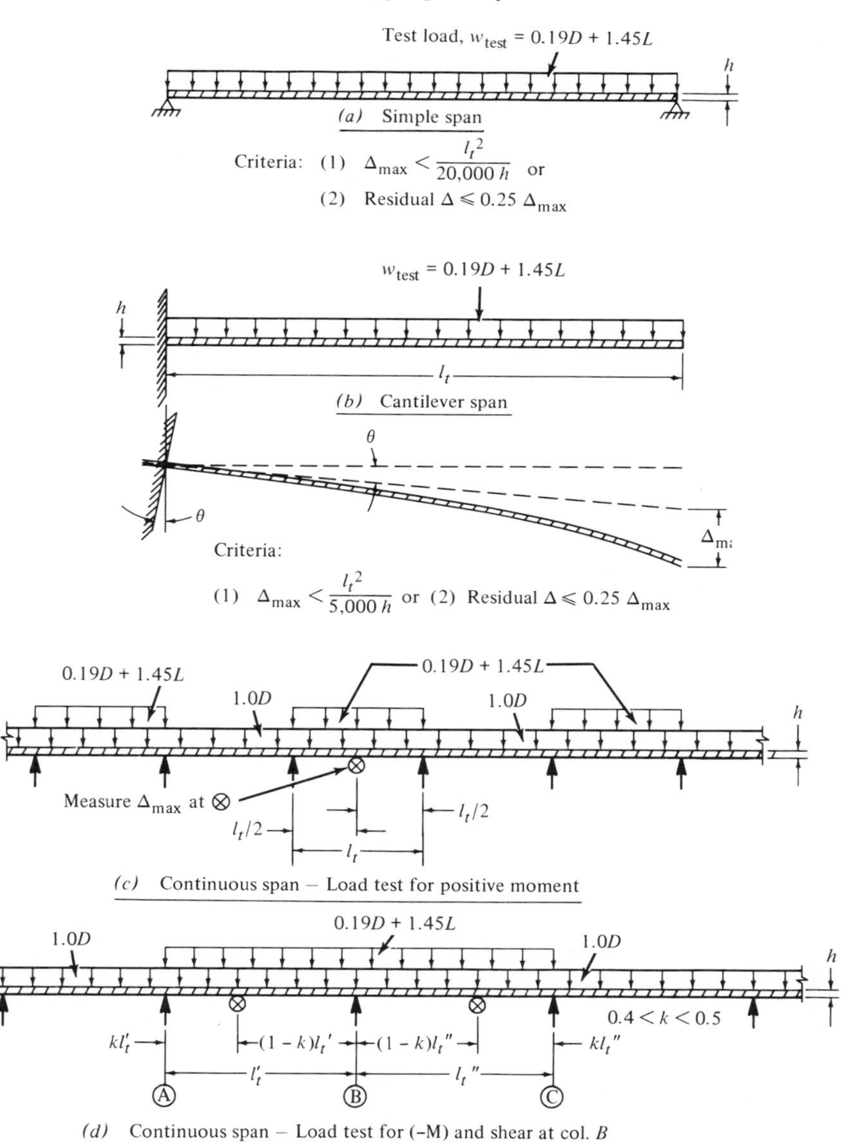

Fig. 17–1 One-way slabs—load tests.

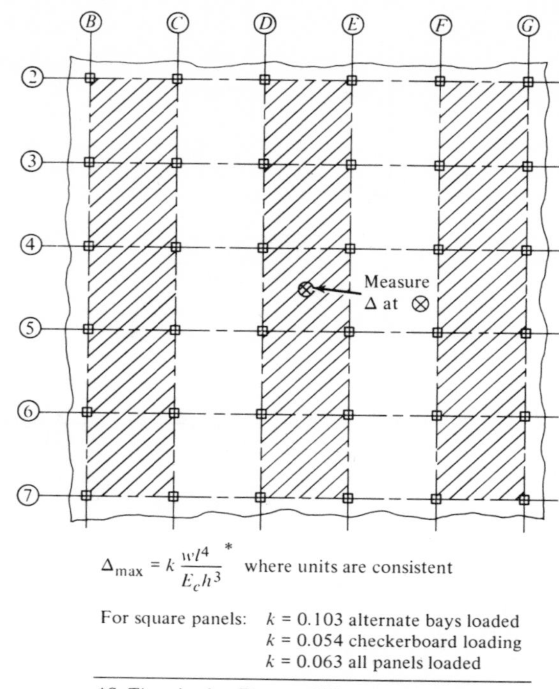

Fig. 17–2 Two-way slabs—loading for Δ_{max} and $(+M_{E\text{-}W})$.

Critical negative moment capacity and shear capacity in two-way construction is developed by a simpler loading pattern (see Fig. 17-3), but again the maximum allowable deflection criteria, if utilized, must be reduced appropriately.

EXAMPLE 1. For the construction shown in Fig. 17-4, assume that all tests indicate concrete strength in the panel *C-D*-3-4 substantially below f_c'. The live load prescribed in the general code for the floor is 100 psf. Compute the test loads applicable for load testing (1) slabs, (2) beams *B*-5, and (3) girder *G*-5. No dead loads other than the weight of the concrete floor are involved.

(1) *Slab.* Live load reduction allowed, $R = 0$. Test load = the total load (Section 20.4.3) minus the dead load already in place. Test load, $w_{test} = (0.85)\ (1.4\ D + 1.7\ L) - 1.0\ D = 0.19\ D + 1.45\ L$

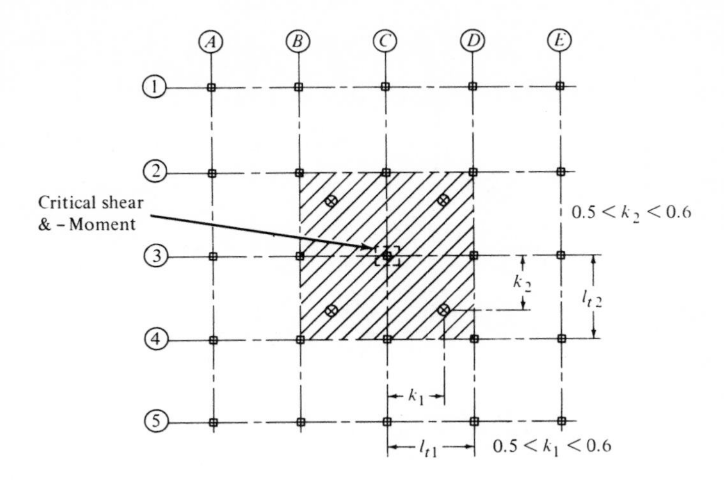

Fig. 17–3 Two-way slabs—loading for critical shear and—moment at col. C3.

(2) *Beam.* Live load reduction allowed by general codes, $R\ \% = 0.08\ A$. (see Fig. 2-1). $R\ \% = (0.08)\ (2 \times \frac{10)}{2}\ (30) = 24\%$ $\quad L\ (\text{test}) = 1.00\ L - 0.24\ L = 0.76\ L$. Test load, $w_{\text{test}} = (0.85)\ [(1.4\ \text{D}) + (1.7)\ (0.76\ \text{L})] - 1.0\text{D} = 0.19\text{D} + 1.10\text{L}$

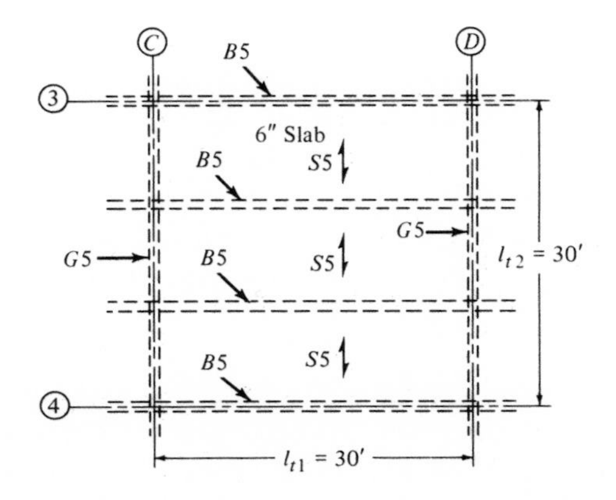

Fig. 17–4 One-way slab, beam, and girder.

(3) *Girder.* Live load reduction allowed, $R\ \% = (0.08\ A) \leqq 100$ Enter Fig. 2-1 with D/L = 0.75) read maximum R = 40.5% L (test) = $1.00\ L - 0.405\ L = 0.595\ L$. *Test load,* $w_{test} = (0.85)\ [(1.4\ D) + (1.7)\ (0.595\ L)] - 1.0\ D = 0.19\ D + 0.86\ L$

appendix 1

CODE NOTATION AND SYMBOLS

a = depth of equivalent rectangular stress block.
a = shear span, distance between concentrated load and face of support.
a = maximum deflection under test load of a member relative to a line joining the ends of the span, or of the free end of a cantilever relative to its support, in.
A = effective tension area of concrete surrounding the main tension reinforcing bars and having the same centroid as that reinforcement, divided by the number of bars, sq. in. When the main reinforcement consists of several bar sizes the number of bars shall be computed as the total steel area divided by the area of the largest bar used.
A = area of the cross section between the tension face and the center of gravity of the gross section (prestressed).
A_a = area of stirrups provided at termination of tension reinforcement, sq in.
A_b = area of an individual bar, sq in.
A_c = area of core of spirally reinforced column measured to the outside diameter of the spiral, sq in.
A_c = area of concrete at the (prestressed) cross section considered.
A_g = gross area of section sq in.
A_l = total area of longitudinal reinforcement to resist torsion, sq in.
A_{ps} = area of prestressed reinforcement in tension zone.
A_s = area of nonprestressed tension reinforcement, sq in.
A_s' = area of compression reinforcement, sq in.
A_t = area of one leg of a closed stirrup resisting torsion within a distance s, sq in.
A_v = area of shear reinforcement within a distance s, or area of shear reinforcement perpendicular to main reinforcement within a distance s for deep beams, sq in.
A_w = area of a deformed wire, sq in.
A_1 = loaded area.
A_2 = maximum area of the portion of the supporting surface that is geometrically similar to and concentric with the loaded area.
b = width of compression face of member.
b_o = periphery of critical section for slabs and footings.
b_w = web width.
B = width of center band across a rectangular two-way footing in the short direction for concentration of short reinforcement (width B = length of short side of footing). See B. Also called b.

c = distance from extreme compression fiber to neutral axis.
c_1 = size of rectangular or equivalent rectangular column, capital, or bracket measured in the direction in which moments are being determined.
c_2 = size of rectangular or equivalent rectangular column, capital, or bracket measured transverse to the direction in which moments are being determined.
C = cross-sectional constant to define the torsional properties.
C_m = a factor relating the actual moment diagram to an equivalent uniform moment diagram.
d = distance from extreme compression fiber to centroid of tension reinforcement, prestressing steel, or to combined centroid when nonprestressing tension reinforcement is included, in.
d' = distance from extreme compression fiber to centroid of compression reinforcement, in.
d_b = nominal diameter of bar, wire, or prestressing strand, in.
d_c = thickness of concrete cover measured from the extreme tension fiber to the center of the bar located closest thereto.
d_p = diameter of the pile at footing base, in.
d_s = distance from centroid of tension reinforcement to the tensile face of the member, in.
D = dead loads, or their related internal moments and forces.
e = eccentricity of design load parallel to axis measured from the centroid of the section. It may be calculated by conventional methods of frame analysis.
e = base of Napierian logarithms.
E = load effects of earthquake, or their related internal moments and forces.
E_c = modulus of elasticity of concrete, psi.
E_{cb} = modulus of elasticity for beam concrete.
E_{cc} = modulus of elasticity for column concrete.
E_{cs} = modulus of elasticity for slab concrete.
EI = flexural stiffness of compression members.
E_s = modulus of elasticity of steel, psi.
f_c' = specified compressive strength of concrete, psi.
$\sqrt{f_c'}$ = square root of specified compressive strength of concrete, psi.
f_{ci}' = compressive strength of concrete at time of initial prestress.
f_{ct} = average splitting tensile strength of lightweight aggregate concrete, psi.
f_d = stress due to dead load, at the extreme fiber of a section at which tensile stresses are caused by applied load, psi.
f_h = tensile stress developed by a standard hook, psi.
f_{pc} = compressive stress in the concrete, after all prestress losses have occurred, at the centroid of the cross section resisting the applied loads or at the junction of the web and flange when the centroid lies in the flange, psi. (In a composite member, f_{pc} will be the resultant compressive stress at the centroid of the composite section, or at the junction of the web and flange when the centroid lies within the flange,

due to both prestress and to bending moments resisted by the precast member acting alone.)

f_{pe} = compressive stress in concrete due to prestress only after all losses, at the extreme fiber of a section at which tensile stresses are caused by applied loads, psi.

f_{ps} = calculated stress in prestressing steel at design load, psi.

f_{pu} = ultimate strength of prestressing steel, psi.

f_{py} = specified yield strength of prestressing steel, psi.

f_r = modulus of rupture of concrete, psi.

f_s = calculated stress in reinforcement at service loads, ksi.

f_{se} = effective stress in prestressing steel, after losses, psi.

f_y = specified yield strength of nonprestressed reinforcement, psi.

F = lateral or vertical pressure of liquids, or their related internal moments and forces.

h = overall thickness of member, in.

h_v = total depth of shearhead cross section.

h_w = total height of wall from its base to its top.

H = lateral earth pressure, or its related internal moments and forces.

I = moment of inertia of section resisting externally applied design loads.

I_b = moment of inertia about centroidal axis of gross section of a beam as defined in Section 13.1.15.

I_c = moment of inertia of gross cross section of columns.

I_{cr} = moment of inertia of cracked section transformed to concrete.

I_e = effective moment of inertia for computation of deflection.

I_g = moment of inertia of gross concrete section about the centroidal axis, neglecting the reinforcement.

I_s = moment of inertia about centroidal axis of gross section of slab = $h^3/12$ times width of slab specified in definitions of α and β_t.

I_{se} = moment of inertia of reinforcement about the centroidal axis of the member cross section.

k = effective length factor for compression members.

k = fraction of effective depth in compression due to flexure by straight-line theory.

K = wobble friction coefficient per foot of prestressing steel.

K_b = flexural stiffness of beam; moment per unit rotation.

K_c = flexural stiffness of column; moment per unit rotation.

K_{ec} = flexural stiffness of an equivalent column; moment per unit rotation.

K_s = flexural stiffness of slab; moment per unit rotation.

K_t = torsional stiffness of torsional member; moment per unit rotation.

l = span length of beam or one-way slab, as defined in Section 8.5.2; clear projection of cantilever, in.

l = length of prestressing steel element from jacking end to any point x.

l_a = additional embedment length at support or at point of inflection, in.

l_c = height of column, center-to-center of floors or roof.

l_c = vertical distance between supports for empirical design of walls.

l_d = development length, in.

l_e = equivalent embedment length, in.

l_n = clear span for positive moment or shear and the average adjacent clear spans for negative moment.
l_n = length of clear span in long direction of two-way construction.
l_n = length of clear span, in the direction moments are being determined, measured face-to-face of supports.
l_t = span of member under load test (the shorter span of flat slabs and of slabs supported on four sides). The span, except as provided in Section 20.4.6(c) is the distance between the centers of the supports or the clear distance between supports plus the depth of the member, whichever is smaller, in.
l_u = unsupported length of compression member.
l_v = length of shearhead arm from centroid of concentrated load or reaction.
l_w = total lengths of wire extending beyond outermost cross wires, for each pair of spliced wires, in.
l_w = horizontal length of wall.
l_1 = length of span in the direction moments are being determined measured center-to-center of supports.
l_2 = length of span transverse to l_1, measured center-to-center of supports.
L = live loads, or their related internal moments and forces.
M_a = maximum moment in member at stage for which deflection is computed.
M_c = moment to be used for design of compression member.
M_{cr} = cracking moment.
M_m = modified bending moment.
M_{max} = maximum bending moment due to externally applied design loads.
M_o = total static design moment.
M_p = required full plastic moment of shearhead cross section.
M_t = theoretical moment strength, in.-lb, of a section = $A_s f_y(d - a/2)$.
M_u = applied design load moment at a section, in.-lb.
M_v = moment resistance contributed by shearhead reinforcement.
M_1 = smaller end moment on compression member from conventional elastic frame analysis, positive if bent in single curvature, negative if bent in double curvature.
M_2 = larger end moment on compression member calculated from conventional elastic frame analysis, always positive.
n = modular ratio = E_s/E_c.
n = number of pairs of cross wires in splice.
n = number of cross wires in anchorage zone of welded deformed wire fabric.
N_c = tensile force in the concrete under load of $D + 1.2$ L.
P_b = axial load capacity at simultaneous assumed ultimate strain of concrete and yielding of tension steel (balanced conditions).
P_c = critical load (buckling).
P_s = steel force at jacking end.
P_u = axial design load in compression member.
P_x = steel force at any point, x (prestressed).

r = radius of gyration of the cross section of a compression member.
s = tie spacing, in.
s = spacing of stirrups, in.
s_w = spacing of deformed wires, in.
T = cumulative effects of temperature, creep, shrinkage, and differential settlement.
T_u = design torsional moment.
U = required strength to resist design loads or their related internal moments and forces.
v_c = nominal permissible shear stress carried by concrete.
v_{ci} = shear stress at diagonal cracking due to all design loads, when such cracking is the result of combined shear and moment.
v_{cw} = shear stress at diagonal cracking due to all design loads, when such cracking is the result of excessive principle tensile stresses in the web.
v_{tc} = nominal permissible torsion stress carried by concrete.
v_{tu} = nominal total design torsion stress.
v_u = nominal total design shear stress.
v_d = shear force at section due to dead load.
v_l = shear force at section occurring simultaneously with M_{max}.
V_p = vertical component of the effective prestress force at the section considered.
V_u = total applied design shear force at section.
w = design load per unit length of beam or per unit area of slab.
w = weight of concrete, lb per cu ft.
w_d = design dead load per unit area.
w_l = design live load per unit area.
W = wind load, or its related internal moment and forces.
x = shorter over-all dimension of a rectangular part of a cross section.
x_1 = shorter center-to-center dimension of a closed rectangular stirrup.
y = longer over-all dimension of a rectangular part of a cross section.
y_t = distance from the centroidal axis of gross section, neglecting the reinforcement, to the extreme fiber in tension.
y_1 = longer center-to-center dimension of a closed rectangular stirrup.
z = a quantity limiting distribution of flexural reinforcement.
α = ratio of flexural stiffness of beam section to the flexural stiffness of a width of slab bounded laterally by the centerline of the adjacent panel, if any, on each side of the beam $= E_{cb}I_b/E_{cs}I_s$.
α = total angular change of prestressing steel profile in radians from jacking end to any point, x.
α_c = ratio of flexural stiffness of the columns above and below the slab to the combined flexural stiffness of the slabs and beams at a joint taken in the direction moments are being determined $= \Sigma K_c/\Sigma(K_s + K_b)$.
α_{ec} = ratio of flexural stiffness of the equivalent column to the combined flexural stiffness of the slabs and beams at a joint taken in the direction moments are being determined $= K_{ec}/\Sigma(K_s + K_b)$.
α_m = average value of α for all beams on the edges of a panel.

α_{min} = minimum α_c to satisfy Section 13.3.6.1(a).
α_t = a coefficient as a function of y_1/x_1.
α_v = ratio of stiffness of shearhead arm to surrounding composite slab section.
α_1 = α in the direction of l_1.
α_2 = α in the direction of l_2.
β = ratio of clear spans in long to short direction of two-way construction.
β = ratio of long side to short side of a footing.
β_a = ratio of dead load per unit area to live load per unit area in each case without load factors.
β_b = ratio of area of bars cut off to total area of bars at the section.
β_d = the ratio of maximum design dead load moment to maximum design total load moment, always positive.
β_s = ratio of length of continuous edges to total perimeter of a slab panel.
β_t = ratio of torsional stiffness of edge beam section to the flexural stiffness of a width of slab equal to the span length of the beam, center-to-center of supports = $E_{cb}C/2E_{cs}I_s$.
β_1 = a factor defined in Section 10.2.7.
δ = moment magnification factor for columns.
δ_s = factor defined by Eq. (13-4). See Section 13.3.6.1.
μ = coefficient of friction. See Section 11.15.
μ = prestress curvature friction coefficient.
ξ = constant for standard hook.
ρ = A_s/bd = ratio of nonprestressed tension reinforcement.
ρ' = A_s'/bd.
ρ_b = reinforcement ratio producing balanced conditions.
ρ_{min} = minimum reinforcement ratio.
ρ_n = the ratio of vertical shear reinforcement area to the gross concrete area of a horizontal section.
ρ_p = A_{ps}/bd; ratio of prestressed reinforcement.
ρ_s = ratio of volume of spiral reinforcement to total volume of core (out-to-out of spirals) of a spirally reinforced concrete column.
ρ_v = $(A_s + A_h)/bd$.
ρ_w = A_s/bd.
ϕ = capacity reduction factor.
ω = $\rho f_y/f_c'$—prestress.
ω' = $\rho' f_y/f_c'$—prestress.
ω_p = $\rho_p f_{ps}/f_c'$—prestress.

ω_w, ω_{pw}, ω_w' = reinforcement indices for flanged sections computed as for ω, ω_p, ω', except that b shall be the web width, and the steel area shall be that required to develop the compressive strength of the web only.

appendix 2

CODE COMMENTARY NOTATION AND SYMBOLS

A_c = $2d(c_1 + c_2 + 2d)$ for interior columns—Chapter 11, Commentary.

A_f = base area of footing—Chapter 15, Commentary.

A_{pf} = $A_{ps} - A_{pw}$—Chapter 18, Commentary.

A_{pw} = $0.85\, f_c'(b - b_w)\, \dfrac{h_f}{f_{ps}}$—Chapter 18, Commentary.

A_{sa} = increased area of bars $= \dfrac{2A_{st}\beta}{\beta + 1}$—Chapter 15, Commentary.

A_{sp} = area of spiral reinforcement—Appendix A, Commentary.

A_{st} = total area of steel in the short direction of footing—Chapter 15, Commentary.

A_1 = effective area of concrete per bar—Chapter 10, Commentary.

A_1 = bearing area of anchor plate of posttensioning steel—Chapter 18, Commentary.

A_2 = maximum area of the portion of the anchorage surface that is geometrically similar to and concentric with the area of the anchor plate of the posttensioning steel—Chapter 18, Commentary.

b_w = wall thickness of box section—Chapter 11, Commentary.

d = diameter—Appendix A, Commentary.

f_{cp} = permissible concrete bearing stress under anchor plate of posttensioning steel—Chapter 18, Commentary.

f_{cr} = required average strength—Chapter 4, Commentary.

f_s = tensile stress in reinforcing steel—Appendix A, Commentary.

f_s' = compressive stress in reinforcing steel—Appendix A, Commentary.

f_t = tensile strength of concrete—Chapter 11, Commentary.

G = modulus of elasticity or rigidity in shear—Chapter 13, Commentary.

h_f = flange thickness—Chapter 18, Commentary.

h_w = web height or projection of beam below slab—Chapter 13, Commentary.

H = horizontal forces—Appendix A, Commentary.

I_{sb} = moment of inertia of the width of slab used for the calculation of I_s but including the contribution of that portion of the beam stem extending above or below the slab; the beam as defined in

Section 13.1.5 does not apply in this calculation—Chapter 13, Commentary.

J_c = $\frac{d(c_1 + d)^3}{6} + \frac{(c_1 + d)d^3}{6} + \frac{d(c_2 + d)(c_1 + d)^2}{2}$ for interior columns—Chapter 11, Commentary.

K_{ta} = increased torsional stiffness, K_t, due to presence of parallel beam —Chapter 13, Commentary.

l = span length—Chapter 11, Commentary.

M = moment—Chapter 11, Commentary.

M_{p1}, M_{p2} = end moments shown on free-body diagram Fig. A-1—Appendix A, Commentary.

M_u = moment capacity of a section—Chapters 9 and 18, Commentary.

P = unfactored design load for a compression member—Chapter 10, Commentary.

P_o = theoretical axial load capacity of a short compression member—Chapter 10, Commentary.

q_s = soil reaction due to factored loading—Chapter 15, Commentary.

u_u = bond stress—Chapter 12, Commentary.

U = factored concentric load to be resisted by the footing—Chapter 15, Commentary.

V = shear—Chapter 11, Commentary.

V = vertical forces—Appendix A, Commentary.

V_c = diagonal cracking shear force—Chapter 11, Commentary.

w = crack width—Chapter 10, Commentary.

α (alpha) = $\left(1 - \frac{1}{1 + \frac{2}{3}\sqrt{\frac{c_1 + d}{c_2 + d}}}\right)$—Chapter 11, Commentary.

β (beta) = ratio of distances to the neutral axis from the extreme tension fiber and from the centroid of the main reinforcement—Chapter 10, Commentary.

γ (gamma) = ratio of distance center-to-center of reinforcement on opposite faces of a column to the overall dimension of the column cross-section $= \frac{h - d' - d_s}{h}$—Chapter 9, Commentary.

ϵ_s (epsilon) = strain in steel—Chapter 10, Commentary.

λ (lambda) = multiplier for long-time deflection—Chapter 9, Commentary.

σ (sigma) = standard deviation of individual strength tests—Chapter 4, Commentary.

ψ (psi) = ratio of sum of stiffnesses, ΣK, of compression members to ΣK of flexural members in a plane at one end of a compression member—Chapter 10, Commentary.

subject index

section index

commentary index